Acylated Anthocyanins

Anthocyanins esterified with one or more acid groups are known as acylated anthocyanins, which exhibit higher stability and antioxidant activity compared to nonacylated ones and are of interest to the food and beverage industry. It is known that the structural differences among anthocyanins will influence their specific roles and effectiveness in the different applications in which they may be used. This book compiles and summarizes data regarding knowledge on acylated anthocyanins, ranging from their physical and chemical particularities to their health effects, including bioaccessibility, bioavailability, and bioactivity, as well as their different technological applications.

Key Features:

- Provides information on the different areas where acylated anthocyanins are being utilized
- Discusses the characteristics of acylated anthocyanins and the effect of their structural features on different properties
- Includes a database of acylated anthocyanins from both nonedible and edible sources

Acylated Anthocyanins

Chemistry, Bioavailability, Bioactivity, and Applications

Edited by
Hélder José Couto Oliveira and Matteo Bordiga

CRC Press
Taylor & Francis Group
Boca Raton London New York

CRC Press is an imprint of the
Taylor & Francis Group, an **informa** business

Designed cover image: Hélder José Couto Oliveira

First edition published 2026
by CRC Press
2385 NW Executive Center Drive, Suite 320, Boca Raton FL 33431

and by CRC Press
4 Park Square, Milton Park, Abingdon, Oxon, OX14 4RN

CRC Press is an imprint of Taylor & Francis Group, LLC

ISBN: 978-1-032-84451-0 (hbk)
ISBN: 978-1-032-85732-9 (pbk)
ISBN: 978-1-003-51959-1 (ebk)

DOI: 10.1201/9781003519591

Typeset in Times
by Apex CoVantage, LLC

Contents

Preface

Color has always held a special place in our connection with nature. It signals ripeness, attracts pollinators, warns predators, and evokes emotion in those who observe it. Think of the deep red of a ripe cherry, the velvety purple of a petunia in full bloom, or the striking blue of a butterfly pea flower steeped in warm water. These colors are not only beautiful but are also biochemical signatures, often shaped by a remarkable group of compounds known as anthocyanins.

Anthocyanins are a consolidated group of natural compounds with more than 700 structures described so far. They are responsible for the bright red, blue, and violet colors of many natural edible and nonedible sources. Their health benefits as well as their physical and chemical properties have been widely explored during the last few years, and applications of these compounds in several technological and health fields are increasing at an exponential rate.

It is known that the structural differences among anthocyanins will influence their specific roles and effectiveness in the different applications in which they may be used. In particular, acylated anthocyanins are described as presenting overall higher stability of different factors that affect the integrity of these compounds, including temperature, pH, oxygen, light, and the enzymatic/microbial machinery *in vivo*, probably due to their higher structural complexity that may promote a higher internal hindrance/protective effect.

The idea for this book was born from a desire to bring together the growing body of knowledge on acylated anthocyanins in a way that is both comprehensive and accessible. It aims to serve as a reference for anyone interested in these compounds, including their physical and chemical particularities; their health effects, including bioaccessibility, bioavailability, and bioactivity; and also their different technological applications.

Organized into four thematic sections and 15 chapters, the book reflects the collaborative spirit of science. Thirty-nine authors from nine countries (Portugal, Italy, China, Turkey, India, Pakistan, Peru, the United States, and Mexico) have contributed their expertise, perspectives, and passion. Their work enriches this volume and highlights the global interest in understanding and harnessing the potential of acylated anthocyanins.

We sincerely thank each of the authors. This book is a celebration of shared knowledge, curiosity, and the dynamic science behind acylated anthocyanins.

Hélder José Couto Oliveira
Matteo Bordiga

Editors

Hélder José Couto Oliveira is a Principal Researcher at LAQV-REQUIMTE. He earned his Ph.D. in chemistry in 2018 and MSc in biochemistry in 2014 from the University of Porto. His field of research concerns food chemistry and biochemistry, essentially food polyphenols and the true absorption of these compounds at different cellular barriers, their metabolization, and their effect as bioactives in different pathological contexts. Recently, he initiated a research line dedicated to botanics and edible flowers, being responsible for different projects within this area. He has published more than 40 research papers in peer-reviewed international journals, participated in several dozens of funded projects and supervised more than 20 students among BSc, MSc, and PhD degrees. From 2018 to 2025, Hélder was an invited professor at Escola Superior de Saúde-IPP and University of Porto.

Dr. Matteo Bordiga is currently Associate Professor of Food Chemistry at the Università del Piemonte Orientale (UPO), Novara, Italy. He earned a PhD in food science and an MS in chemistry and pharmaceutical technologies from the same university. His main research activities concern food chemistry and investigating different classes of bioactive compounds from analytical, technological, and nutritional points of view. More recently, he also moved his research interests to wine chemistry, focusing his attention on the entire production process also with a view to valorizing by-products. He has published more than 70 research papers in peer-reviewed international and national journals. He is currently the editor of *LWT – Food Science and Technology* (Q1; Elsevier) and associate editor of *Food Science and Nutrition* (Q2; Wiley) and *Quality Assurance and Safety of Crops & Foods* (Q1; Codon Publications). He has edited *Valorization of Wine Making By-Products* (CRC Press, 2015); *Post-Fermentation and -Distillation Technology: Stabilization, Aging, and Spoilage* (CRC Press, 2018); *Food Aroma Evolution: During Food Processing, Cooking, and Aging* (CRC Press, 2019); *Concise Encyclopedia of Science and Technology of Wine* (CRC Press, 2021); *Edible Insects Processing for Food and Feed: From Startups to Mass Production* (CRC Press, 2023); *Flavoromics: An Integrated Approach to Flavor and Sensory Assessment* (CRC Press, 2023); and *How Technological Advances Change Human Food* (CRC Press, 2025).

Contributors

Asma Ayub
Department of Botany
University of Agriculture
Faisalabad, Pakistan

Nataly Bernuy-Osorio
Department of Nutrition
Faculty of Animal Science
National Agrarian University La Molina
Lima, Peru

Raffaella Colombo
Department of Drug Sciences
University of Pavia
Pavia, Italy

Patrícia Correia
Faculty of Sciences
University of Porto
Porto, Portugal

Luis Cruz
Faculty of Sciences
University of Porto
Porto, Portugal

Victor de Freitas
Faculty of Sciences
University of Porto
Porto, Portugal

Vincenzo Disca
Dipartimento di Scienze del Farmaco
Universita degli Studi del Piemonte Orientale
Novara, Italy

Dureshahwar
Department of Zoology, Wildlife and Fisheries
University of Agriculture
Faisalabad, Pakistan

Ana Faria
Nova Medical School
Universidade Nova de Lisbon
Lisbon, Portugal

Ana Fernandes
Faculty of Sciences
University of Porto
Porto, Portugal

Iva Fernandes
Faculty of Sciences
University of Porto
Porto, Portugal

Gamze Güçlü
Department of Food Engineering
Faculty of Agriculture
Cukurova University
Adana, Turkey

Sheyla S. Gutierrez-Asencios
Doctoral Program in Nutrition
National Agrarian University La Molina
Lima, Peru

Muhammad Azlan Habib
Department of Zoology
Government College University
Lahore, Pakistan

Jingren He
School of Modern Industry for Selenium Science and Engineering
Wuhan Polytechnic University
Wuhan, China

Deepika Kaushik
Department of Biotechnology
Faculty of Applied Sciences and Biotechnology
Shoolini University
Solan, India

Hasim Kelebek
Department of Food Engineering
Faculty of Engineering
Adana Alparslan Turkes Science and Technology University
Adana, Turkey

Ozlem Kilic-Buyukkurt
Department of Food Technology
Kadirli Applied Sciences School, Osmaniye Korkut Ata University
Osmaniye, Turkiye

Mukul Kumar
Department of Food Technology and Nutrition
Lovely Professional University
Punjab, India

Waqar Majeed
Department of Zoology, Wildlife and Fisheries
University of Agriculture
Faisalabad, Pakistan

Nuno Mateus
Chemistry and Biochemistry Department
Faculty of Sciences, University of Porto
Porto, Portugal

Ana Rita Monteiro
Chemistry and Biochemistry Department
Faculty of Sciences, University of Porto
Porto, Portugal

Ammu P. Nair
Department of Food Technology and Nutrition
Lovely Professional University
Punjab, India

Muhammad Naveed
Department of Zoology, Wildlife and Fisheries
University of Agriculture
Faisalabad, Pakistan

Joana Oliveira
Chemistry and Biochemistry Department
Faculty of Sciences, University of Porto
Porto, Portugal

Emel Oz
Department of Food Engineering
Faculty of Agriculture, Ataturk University
Erzurum, Turkey

Fatih Oz
Department of Food Engineering
Faculty of Agriculture, Ataturk University
Erzurum, Turkey

Adele Papetti
Department of Drug Sciences
University of Pavia
Pavia, Italy

Saroj Paudel
Department of Nutrition, Dietetics, and Food Sciences
Utah State University
Logan, USA

Ana Rita Pereira
Chemistry and Biochemistry Department
Faculty of Sciences, University of Porto
Porto, Portugal

Joana Pico
Department of Nutrition, Dietetics, and Food Sciences
Utah State University
Logan, USA

Fernando Pina
Departamento de Quimica
Faculdade de Ciencias e Tecnologia, Universidade Nova de Lisboa
Lisbon, Portugal

Erika Salas
Facultad de Ciencias Quimicas
Universidad Autonoma de Chihuahua
Chihuahua, Mexico

Serkan Selli
Department of Food Engineering
Faculty of Agriculture, Cukurova University
Adana, Turkey

Wen Tao
Chemistry and Biochemistry Department
Faculty of Sciences, University of Porto
Porto, Portugal

Margarida Teixeira
Chemistry and Biochemistry Department
Faculty of Sciences, University of Porto
Porto, Portugal

Wilson Vasquez-Rojas
Doctoral Program in Nutrition
National Agrarian University La Molina
Lima, Peru

Muci Wu
School of Modern Industry for Selenium Science and Engineering
Wuhan Polytechnic University
Wuhan, China

Rui Zhang
School of Modern Industry for Selenium Science and Engineering
Wuhan Polytechnic University
Wuhan, China

PART I

The Chemistry of Acylated Anthocyanins

Anthocyanins

1

Erika Salas, Matteo Bordiga, and
Hélder José Couto Oliveira

1.1 INTRODUCTION TO ANTHOCYANINS

Natural compounds have always played a pivotal role in human societies, driving scientific advancements and offering solutions for health, nutrition, and medicine. Among these bioactive molecules, polyphenols stand out due to their remarkable antioxidant, anti-inflammatory, and disease-preventive properties [1, 2]. Found abundantly in fruits, vegetables, and medicinal plants, polyphenols contribute not only to the vibrant colors and flavors of our food but also to their protective effects against chronic diseases such as cardiovascular disorders, diabetes, and cancer.

Within the vast family of polyphenols, anthocyanins, pigments responsible for the red, purple, and blue hues in berries, grapes, and other plants, have garnered significant scientific interest. Beyond their aesthetic appeal, anthocyanins exhibit potent health benefits, including neuroprotective effects, improved cardiovascular health, and antiaging properties. Their ability to modulate oxidative stress and inflammation makes them promising candidates for nutraceuticals and functional foods.

Historically, anthocyanins were likely utilized long before they were understood. Various cultures have used brightly colored fruits and plants for food, medicine, and dyes, recognizing their vibrant colors but not knowing their chemical composition. The use of berries and flowers in cooking and traditional medicine dates back centuries. For instance, Egyptians and Mesopotamians used plant-based dyes from madder root (*Rubia tinctorum*) and grapes in textiles and cosmetics [3]. Romans and Greeks used crushed berries and purple cabbage as pH indicators in rudimentary chemical tests [3]. Chinese and Ayurvedic traditional medicine systems utilized anthocyanin-rich plants (e.g., hibiscus and black rice) for treating inflammation and circulatory disorders [4].

The chemical structure of anthocyanins began to be studied in the 19th century. The word "anthocyanin" was first used in the 1830s by Ludwig Marquart, derived from the Greek words *anthos* (flower) and *kyanos* (dark blue). Nobel Prize-winning chemist Richard Willstätter identified the first anthocyanin structures (e.g., cyanidin in roses) and linked their colors to pH-dependent molecular changes [5]. Prior to that, pigmented extracts from blue cornflowers (*Centaurea cyanus*) were isolated. The advances in chromatographic techniques such as paper chromatography enabled the separation and identification of over 700 anthocyanins in nature [6–9].

As research continues to uncover the mechanisms behind these natural compounds, the importance of polyphenols, particularly anthocyanins, becomes increasingly evident. By harnessing their potential, we can develop innovative strategies for disease prevention and health promotion, bridging traditional knowledge with modern science for a healthier future.

These compounds also play crucial ecological roles, attracting pollinators and seed dispersers while also providing photoprotection and antioxidant defense in plants. Flowers like petunia and ipomoea use

DOI: 10.1201/9781003519591-2

anthocyanin-derived colors to signal to bees and butterflies [10–12]. In plants growing in high-light environments (e.g., alpine plants), anthocyanins act as "sunscreens," absorbing UV-B and excess blue/green wavelengths to prevent oxidative damage [13, 14]. Anthocyanin accumulation in leaves (e.g., red maple in autumn) correlates with cold tolerance and herbivore deterrence [15].

1.2 CHEMICAL STRUCTURE AND CLASSIFICATION OF ANTHOCYANINS

The chemical structure of anthocyanins is influenced by several factors, including pH, glycosylation and acylation patterns, and the interactions with co-pigments and metal ions. Consequently, their stability and color expression are directly influenced. As water-soluble natural compounds, anthocyanins are formed through the glycosylation of anthocyanidins, a process where sugar molecules bind to the anthocyanidin core *via* glycosidic bonds. The glycosylation reaction occurs because anthocyanidins – the aglycone (non-sugar) forms of anthocyanins – are chemically unstable in their free state [16]. Anthocyanidins consist of a 2-phenylbenzopyrylium (flavylium) cation and a positively charged oxygen-containing heterocycle, with varying hydroxyl (-OH) and methoxyl (-OCH_3) substitutions (Figure 1.1). Since the substituents can appear at the C3, C5, C7, C3′, C4′, and C5′ positions of the flavylium ion, the types of anthocyanidins can be diverse. Up to now, more than 700 kinds of anthocyanins based on six main types of anthocyanidins have been identified and registered in the literature [8, 9]. Table 1.1 shows the substitution patterns of the flavylium cation to form the most common anthocyanidins that originate anthocyanins. These substitutions determine the type of anthocyanidin (e.g., cyanidin, delphinidin, and malvidin) formed and, consequently, the color properties of the resulting anthocyanin.

Anthocyanin compounds typically contain hexoses (such as glucose or galactose) or pentoses (including rhamnose, xylose, and arabinose) attached to the aglycone core as mono-, di-, or tri-glycosides [16]. During anthocyanin biosynthesis (discussed later), sugar units (e.g., glucose, galactose, rhamnose, xylose, or arabinose) are primarily linked to the aglycone at the C3 position (via the C3-OH group on ring C), as glycosylation at this site enhances the molecular stability previously referred [17]. While monoglycosylated anthocyanins represent the simplest structural form, additional sugars may attach to different positions. These sugars are more likely to be attached to C3, C5, and C7, and less frequently to positions C3' and C5' [18]. Anthocyanins can also undergo acylation, where aromatic acids (e.g., *p*-coumaric, sinapic, caffeic, and ferulic acids) or aliphatic acids (e.g., malonic, acetic, and malic acids) bind to sugar moieties, typically at the C3 or C5 position of the flavonoid skeleton [19]. These acyl substituents are usually esterified to sugars at the 6-OH or less frequently at the 4-OH [19]. Apart from these, a unique structural variation occurs in 3,5-*O*-diglycosides, where an acyl group bridges two sugar units, forming a cyclic structure – a rare feature observed in certain Dianthus species (e.g., *Dianthus caryophyllus*)

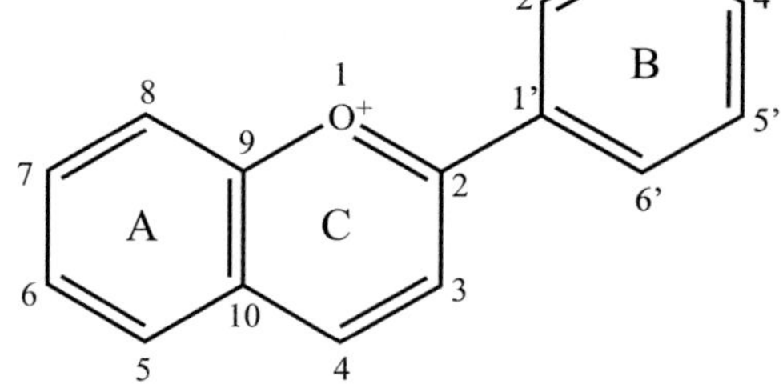

FIGURE 1.1 Chemical structure of 2-phenylbenzopyrylium cation (flavylium ion).

TABLE 1.1 Major Anthocyanidins That Originate Anthocyanins

ANTHOCYANIDIN	*SUBSTITUENTS (C1′–C5′)*	*COMMON SOURCES*	*COLOR RANGE*
Cyanidin	–OH, –H, –OH, –OH, –H	Berries, cherries	Red–orange
Delphinidin	–OH, –OH, –OH, –OH, –H	Blueberries, grapes	Blue–purple
Pelargonidin	–H, –H, –OH, –OH, –H	Strawberries, radishes	Orange–red
Peonidin	$-OCH_3$, –H, –OH, –OH, –H	Cranberries, plums	Red–purple
Petunidin	$-OCH_3$, –OH, –OH, –OH, –H	Eggplants, black currants	Purple
Malvidin	$-OCH_3$, $-OCH_3$, –OH, –OH, –H	Red wine, grapes	Purple–blue

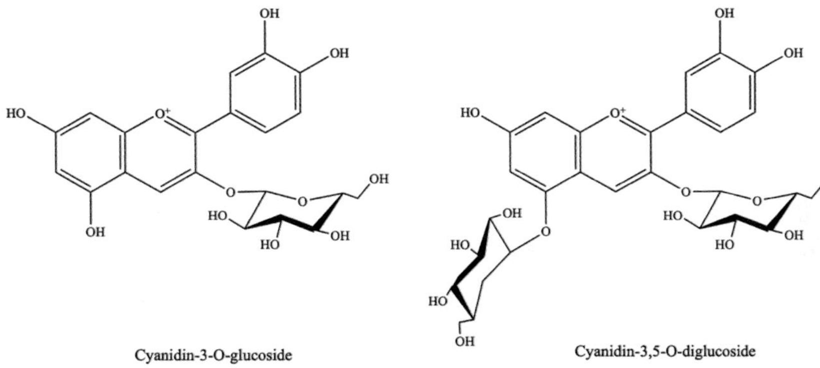

FIGURE 1.2 Mono- and polyglycosylated anthocyanins example.

(as in 3,5-*O*-diglycosides), and the acyl group is connected to both, forming a cyclic structure [20, 21]. All of these non-monoglycosylated (i.e., di- or tri-glycosylated) anthocyanins and their acylglycosylated or cyclic forms can be defined as polyglycosylated anthocyanins [7]. Figure 1.2 shows an example of a mono- and a polyglycosylated anthocyanin. Given the great number of possibilities, acyl-polyglycosylated anthocyanins are the main group of anthocyanins present in nature and will be discussed in depth in the next chapters of this book.

The inherent structural characteristics of anthocyanins render these pigments particularly susceptible to molecular transformations, with pH serving as the primary determinant of their stability. Under strongly acidic conditions (pH 1), anthocyanins exist in their most stable configuration as the flavylium cation form, characterized by a positively charged pyrylium ring. As pH increases, two distinct chemical processes occur simultaneously but at different rates: (1) a rapid proton-transfer reaction leading to the formation of quinoidal bases through deprotonation and (2) a slower hydration process where water molecules attack the pyran ring nucleophilically, producing hemiketal structures. These hemiketal forms become increasingly prevalent in weakly acidic environments, typically resulting in significant reduction in color vibrancy [22]. In addition to these transformations, the pyran ring can undergo cleavage through tautomerization. This process converts the hemiketal forms first into *cis*-chalcones, followed by their *trans*-chalcone isomers. The rate of this isomerization varies depending on the specific substitution pattern of the flavylium cation's core structure [22]. Under alkaline conditions (pH >7), anthocyanins undergo significant structural changes through two primary pathways: either complete molecular degradation or deprotonation reactions that generate corresponding anionic base forms. The specific chemical equilibrium established between these forms depends largely on the substitution patterns present in each anthocyanin's molecular structure [23]. Figure 1.3 summarizes the structural changes suffered by anthocyanins according to the pH environment. Co-pigmentation is also among the main factors affecting the

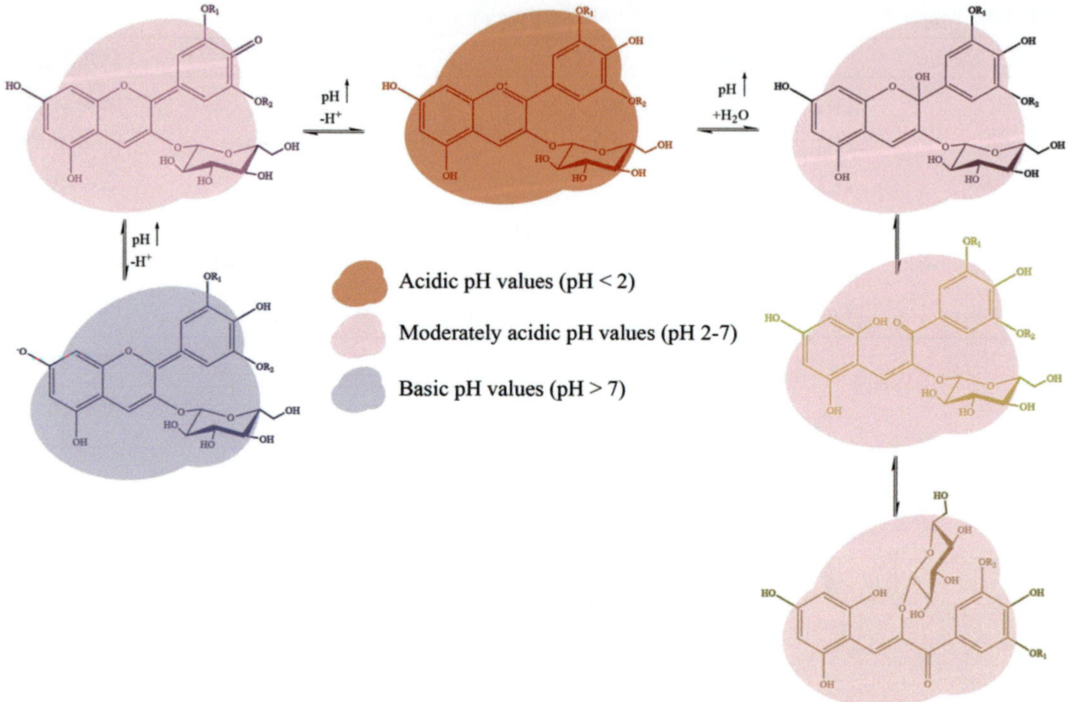

FIGURE 1.3 Structural dependence of anthocyanins on pH.

stability of anthocyanins. Such phenomenon occurs when colorless organic molecules form non-covalent complexes with anthocyanins. The process usually occurs through multiple molecular interactions that collectively enhance both color intensity and molecular durability. These include π–π stacking between aromatic ring systems, hydrogen bonding with phenolic groups, and van der Waals forces in hydrophobic regions [24, 25]. Such interactions create a protective microenvironment around the flavylium cation, particularly important at pH ranges (3–6) where anthocyanins are most vulnerable to nucleophilic attacks by water molecules [26]. The diversity of natural co-pigments contributes to the wide variability observed in anthocyanin stability across different plant species. Major classes of co-pigments include flavonoids such as quercetin and rutin; phenolic acids such as caffeic and chlorogenic acids; and other compounds including certain alkaloids and amino acids. Among these, flavonoids exhibit particularly strong stabilizing effects, with studies by Mazza and Brouillard showing that optimal molar ratios (typically 1:10 anthocyanin to co-pigment) can produce up to 300% increases in color intensity [26].

From biological and bioactive perspectives, co-pigmentation also serves multiple essential functions in plants and animals, such as the modulation of color for attracting pollinators and protecting from UV-Vis damage, or enhancing the overall antioxidant capacity of pigmented tissues [27].

Besides pH and co-pigmention, anthocyanins are also susceptible to other factors including temperature, light, oxygen, solvents, the presence of enzymes, other polyphenols, and metallic ions [28]. For instance, thermal degradation of most anthocyanins follows first-order kinetics, with degradation rates doubling for every 10°C increase [29, 30], and light exposure can induce photo-oxidation through radical formation, accelerating degradation by two to five times compared to dark storage [31], while the presence of enzymes, such as polyphenol oxidases and peroxidases, can degrade anthocyanins within minutes [32, 33]. On the other hand, metallic ions, particularly Fe^{3+}, Al^{3+}, and Cu^{2+}, can either stabilize anthocyanins through complex formation or accelerate degradation *via* redox reactions, depending on their concentration and pH [34, 35].

1.3 BIOLOGICAL FUNCTIONS OF ANTHOCYANINS IN PLANTS

1.3.1 Role of Anthocyanins in Plant Coloration: Attraction of Pollinators and Seed Dispersers

Anthocyanins are a group of flavonoid pigments responsible for the red, blue, and purple hues in plant organs, including flowers and fruits. These pigments play a critical role in plant reproductive success by serving as visual signals that attract pollinators and seed dispersers [36]. The ecological significance of anthocyanins in these interactions has been widely studied, revealing their role in facilitating plant–animal mutualisms. Flowers rely on their pigmentation to attract pollinators such as bees, butterflies, birds, and bats. The ability of pollinators to perceive flower color varies based on their visual systems. Bees, for instance, have trichromatic vision, sensitive to ultraviolet (UV), blue, and green light, making them particularly attracted to blue and purple flowers, which often owe their coloration to anthocyanins. Birds, particularly hummingbirds, prefer red flowers, a coloration frequently attributed to anthocyanin accumulation [36].

Anthocyanin accumulation is often linked with UV reflectance, which further enhances floral visibility to pollinators. UV-reflecting patterns serve as nectar guides, directing pollinators toward reproductive structures. For example, a study has shown that the removal of UV reflectance significantly reduced flower visitation by pollinators. This suggests that anthocyanin-based pigmentation, coupled with UV reflection, enhances floral attractiveness and reproductive efficiency. Beyond attraction, anthocyanins may also influence pollinator behavior by acting as indicators of floral age and nectar reward. Flowers with fading anthocyanin pigmentation may signal to pollinators that they have already been visited and have reduced nectar availability, thereby optimizing pollination efficiency [37].

In addition to pollination, anthocyanins play a crucial role in fruit and seed dispersal by making fruits visually conspicuous to dispersers. The evolution of fleshy fruits with bright colors is considered an adaptation to attract frugivorous animals, including birds and mammals, that facilitate seed dispersal. Red and black are the most common colors among bird-dispersed fruits. Birds, possessing tetrachromatic vision, can easily detect red fruits against a green foliage background, increasing the chances of fruit removal and seed dispersal [38]. Conversely, black fruits, which contain high levels of anthocyanins, provide a strong achromatic contrast that remains visible under varying light conditions. Interestingly, blue fruits are less common, despite their conspicuousness to birds. One hypothesis suggests that blue fruits may also be attractive to undesired dispersers such as insects and dichromatic mammals, leading to their relative scarcity. This illustrates the role of selective pressures in shaping fruit color evolution. Anthocyanins also contribute to seed dispersal by influencing fruit maturation and signaling ripeness. Many fruits undergo a color change from green to red, purple, or black as they ripen, which signals to dispersers that they are ready for consumption. This color transition, controlled by anthocyanin biosynthesis, ensures that fruits are consumed at an optimal stage for seed dispersal [39].

1.3.2 Anthocyanins in Multispecies Interactions

Plants often use combinations of color traits to maximize their chances of successful pollination and seed dispersal while avoiding herbivory. Some species display ripe black fruits alongside unripe red fruits or red accessory structures to enhance visual attraction while ensuring that only ripe seeds are dispersed. Moreover, UV-reflecting epicuticular wax layers on some fruits further improve their conspicuousness to dispersers [40]. Thus, anthocyanins serve a dual function in plant–animal interactions: they enhance reproductive success by attracting pollinators and seed dispersers while potentially deterring herbivores. Their widespread occurrence in flowers and fruits underscores their ecological importance in shaping plant reproductive strategies.

1.3.3 Protective Functions Against UV Radiation, Oxidative Stress, and Herbivores

Anthocyanins, a subgroup of flavonoids, play a crucial role in protecting plants against various environmental stressors, including UV radiation, oxidative stress, and herbivore attack. Their accumulation in plant tissues serves as a multifunctional defense mechanism that ensures survival and adaptability in dynamic ecological settings. Exposure to UV radiation, particularly UV-B (280–320 nm), can have detrimental effects on plant physiology by damaging DNA, proteins, and membrane lipids. To mitigate this impact, plants synthesize UV-absorbing phenolic compounds, including anthocyanins, which accumulate in epidermal tissues and function as natural sunscreens [41, 42]. These pigments absorb excess radiation, thereby reducing the penetration of harmful UV rays into deeper cellular structures. Moreover, acylated anthocyanins, which contain phenylpropanoid modifications, exhibit enhanced UV absorption properties, providing an additional protective layer against photodamage. Studies have demonstrated that red-leafed plant varieties, which contain higher anthocyanin concentrations, retain greater photosynthetic efficiency following UV exposure compared to their green-leafed counterparts. The "leaf-thickness hypothesis" suggests that thicker leaves benefit more from anthocyanin accumulation compared to thinner ones. However, this is not a universal rule, as some species exhibit anthocyanin-based UV protection regardless of leaf thickness. Beyond UV protection, anthocyanins are powerful antioxidants that play a critical role in neutralizing reactive oxygen species (ROS), which accumulate during abiotic stress conditions such as high light intensity, drought, and extreme temperatures [42]. The ability of anthocyanins to scavenge ROS helps prevent oxidative damage to cellular macromolecules, including lipids, proteins, and nucleic acids. The structural diversity of anthocyanins, particularly their hydroxylation and glycosylation patterns, influences their radical-scavenging potential. For instance, B-ring ortho-dihydroxy-substituted anthocyanins exhibit superior antioxidant activity compared to their monohydroxy counterparts. Furthermore, the localization of anthocyanins in vacuoles allows them to act as modulators of ROS signaling cascades, which are integral to stress response mechanisms. The presence of anthocyanins in plant tissues is often correlated with increased activity of enzymatic antioxidants such as superoxide dismutase (SOD) and catalase (CAT), further enhancing their protective function. In addition to their photoprotective and antioxidant roles, anthocyanins contribute to plant defense against herbivores through both direct and indirect mechanisms. The presence of anthocyanins in leaves, stems, and fruits can reduce herbivore preference due to their bitter taste and potential toxicity. Many herbivorous insects and mammals avoid feeding on anthocyanin-rich tissues, as these compounds interfere with digestion and nutrient absorption. Moreover, anthocyanins often co-occur with other secondary metabolites, such as tannins and alkaloids, which enhance plant resistance to herbivory [43]. In some cases, anthocyanin accumulation is inducible in response to herbivore attack, serving as an adaptive mechanism that strengthens plant defenses when needed. Some studies suggest that anthocyanins may act as feeding deterrents by mimicking toxic compounds, further reducing herbivore damage. Another key aspect of anthocyanin-mediated defense is their role in indirect plant protection through signaling interactions with natural enemies of herbivores. Certain anthocyanin-rich plant species attract predatory insects that target herbivorous pests, thereby reducing herbivore pressure on the plant. Additionally, anthocyanins can function in plant–plant signaling, warning neighboring individuals of imminent herbivore threats and triggering defensive responses in nearby plants. In some cases, anthocyanins contribute to camouflage or mimicry strategies that make plants less conspicuous to herbivores, thus reducing the likelihood of attack.

1.4 ANTHOCYANINS IN THE HUMAN DIET AND HEALTH

1.4.1 Sources of Anthocyanins in the Human Diet (Fruits, Vegetables, and Beverages)

Anthocyanins are an important class of flavonoid pigments that contribute to the vibrant red, blue, and purple colors of many plant-based foods. Their presence in the human diet is significant due to their potential health benefits, including antioxidant, anti-inflammatory, and cardioprotective properties. The primary dietary sources of anthocyanins include fruits, vegetables, beverages, and flowers, with considerable variation in their composition and concentration among different plant species and cultivars [16, 44].

Structurally, anthocyanins belong to the flavonoid group of phenolic compounds and are characterized by a highly conjugated 2-phenylbenzopyran cation, forming a C6–C3–C6 skeleton. This basic structure leads to the formation of various anthocyanidins, the six most common being cyanidin, pelargonidin, delphinidin, peonidin, petunidin, and malvidin. Free anthocyanidins are rare in nature; they are typically present as glycosides, in which a sugar moiety such as glucose, galactose, arabinose, or rhamnose is attached through glycosidic bonds. Acylated anthocyanins, formed *via* esterification with organic acids, exhibit enhanced stability and resistance to degradation from environmental factors such as light and temperature [45].

Fruits are the most abundant source of anthocyanins, particularly berries such as blueberries, blackberries, raspberries, and strawberries. The total anthocyanin content in these fruits can range widely, with bilberries containing up to 698 mg/100 g fresh weight (FW) and black currants between 130 and 476 mg/100 g FW. Other anthocyanin-rich fruits include grapes, which contain primarily malvidin glycosides, with red grape varieties showing anthocyanin concentrations up to 250 mg/100 g FW, while red wines derived from these grapes can reach approximately 500 mg/L. Cherries, pomegranates, plums, and black goji berries also contribute substantial amounts of anthocyanins, with chokeberries showing one of the highest recorded levels, reaching 1,480 mg/100 g FW [41, 46].

Vegetables, although generally containing lower anthocyanin levels than fruits, still represent an important dietary source. Red cabbage is particularly rich, with reported concentrations reaching up to 629 mg/100 g dry weight (DW), depending on cultivar and growing conditions. Other anthocyanin-containing vegetables include purple sweet potatoes, which exhibit levels between 87 and 168 mg/100 g DW, and violet cauliflower, where concentrations range from 1.81 to 73.9 mg/100 g FW. Additionally, anthocyanins have been identified in radishes, eggplants, and various *Brassica* species such as kale and mustard greens, with their profiles differing based on environmental factors and genetic variability [7, 47].

Beverages derived from anthocyanin-rich plant sources also contribute to dietary intake. Red wine produced from grape varieties high in anthocyanins is among the most well-documented sources. Other anthocyanin-containing beverages include fruit juices made from blueberries, cherries, and pomegranates, which can retain high levels of these pigments when minimally processed. Herbal teas, particularly those made from hibiscus and butterfly pea flowers, also provide significant anthocyanin content.

Edible flowers represent a lesser-known but valuable source of anthocyanins in the diet. Common examples include hibiscus, which is rich in delphinidin derivatives, as well as violets, pansies, and roses, which contain various glycosylated anthocyanins. Edible flowers are increasingly used in culinary applications, including salads, desserts, and beverages, contributing both aesthetic appeal and potential health benefits. Some flowers, such as *Bauhinia purpurea* and *Pelargonium* spp., have been identified as particularly rich in cyanidin and malvidin derivatives [7, 41, 48].

The anthocyanin content of plant foods is influenced by multiple factors, including genetic background, growing conditions, and postharvest storage. Temperature, light exposure, and agricultural practices all contribute to variations in anthocyanin accumulation. Additionally, food processing and cooking methods can affect anthocyanin stability, with some studies showing reductions in anthocyanin content

TABLE 1.2 Main Plant Sources of Anthocyanins

FAMILY	*GENUS*	*SPECIES*	*CONCENTRATION (MG/100 G)*	*REFERENCES*
Rosaceae	*Rubus* L.	*Rubus fruticosus* Pollich	100–400	[49]
	Crataegus L.	*Crataegus pinnatifida* Bunge	1,300	[50]
	Rosa L.	*Rosa* ssp.	190–220	[51]
	Amygdalus L.	*Amygdalus persica* L.	10,500–21,600	[52]
Rhododendron	*Vaccinium*	*Vaccinium myrtillus* Linn.	300	[53]
	Rhododendron L.	*Rhododendron simsii* Planch.	260–950	[54]
Brassicaceae	*Brassica* L.	*Brassica oleracea* var. *capitata rubra*	180	[55]
		Brassica rapa var. *chinensis*	310	[56]
		Brassica juncea var. *tumida*	190	[57]
Compositae	*Chrysanthemum* L.	*Chrysanthemum* x *morifolium* Ramat.	4–1,100	[58]
	Cremanthodium Benth.	Red chicory	40	[59]
Caprifoliaceae	*Sambucus* L.	*Sambucus williamsii* Hance	63–430	[60]
Solanaceae	*Petunia* Juss.	*Petunia hybrida* (J. D. Hooker) *Vilmorin*	300	[61]
	Lycium L.	*Lycium ruthenicum* Murr.	560	[62]
Lamiaceae	*Thymus* L.	*Thymus mongolicus* Ronn.	0.1–14	[63]
Poaceae	*Oryza* L.	*Oryza sativa* L.	0.7–330	[64]
		Purple rice	26–250	[65]
	Zea L.	*Zea mays* L.	0.7–326	[64]
	Triticum L.	*Triticum aestivum* L.	0.7–328	[64]
Saxifragaceae	*Ribes*	*Ribes nigrum* L.	128–410	[66]
Grossulariaceae	*Ribes*	*Ribes rubrum* L.	7.5–7.8	[49]
Boraginaceae	*Borago* L.	*Borago officinalis* Linn.	7–260	[67]
Dioscoreaceae	*Dioscorea*	*Solanum tuberosum*	32–1,400	[68]
Moraceae	*Morus* Linn.	*Morus alba* L.	120	[69]
Vitiaceae	*Vitis* L.	*Vitis vinifera* L.	28–300	[70]
Myrtaceae	*Syzygium* Gaertn.	*Syzygium jambos* (L.) Alston	764	[71]
Garciniaaceae	*Garcinia* L.	*Garcinia mangostana* L.	7,370	[72]

due to thermal degradation, while others report increased bioavailability following certain treatments [41]. Table 1.2 summarizes some of the main plant sources of anthocyanins and their amount.

1.4.2 Absorption, Metabolism, and Bioavailability of Anthocyanins in Humans

The absorption, metabolism, and bioavailability of anthocyanins in humans are complex and influenced by multiple factors. Anthocyanins are considered to have relatively low bioavailability due to their rapid degradation in the gastrointestinal tract and extensive metabolism in the body. Studies have shown that anthocyanins are stable in the acidic conditions of the stomach, but their recovery after pancreatic digestion is significantly reduced, often below 30%. The bioavailability of anthocyanins depends on their chemical structure, particularly the degree of glycosylation and acylation. Nonacylated anthocyanins have been reported to be more efficiently absorbed than their acylated counterparts, as acylation can hinder their uptake through the intestinal epithelium [73, 74].

Anthocyanin absorption begins in the stomach and continues in the small intestine, where they are transported across gastric cells and enterocytes *via* glucose transporters, and sodium-dependent glucose transporters (SGLT1) and bilitranslocase, respectively. Their absorption at the stomach level in reported as being very low, so they proceed to the small intestine [75–77]. However, due to their instability in neutral pH conditions and susceptibility to enzymatic degradation, only a small proportion of intact anthocyanins enter systemic circulation. A significant portion undergoes modification by phase II metabolic enzymes, including UDP-glucuronosyltransferases, sulfotransferases, and catechol-*O*-methyltransferases, leading to the formation of glucuronidated, sulfated, and methylated derivatives. These metabolites exhibit altered biological activity and enhanced solubility, facilitating their excretion *via* urine and bile.

Studies have shown that anthocyanins exhibit a plasma half-life of approximately 2–3.3 h, with peak concentrations occurring between 1.5 and 4 h after ingestion. Despite relatively high plasma levels shortly after consumption, anthocyanins are rapidly cleared from circulation, contributing to their low systemic bioavailability. However, their metabolites may persist for longer durations, extending their physiological impact. Urinary excretion of anthocyanins is generally low, typically ranging between 0.004% and 0.2% of the ingested dose, suggesting extensive metabolism and tissue distribution [9, 74].

A significant proportion of unabsorbed anthocyanins reaches the colon, where they undergo microbial metabolism. The gut microbiota plays a crucial role in the degradation and transformation of anthocyanins into smaller phenolic acids, such as protocatechuic acid and vanillic acid, which can be absorbed and contribute to biological effects. Bacterial enzymes such as β-glucosidases hydrolyze anthocyanins into their aglycone forms, facilitating further breakdown into bioavailable phenolic metabolites. The composition of the gut microbiome strongly influences anthocyanin metabolism, and individual variations in microbial populations may impact the extent of anthocyanin bioavailability. The efficiency of anthocyanin absorption is also modulated by dietary components. The presence of alcohol has been reported to enhance anthocyanin absorption, likely by increasing membrane permeability, while high sugar intake may reduce bioavailability by competing for transport mechanisms. Additionally, food matrix effects play a role, as anthocyanins consumed in whole foods, such as berries, may exhibit different bioavailability compared to anthocyanins extracted and administered in isolation. Certain macronutrients, including dietary lipids and proteins, have been suggested to affect anthocyanin absorption by altering gastrointestinal conditions and interacting with anthocyanin molecules [78].

Anthocyanin bioavailability may vary based on genetic factors. Polymorphisms in genes encoding metabolic enzymes and transport proteins can affect individual differences in anthocyanin metabolism and systemic availability. Moreover, long-term consumption of anthocyanin-rich diets may influence gut microbiota composition, potentially enhancing microbe-mediated metabolism and increasing anthocyanin-derived metabolites in circulation.

1.4.3 Health Benefits: Antioxidant Activity, Anti-Inflammatory Properties, Cardiovascular Health, and Cognitive Function

Anthocyanins have gained considerable attention due to their wide-ranging health benefits, which include antioxidant activity, anti-inflammatory properties, cardiovascular protection, and cognitive function support. These naturally occurring pigments, found in various fruits, vegetables, and flowers, play a vital role in mitigating oxidative stress, reducing inflammation, and promoting overall well-being. Anthocyanins exhibit potent antioxidant activity, primarily through their ability to scavenge free radicals and reduce oxidative damage at the cellular level. Their structural composition, particularly the presence of hydroxyl groups, enables them to neutralize ROS, thereby protecting lipids, proteins, and DNA from oxidative harm. The antioxidant capacity of anthocyanins has been extensively studied, with evidence showing that acylated anthocyanins, such as those found in red cabbage, possess enhanced radical-scavenging abilities due to their specific chemical configurations. Additionally, anthocyanins contribute to cellular antioxidant defenses by modulating endogenous antioxidant enzymes, such as CAT and SOD, further enhancing their protective effects [9, 78–80].

Beyond their antioxidant properties, anthocyanins possess strong anti-inflammatory activity, which is crucial in preventing chronic diseases associated with inflammation. These bioactive compounds inhibit key pro-inflammatory mediators, including cytokines, cyclooxygenase-2 (COX-2), and nuclear factor kappa-B (NF-κB), thereby reducing inflammatory responses in various tissues. Studies have demonstrated that dietary anthocyanins can significantly suppress inflammation-induced oxidative stress and improve immune system regulation. Furthermore, anthocyanin-rich extracts have been shown to reduce inflammation in models of colitis and arthritis, suggesting their potential therapeutic role in inflammatory disorders. Clinical trials have supported these findings, with studies indicating that anthocyanin supplementation can significantly decrease inflammatory markers such as C-reactive protein and interleukin-6 in patients with metabolic syndrome [80].

Cardiovascular health is another major area where anthocyanins exert beneficial effects. These flavonoid compounds help maintain blood vessel integrity by promoting endothelial function and reducing arterial stiffness. Regular consumption of anthocyanin-rich foods has been associated with improved blood pressure regulation and a reduced risk of atherosclerosis by inhibiting the oxidation of low-density lipoproteins (LDL) and mitigating vascular inflammation. Clinical studies have reported that individuals with higher anthocyanin intake exhibit improved lipid profiles, with reductions in total cholesterol and triglycerides. Additionally, epidemiological studies suggest that individuals with higher dietary anthocyanin intake experience a significantly lower incidence of myocardial infarction and coronary artery disease [79, 80].

Cognitive function is another area where anthocyanins demonstrate promising health benefits. Research has shown that these compounds can cross the blood–brain barrier and exert neuroprotective effects by reducing oxidative stress and neuroinflammation, both of which are implicated in neurodegenerative diseases [80]. Animal and human studies have reported improvements in memory, learning, and overall cognitive function following dietary supplementation with anthocyanins. Clinical trials have indicated that anthocyanin-rich interventions, such as blueberry or cherry juice supplementation, enhance neuronal signaling, increase brain blood flow, and improve cognitive performance in aging populations. Longitudinal studies suggest that higher anthocyanin intake is associated with a slower rate of cognitive decline and a reduced risk of Alzheimer's disease in older adults.

1.5 STABILITY OF ANTHOCYANINS

The chemical structure of anthocyanins plays a pivotal role in determining their stability. As stated before, anthocyanins consist of an anthocyanidin aglycone linked to one or more sugar moieties, and often acylated with aromatic or aliphatic acids. The nature and position of substituents on the B-ring of the anthocyanidin core influence their susceptibility to degradation. Generally, methoxylated anthocyanidins (e.g., malvidin) exhibit greater stability than their hydroxylated counterparts (e.g., delphinidin). Also, glycosylation enhances water solubility and stabilizes the flavylium ion, whereas acylation confers further stabilization through intramolecular co-pigmentation and steric protection. Acylated anthocyanins, particularly those from purple sweet potato or red cabbage, are notably more resistant to thermal and oxidative degradation, as we will see in detail along this book.

Nevertheless, some key factors are well known to influence the overall stability of anthocyanins [22]. Anthocyanins are highly sensitive to pH variations, which modulate their structural forms and consequently their color and stability. In highly acidic environments (pH 1–3), anthocyanins exist predominantly in the flavylium cation form, exhibiting red coloration and relatively high stability. At moderately acidic to neutral pH (pH 4–6), structural transformations to hemiketal and chalcone forms occur, leading to discoloration and rapid degradation. At alkaline pH, anthocyanins form quinonoidal bases and anionic species, which are unstable and prone to further degradation. Thermal processing also is known to accelerate anthocyanin degradation through hydrolysis, cleavage of glycosidic bonds, and oxidation [81]. The extent of thermal degradation is influenced by the structure of the anthocyanin, the food matrix, and processing parameters such as time, temperature, and oxygen exposure.

Exposure to light, particularly UV radiation, leads to photodegradation of anthocyanins, resulting in color loss and reduced functionality [82]. Similarly, oxidative degradation in the presence of oxygen and pro-oxidants (e.g., metal ions or ascorbic acid) contributes to anthocyanin instability *via* mechanisms such as hydroxylation and ring cleavage. Polyphenol oxidase (PPO) and peroxidase (POD), enzymes present in plant tissues and during processing, catalyze oxidative reactions that lead to the breakdown of anthocyanins. These enzymatic reactions are particularly relevant in minimally processed fruits and vegetables and during storage, necessitating the use of enzyme inactivation steps such as blanching [83].

Anthocyanins can also form non-covalent complexes with other phenolic compounds, polysaccharides, proteins, and metal ions, which modulate their stability and color expression [82, 84]. Besides, co-pigmentation with flavonoids or phenolic acids stabilizes the chromophore *via* π–π stacking interactions, enhancing both color intensity and resistance to degradation. Interactions with dietary components or food matrix constituents may either protect anthocyanins or promote their breakdown, depending on the physicochemical environment. Solid and semisolid matrices typically provide better protection against degradation compared to aqueous solutions [85–87]. Moreover, encapsulation in biopolymers, liposomes, or emulsions has been shown to significantly enhance anthocyanin stability [88–91]. Finally, in the gastrointestinal tract, anthocyanins exhibit moderate stability in the acidic gastric environment but are rapidly degraded in the neutral to alkaline conditions of the small intestine, often through microbial metabolism. This instability limits their bioavailability, although their degradation products may still exert biological activity [92].

Overall, anthocyanins face significant challenges to maintaining the stability of their original chemical structure. However, as we will see along this book, acylated anthocyanins, as a reflex of plant biosynthetic pathways, acquired superior resistance to the majority of these factors, mainly due to their more complex structures and the possibility of protecting themselves throughout different intra- and intermolecular interactions. Therefore, we are about to embark on a journey exploring the different aspects of these particular kinds of anthocyanins.

REFERENCES

1. Cory, H., et al., The role of polyphenols in human health and food systems: a mini-review. *Frontiers in Nutrition*, 2018. **5**. https://doi.org/10.3389/fnut.2018.00087.
2. Pandey, K.B. and S.I. Rizvi, Plant polyphenols as dietary antioxidants in human health and disease. *Oxidative Medicine and Cellular Longevity*, 2009. **2**(5): p. 270–278.
3. Rackham, H., *Pliny the Elder: Naturalis Historia*. Harvard University Press, 1938. Book XXI, Chapter 22.
4. Wilms, S., *Shén Nóng Běncǎo Jīng: The Divine Farmer's Classic of Materia Medica*. Happy Goat Productions, 2017. 3rd Edition.
5. Goto, T. and T. Kondo, Structure and molecular stacking of anthocyanins – flower color variation. *Angewandte Chemie International Edition in English*, 1991. **30**(1): p. 17–33. https://doi.org/10.1002/anie.199100171.
6. Harborne, J.B., The chromatographic identification of anthocyanin pigments. *Journal of Chromatography A*, 1958. **1**: p. 473–488. https://doi.org/10.1016/S0021-9673(00)93449-6.
7. He, J., et al., Dietary polyglycosylated anthocyanins, the smart option? A comprehensive review on their health benefits and technological applications. *Comprehensive Reviews in Food Science and Food Safety*, 2022. **21**(4): p. 3096–3128. https://doi.org/10.1111/1541-4337.12970.
8. Smeriglio, A., et al., Chemistry, pharmacology and health benefits of anthocyanins. *Phytotherapy Research*, 2016. **30**(8): p. 1265–1286. https://doi.org/10.1002/ptr.5642.

9. Gowd, V., Z. Jia, and W. Chen, Anthocyanins as promising molecules and dietary bioactive components against diabetes – a review of recent advances. *Trends in Food Science & Technology*, 2017. **68**: p. 1–13.
10. Gould, K.S. and C. Lister, Chapter 8: Flavonoid functions in plants. In *Flavonoids: Chemistry, Biochemistry, and Applications*. Eds. Ø.M. Andersen and K.R. Markham. CRC Press, 2006: p. 397–441. https://doi.org/10.1201/9781420039443
11. Zufall, R.A. and M.D. Rausher, Genetic changes associated with floral adaptation restrict future evolutionary potential. *Nature*, 2004. **428**(6985): p. 847–850. https://doi.org/10.1038/nature02489.
12. Giurfa, M., et al., Colour preferences of flower-naive honeybees. *Journal of Comparative Physiology A*, 1995. **177**(3): p. 247–259. https://doi.org/10.1007/BF00192415.
13. Gould, K.S., J. McKelvie, and K.R. Markham, Do anthocyanins function as antioxidants in leaves? Imaging of H_2O_2 in red and green leaves after mechanical injury. *Plant, Cell & Environment*, 2002. **25**(10): p. 1261–1269. https://doi.org/10.1046/j.1365-3040.2002.00905.x.
14. Steyn, W.J., et al., Anthocyanins in vegetative tissues: a proposed unified function in photoprotection. *New Phytologist*, 2002. **155**(3): p. 349–361. https://doi.org/10.1046/j.1469-8137.2002.00482.x.
15. Schaberg, P.G., et al., Factors influencing red expression in autumn foliage of sugar maple trees. *Tree Physiology*, 2003. **23**(5): p. 325–333. https://doi.org/10.1093/treephys/23.5.325.
16. Khoo, H.E., et al., Anthocyanidins and anthocyanins: colored pigments as food, pharmaceutical ingredients, and the potential health benefits. *Food & Nutrition Research*, 2017. **61**(1): p. 1361779. https://doi.org/10.1080/16546628.2017.1361779.
17. Wang, L.-S. and G.D. Stoner, Anthocyanins and their role in cancer prevention. *Cancer Letters*, 2008. **269**(2): p. 281–290. https://doi.org/10.1016/j.canlet.2008.05.020.
18. de Pascual-Teresa, S. and M. Sanchez-Ballesta, Anthocyanins: from plant to health. *Phytochemistry Reviews*, 2008. **7**: p. 281–299. https://doi.org/10.1007/s11101-007-9074-0.
19. Giusti, M.M. and R.E. Wrolstad, Acylated anthocyanins from edible sources and their applications in food systems. *Biochemical Engineering Journal*, 2003. **14**(3): p. 217–225. https://doi.org/10.1016/S1369-703X(02)00221-8.
20. Okamura, M., et al., Crossbreeding of a metallic color carnation and diversification of the peculiar coloration by ion-beam irradiation. *Euphytica*, 2013. **191**: p. 45–56. https://doi.org/10.1007/s10681-012-0859-x.
21. Sasaki, N., et al., Recent advances in understanding the anthocyanin modification steps in carnation flowers. *Scientia Horticulturae*, 2013. **163**: p. 37–45. https://doi.org/10.1016/j.scienta.2013.07.029.
22. Pina, F., J. Oliveira, and V. de Freitas, Anthocyanins and derivatives are more than flavylium cations. *Tetrahedron*, 2015. **71**(20): p. 14. https://doi.org/10.1016/j.tet.2014.09.051.
23. Pina, F., Anthocyanins and related compounds: detecting the change of regime between rate control by hydration or tautomerization. *Dyes and Pigments*, 2014. **102**: p. 308–314. http://dx.doi.org/10.1016/j.dyepig.2013.10.033.
24. Zhao, X., et al., Intermolecular copigmentation between anthocyanidin-3,5-O-diglucosides and three phenolic compounds: insights from experimental and theoretical studies. *Food Chemistry Advances*, 2022. **1**: p. 100111. https://doi.org/10.1016/j.focha.2022.100111.
25. Trouillas, P., et al., Stabilizing and modulating color by copigmentation: insights from theory and experiment. *Chemical Reviews*, 2016. **116**(9): p. 4937–4982. https://doi.org/10.1021/acs.chemrev.5b00507.
26. Mazza, G. and R. Brouillard, The mechanism of co-pigmentation of anthocyanins in aqueous solutions. *Phytochemistry*, 1990. **29**(4): p. 6. https://doi.org/10.1016/0031-9422(90)85411-8.
27. Fernández-López, J.A., V. Fernández-Lledó, and J.M. Angosto, New insights into red plant pigments: more than just natural colorants. *RSC Advances*, 2020. **10**(41): p. 24669–24682. https://doi.org/10.1039/D0RA03514A.
28. Rein, M., *Copigmentation Reactions and Color Stability of Berry Anthocyanins*. Helsinki: Uniservity of Helsinki, 2005: p. 10–14.
29. Ionescu, C., A. Samide, and C. Tigae, Trend in detection of anthocyanins from fresh fruits and the influence of some factors on their stability impacting human health: kinetic study assisted by UV-Vis spectrophotometry. *Antioxidants (Basel)*, 2025. **14**(2). https://doi.org/10.3390/antiox14020227.
30. Sadilova, E., F.C. Stintzing, and R. Carle, Thermal degradation of acylated and nonacylated anthocyanins. *Journal of Food Science*, 2006. **71**(8): p. C504–C512. https://doi.org/10.1111/j.1750-3841.2006.00148.x.
31. Bąkowska, A., A.Z. Kucharska, and J. Oszmiański, The effects of heating, UV irradiation, and storage on stability of the anthocyanin–polyphenol copigment complex. *Food Chemistry*, 2003. **81**(3): p. 349–355. https://doi.org/10.1016/S0308-8146(02)00429-6.
32. Araya, V., et al. In silico analysis of the molecular interaction between anthocyanase, peroxidase and polyphenol oxidase with anthocyanins found in cranberries. *International Journal of Molecular Sciences*, 2024. **25**. https://doi.org/10.3390/ijms251910437.
33. Ziabakhsh Deylami, M., et al., Effect of blanching on enzyme activity, color changes, anthocyanin stability and extractability of mangosteen pericarp: a kinetic study. *Journal of Food Engineering*, 2016. **178**: p. 12–19. https://doi.org/10.1016/j.jfoodeng.2016.01.001.

34. Sigurdson, G.T., et al., Evaluating the role of metal ions in the bathochromic and hyperchromic responses of cyanidin derivatives in acidic and alkaline pH. *Food Chemistry*, 2016. **208**: p. 26–34. https://doi.org/10.1016/j.foodchem.2016.03.109.
35. Estévez, L., et al., First characterization of the formation of anthocyanin-Ge and anthocyanin-B complexes through UV–Vis spectroscopy and density functional theory quantum chemical calculations. *Journal of Agricultural and Food Chemistry*, 2021. **69**(4): p. 1272–1282. https://doi.org/10.1021/acs.jafc.0c06827.
36. Yan, J., et al., Pollinator responses to floral colour change, nectar and scent promote reproductive fitness in *Quisqualis indica* (Combretaceae). *Scientific Reports*, 2016. **6**(1): p. 24408. https://doi.org/10.1038/srep24408.
37. Trunschke, J., et al., Flower color evolution and the evidence of pollinator-mediated selection. *Frontiers in Plant Science*, 2021. **12**. https://doi.org/10.3389/fpls.2021.617851.
38. Galetti, M., et al., Functional extinction of birds drives rapid evolutionary changes in seed size. *Science*, 2013. **340**(6136): p. 1086–1090. https://doi.org/10.1126/science.1233774.
39. Schaefer, H.M., A. Valido, and P. Jordano, Birds see the true colours of fruits to live off the fat of the land. *Proceedings of the Royal Society B: Biological Sciences*, 2014. **281**(1777): p. 20132516. https://doi.org/10.1098/rspb.2013.2516.
40. Pereira, A.C., et al., Flower color change accelerated by bee pollination in *Tibouchina* (Melastomataceae). *Flora – Morphology, Distribution, Functional Ecology of Plants*, 2011. **206**(5): p. 491–497. https://doi.org/10.1016/j.flora.2011.01.004.
41. Castañeda-Ovando, A., et al., Chemical studies of anthocyanins: a review. *Food Chemistry*, 2009. **113**(4): p. 859–871. https://doi.org/10.1016/j.foodchem.2008.09.001.
42. Ahmed, N.U., et al., Anthocyanin biosynthesis for cold and freezing stress tolerance and desirable color in *Brassica rapa*. *Functional & Integrative Genomics*, 2015. **15**(4): p. 383–394. https://doi.org/10.1007/s10142-014-0427-7.
43. Qiu, Z., et al., The Tomato Hoffman's anthocyaninless gene encodes a bHLH transcription factor involved in anthocyanin biosynthesis that is developmentally regulated and induced by low temperatures. *PLoS One*, 2016. **11**(3): p. e0151067. https://doi.org/10.1371/journal.pone.0151067.
44. Vidana Gamage, G.C., Y.Y. Lim, and W.S. Choo, Sources and relative stabilities of acylated and nonacylated anthocyanins in beverage systems. *Journal of Food Science and Technology*, 2022. **59**(3): p. 831–845. https://doi.org/10.1007/s13197-021-05054-z.
45. You, Q., et al., Comparison of anthocyanins and phenolics in organically and conventionally grown blueberries in selected cultivars. *Food Chemistry*, 2011. **125**(1): p. 201–208. https://doi.org/10.1016/j.foodchem.2010.08.063.
46. Zhao, C.L., et al., Stability-increasing effects of anthocyanin glycosyl acylation. *Food Chemistry*, 2017. **214**: p. 119–128. https://doi.org/10.1016/j.foodchem.2016.07.073.
47. Shi, J., et al., Insight into the pigmented anthocyanins and the major potential co-pigmented flavonoids in purple-coloured leaf teas. *Food Chemistry*, 2021. **363**: p. 130278. https://doi.org/10.1016/j.foodchem.2021.130278.
48. Teixeira, M., et al., Anthocyanin-rich edible flowers, current understanding of a potential new trend in dietary patterns. *Trends in Food Science & Technology*, 2023. **138**: p. 708–725. https://doi.org/10.1016/j.tifs.2023.07.010.
49. Pantelidis, G.E., et al., Antioxidant capacity, phenol, anthocyanin and ascorbic acid contents in raspberries, blackberries, red currants, gooseberries and Cornelian cherries. *Food Chemistry*, 2007. **102**(3): p. 777–783. https://doi.org/10.1016/j.foodchem.2006.06.021.
50. Liu, S., et al., Effects of pretreatments on anthocyanin composition, phenolics contents and antioxidant capacities during fermentation of hawthorn (*Crataegus pinnatifida*) drink. *Food Chemistry*, 2016. **212**: p. 87–95. https://doi.org/10.1016/j.foodchem.2016.05.146.
51. Qiu, L., et al., Effects of infrared freeze drying on volatile profile, FTIR molecular structure profile and nutritional properties of edible rose flower (*Rosa rugosa* flower). *Journal of the Science of Food and Agriculture*, 2020. **100**(13): p. 4791–4800. https://doi.org/10.1002/jsfa.10538.
52. Cao, K., et al., Transcriptome analysis reveals novel genes involved in anthocyanin biosynthesis in the flesh of peach. *Plant Physiology and Biochemistry*, 2018. **123**: p. 94–102. https://doi.org/10.1016/j.plaphy.2017.12.005.
53. Prior, R.L., et al., Antioxidant capacity as influenced by total phenolic and anthocyanin content, maturity, and variety of vaccinium species. *Journal of Agricultural and Food Chemistry*, 1998. **46**(7): p. 2686–2693. https://doi.org/10.1021/jf980145d.
54. Du, H., et al., Characterisation of flower colouration in 30 Rhododendron species via anthocyanin and flavonol identification and quantitative traits. *Plant Biology (Stuttgart)*, 2018. **20**(1): p. 121–129. https://doi.org/10.1111/plb.12649.
55. He, Q., Z. Zhang, and L. Zhang, Anthocyanin accumulation, antioxidant ability and stability, and a transcriptional analysis of anthocyanin biosynthesis in purple heading Chinese cabbage (*Brassica rapa* L. ssp. *pekinensis*). *Journal of Agricultural and Food Chemistry*, 2016. **64**(1): p. 132–145. https://doi.org/10.1021/acs.jafc.5b04674.

56. Zhang, Y., et al., Anthocyanin accumulation and transcriptional regulation of anthocyanin biosynthesis in purple bok choy (*Brassica rapa* var. *chinensis*). *Journal of Agricultural and Food Chemistry*, 2014. **62**(51): p. 12366–12376. https://doi.org/10.1021/jf503453e.
57. Xie, Q., et al., Accumulation and molecular regulation of anthocyanin in purple tumorous stem mustard (*Brassica juncea* var. *tumida* Tsen et Lee). *Journal of Agricultural and Food Chemistry*, 2014. **62**(31): p. 7813–7821. https://doi.org/10.1021/jf501790a.
58. Park, C.H., et al., Anthocyanin and carotenoid contents in different cultivars of chrysanthemum (*Dendranthema grandiflorum* Ramat.) flower. *Molecules*, 2015. **20**(6): p. 11090–11102. https://doi.org/10.3390/molecules200611090.
59. Frond, A.D., et al., Phytochemical characterization of five edible purple-reddish vegetables: anthocyanins, flavonoids, and phenolic acid derivatives. *Molecules*, 2019. **24**. https://doi.org/10.3390/molecules24081536.
60. Stuppner, S., et al., Near-infrared spectroscopy as a rapid screening method for the determination of total anthocyanin content in sambucus fructus. *Sensors*, 2020. **20**. https://doi.org/10.3390/s20174983.
61. Albert, N.W., et al., Light-induced vegetative anthocyanin pigmentation in *Petunia*. *Journal of Experimental Botany*, 2009. **60**(7): p. 2191–2022. https://doi.org/10.1093/jxb/erp097.
62. Qin, X., et al., Maillard-reacted whey protein isolates enhance thermal stability of anthocyanins over a wide pH range. *Journal of Agricultural and Food Chemistry*, 2018. **66**(36): p. 9556–9564. https://doi.org/10.1021/acs.jafc.8b03125.
63. Rababah, T.M., et al., Optimization of extraction conditions of total phenolics, antioxidant activities, and anthocyanin of oregano, thyme, terebinth, and pomegranate. *Journal of Food Science*, 2010. **75**(7): p. C626–C632. https://doi.org/10.1111/j.1750-3841.2010.01756.x.
64. Abdel-Aal, E.-S.M., J.C. Young, and I. Rabalski, Anthocyanin composition in black, blue, pink, purple, and red cereal grains. *Journal of Agricultural and Food Chemistry*, 2006. **54**(13): p. 4696–4704. https://doi.org/10.1021/jf0606609.
65. Yamuangmorn, S. and U.T.C. Prom, The potential of high-anthocyanin purple rice as a functional ingredient in human health. *Antioxidants (Basel)*, 2021. **10**(6). https://doi.org/10.3390/antiox10060833.
66. Moyer, R.A., et al., Anthocyanins, phenolics, and antioxidant capacity in diverse small fruits: *Vaccinium*, *Rubus*, and *Ribes*. *Journal of Agricultural and Food Chemistry*, 2002. **50**(3): p. 519–525. https://doi.org/10.1021/jf011062r.
67. Zannou, O., et al., Extraction of anthocyanins from borage (*Echium amoenum*) flowers using choline chloride and a glycerol-based, deep eutectic solvent: optimization, antioxidant activity, and in vitro bioavailability. *Molecules*, 2021. **27**(1). https://doi.org/10.3390/molecules27010134.
68. Xu, J., et al., Characterisation and stability of anthocyanins in purple-fleshed sweet potato P40. *Food Chemistry*, 2015. **186**: p. 90–96. https://doi.org/10.1016/j.foodchem.2014.08.123.
69. Xie, J., et al., Green extraction of mulberry anthocyanin with improved stability using β-cyclodextrin. *Journal of the Science of Food and Agriculture*, 2019. **99**(5): p. 2494–2503. https://doi.org/10.1002/jsfa.9459.
70. Camire, M.E., et al., Blueberry and grape anthocyanins as breakfast cereal colorants. *Journal of Food Science*, 2002. **67**(1): p. 438–441. https://doi.org/10.1111/j.1365-2621.2002.tb11425.x.
71. Chaudhary, B. and K. Mukhopadhyay, Solvent optimization for anthocyanin extraction from *Syzygium cumini* L. Skeels using response surface methodology. *International Journal of Food Sciences and Nutrition*, 2013. **64**(3): p. 363–371. https://doi.org/10.3109/09637486.2012.738647.
72. Piacham, T., C. Isarankura-Na-Ayudhya, and V. Prachayasittikul, Quercetin-imprinted polymer for anthocyanin extraction from mangosteen pericarp. *Materials Science and Engineering: C*, 2015. **51**: p. 127–131. https://doi.org/10.1016/j.msec.2015.02.051.
73. Kong, J.M., et al., Analysis and biological activities of anthocyanins. *Phytochemistry*, 2003. **64**(5): p. 923–933.
74. Burton-Freeman, B., et al., A selective role of dietary anthocyanins and flavan-3-ols in reducing the risk of type 2 diabetes mellitus: a review of recent evidence. *Nutrients*, 2019. **11**(4). https://doi.org/10.3390/nu11040841.
75. Oliveira, H., et al., Experimental and theoretical data on the mechanism by which red wine anthocyanins are transported through a human MKN-28 gastric cell model. *Journal of Agricultural and Food Chemistry*, 2015. **63**(35): p. 7685–7692. https://doi.org/10.1021/acs.jafc.5b00412.
76. Oliveira, H., et al., GLUT1 and GLUT3 involvement in anthocyanin gastric transport-nanobased targeted approach. *Scientific Reports*, 2019. **9**(789): p. 1–14. https://doi.org/10.1038/s41598-018-37283-2.
77. Oliveira, H., et al., Comparison of the in vitro gastrointestinal bioavailability of acylated and non-acylated anthocyanins: purple-fleshed sweet potato vs red wine. *Food Chemistry*, 2019. **276**: p. 410–418. https://doi.org/10.1016/j.foodchem.2018.09.159.
78. Reis, J.F., et al., Action mechanism and cardiovascular effect of anthocyanins: a systematic review of animal and human studies. *Journal of Translational Medicine*, 2016. **14**: p. 315.
79. Manolescu, B.N., et al., Dietary anthocyanins and stroke: a review of pharmacokinetic and pharmacodynamic studies. *Nutrients*, 2019. **11**(7). https://doi.org/10.3390/nu11071479.

80. Fernandes, I., et al., Bioavailability of anthocyanins and derivatives. *Journal of Functional Foods*, 2014. **7**(Supplement C): p. 54–66. https://doi.org/10.1016/j.jff.2013.05.010.
81. Patras, A., et al., Effect of thermal processing on anthocyanin stability in foods; mechanisms and kinetics of degradation. *Trends in Food Science & Technology*, 2010. **21**(1): p. 3–11. https://doi.org/10.1016/j.tifs.2009.07.004.
82. Xue, H., et al., Factors affecting the stability of anthocyanins and strategies for improving their stability: a review. *Food Chemistry: X*, 2024. **24**: p. 101883. https://doi.org/10.1016/j.fochx.2024.101883.
83. Zhang, Z., et al., Role of peroxidase in anthocyanin degradation in litchi fruit pericarp. *Food Chemistry*, 2005. **90**(1): p. 47–52. https://doi.org/10.1016/j.foodchem.2004.03.023.
84. Gençdağ, E., et al., Copigmentation and stabilization of anthocyanins using organic molecules and encapsulation techniques. *Current Plant Biology*, 2022. **29**: p. 100238. https://doi.org/10.1016/j.cpb.2022.100238.
85. Fernandes, A., et al., Impact of grape pectic polysaccharides on anthocyanins thermostability. *Carbohydrate Polymers*, 2020. **239**: p. 116240. https://doi.org/10.1016/j.carbpol.2020.116240.
86. Teixeira, M., et al. First insights on the bioaccessibility and absorption of anthocyanins from edible flowers: wild pansy, cosmos, and cornflower. *Pharmaceuticals*, 2024. **17**. https://doi.org/10.3390/ph17020191.
87. Li, A., et al., Research advances of purple sweet potato anthocyanins: extraction, identification, stability, bioactivity, application, and biotransformation. *Molecules*, 2019. **24**: p. 3816. https://doi.org/10.3390/molecules24213816
88. Machado, M.H., et al., Microencapsulation by spray drying of red cabbage anthocyanin-rich extract for the production of a natural food colorant. *Biocatalysis and Agricultural Biotechnology*, 2022. **39**: p. 102287. https://doi.org/10.1016/j.bcab.2022.102287.
89. Zhang, R., et al., Preparation of nanoliposomes loaded with anthocyanins from grape skin extracts: stability, gastric absorption and antiproliferative properties. *Food & Function*, 2022. **13**(21): p. 10912–10922. https://doi.org/10.1039/D2FO02008D.
90. Tao, W., et al., Tailored zein-polysaccharide nanoparticles for anthocyanin encapsulation: insights into preparation and characterization. *Food Hydrocolloids*, 2025. **166**: p. 111365. https://doi.org/10.1016/j.foodhyd.2025.111365.
91. Jaouhari, Y., et al., Encapsulation of anthocyanins: a key for improvement of the gut-metabolism health axis? *Food Research International*, 2025. **221**: p. 117380. https://doi.org/10.1016/j.foodres.2025.117380.
92. Dima, C., et al., Bioavailability and bioaccessibility of food bioactive compounds; overview and assessment by in vitro methods. *Comprehensive Reviews in Food Science and Food Safety*, 2020. **19**(6): p. 2862–2884. https://doi.org/10.1111/1541-4337.12623.

Biosynthesis of Acylated Anthocyanins

2

Wilson Vasquez-Rojas, Sheyla S. Gutierrez-Asencios, Matteo Bordiga, Nataly Bernuy-Osorio, and Vincenzo Disca

2.1 INTRODUCTION

Anthocyanins are phytochemicals that have garnered significant interest from the scientific and industry sectors due to their properties linked to their color and potential biological properties. Extensive studies have reported their potential application as food colorants, food packaging materials, and health-promoting foods (Mozos *et al.*, 2021; Sigurdson *et al.*, 2017; M. Xu et al., 2024). Anthocyanins constitute a group of compounds with high structural versatility, determined by variations in the number and position of hydroxyl, methoxyl, glycosyl, and acyl substituents. These structural modifications determine their physicochemical properties, and their absorption and metabolism processes. Although their recognized biological effects, color intensity, and natural origin support their incorporation into the food industry as bioactive compounds and natural colorants, their large-scale application is restricted by the chemical instability that characterizes these molecules (Jokioja *et al.*, 2021). Anthocyanins can be affected by several factors (such as pH, temperature, and light exposure), which can promote their instability, making them reactive (even more so than other flavonoids) due to their heterocyclic ring C (Dangles and Fenger, 2018).

Acylated anthocyanins represent a promising alternative to nonacylated anthocyanins, as they provide greater structural stabilization, which translates into greater color stability, modifications in absorption, changes in bioavailability, greater *in vivo* stability, and variations in their degradation at the colonic level (Jokioja *et al.*, 2021). This property occurs mainly through the following mechanisms: physicochemical (related to low polarity and steric hindrance of acyl groups), stereochemical (intramolecular co-pigmentation, a portion of the same anthocyanin molecule acts as a co-pigment), photochemical (absorption and transfer of light energy from acyl groups with conjugated double bonds), biochemical (combination of membrane transporters induced by acyl groups and inhibitory effects of acylated anthocyanins on digestive enzymes in animals), or environmental (acidifying capacity of acyl groups possessing a free carboxyl group) (Zhao *et al.*, 2017). In addition, acylation with different phenolic acids provides specific characteristics and advantages related to tolerance to pH variations, thermal resistance, light sensitivity, and overall stability of the compound (J. Xu *et al.*, 2015).

DOI: 10.1201/9781003519591-3

Acylated anthocyanins occur naturally in numerous plant species and are found in all plant organs, including roots, stems, flowers, and fruits (Luo *et al.*, 2022). Potential sources that naturally contain acylated anthocyanins include red radish, purple carrot, red cabbage, purple potato, black carrot, and purple sweet potato; among these, the last one mostly contains peonidin and cyanidin (acylated with caffeic acid, ferulic acid, and *p*-hydroxybenzoic acid) (Jokioja *et al.*, 2021; Montilla *et al.*, 2011; Truong *et al.*, 2010). Acylated anthocyanins from purple sweet potato can provide stronger dyeing power to beverages and other foods, and better stability than other anthocyanin pigments from blackberry, grape peel, etc. (Li *et al.*, 2019). Also, they are considered a model food with a high content of acylated anthocyanins, which enhances their industrialization and the nutritional relevance of these compounds. On the other hand, red cabbage, also known as purple cabbage, is an edible source of natural pigments with a high anthocyanin content (≥10 g/kg dry matter), mainly containing cyanidin-3-diglucoside-5-glucoside derivatives in nonacylated or acylated forms. The acyl groups contained in this source are aromatic acids, mainly *p*-coumaric, sinapic, and ferulic acids (Ghareaghajlou *et al.*, 2021).

In order to better utilize acylated anthocyanins, there is a need for a comprehensive understanding of their formation from the perspectives of biosynthesis and chemical synthesis; therefore, this and other related topics are discussed in this book.

2.2 CORE BIOSYNTHETIC PATHWAYS OF ANTHOCYANINS

Anthocyanins are synthesized in the cytosol of plant cells through the phenylpropanoid pathway, a flavonoid biosynthetic pathway, whose central pathway includes key steps and enzymes leading to the formation of six common anthocyanidins, which are unmodified aglycone forms from which more than 90% of natural anthocyanins are derived following glycosylation (e.g., cyanidin-3-*O*-glucoside), acylation (e.g., malvidin-3-*O*-(6-acetylglucoside)), or a combination of both. The common anthocyanidins include pelargonidin (orange-red), cyanidin (purple-red), delphinidin (violet-blue), peonidin (red), petunidin (blue-purple), and malvidin (purple-blue, very stable, typical of grapes and red wine) (Sun *et al.*, 2021). The central biosynthetic pathway of anthocyanins comprises the several key steps and involved enzymes, as shown in Figure 2.1.

2.2.1 Phenylpropanoid Pathway

Phenylpropanoids are plant secondary metabolites, which include flavonoids (anthocyanin, proanthocyanidin, flavonols, etc.), phenolic acids, stilbenes, and coumarins. Their bioconversion begins in the phenylpropanoid pathway with amino acid phenylalanine as the precursor, a product of the shikimate pathway (Kumar *et al.*, 2023). In plants, phenylalanine is considered a "bridge" due to its ability to connect primary metabolism with secondary phenolic metabolism (Kumari *et al.*, 2023).

The enzyme phenylalanine ammonia-lyase (PAL) removes the amino group from phenylalanine, forming cinnamic acid. This initial step in the phenylpropanoid pathway is considered a rate-limiting step in the flow toward flavonoids and anthocyanins, thus regulating the amount of phenolic compounds produced because PAL activity responds to environmental signals such as light, stress, and wounds (Kaur *et al.*, 2025). Cinnamic acid is subsequently hydroxylated in the endoplasmic reticulum by cinnamate 4-hydroxylase (C4H), a cytochrome P450-dependent monooxygenase, which introduces a hydroxyl group at position 4 of the aromatic ring to produce *p*-coumarate (Anwar *et al.*, 2021).

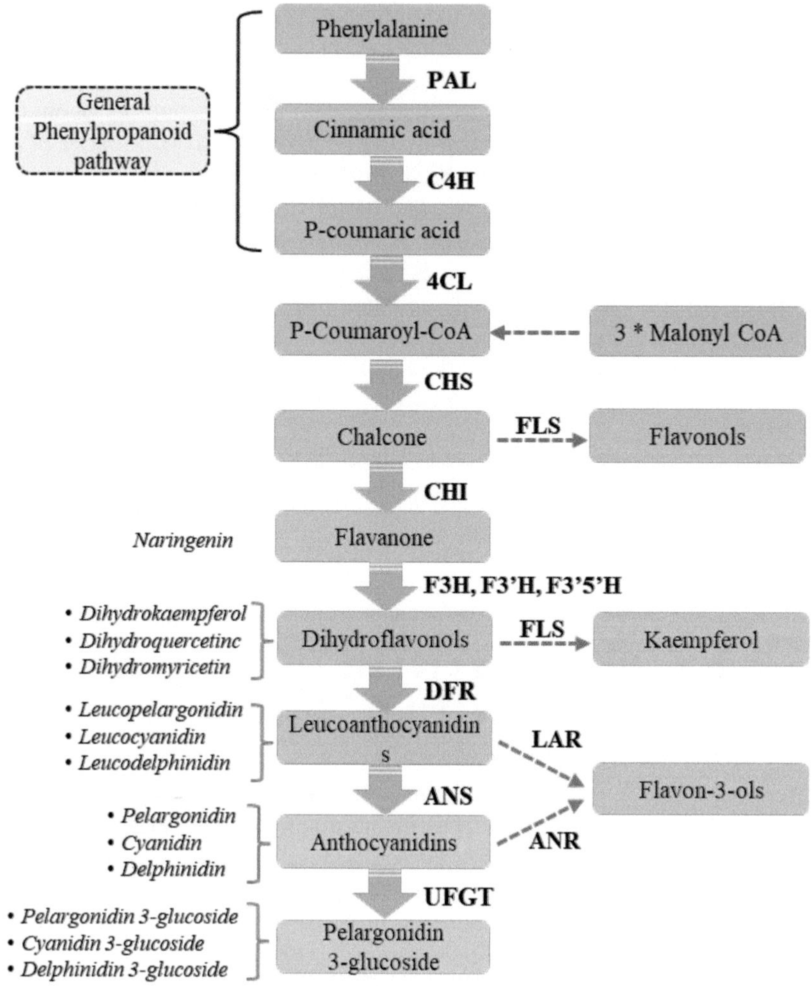

FIGURE 2.1 Biosynthetic pathways of anthocyanins. PAL, phenylalanine ammonia-lyase; C4H, cinnamate 4-hydroxylase; 4CL, *p*-coumaroyl coenzyme A ligase; CHS, chalcone synthase; CHI, chalcone isomerase; F3H, flavanone 3-hydroxylase; F3'H, flavonoid 3'-hydroxylase; F3'5'H, flavonoid 3',5'-hydroxylase; DFR, dihydroflavonol 4-reductase; FLS, flavonol synthase; ANS, anthocyanidin synthase; LAR, leucoanthocyanidin reductase; ANR, anthocyanin reductase; UFGT, flavonoid 3-*O*-glucosyltransferase; asterisk indicates multiplication.

2.2.2 Flavonoid Skeleton Formation (CHS, CHI → Naringenin)

The interest is to form the basic skeleton of flavonoids – this is a key step in the biosynthesis of anthocyanins and other phenolic compounds, where the substrates (*p*-coumaroyl-CoA from the phenylpropanoid pathway plus 3-malonyl-CoA units) and two essential enzymes participate. Chalcone synthase (CHS) catalyzes the synthesis of chalcone by polyketide reactions to form naringenin chalcone as an immediate precursor; this CHS has multiple isoforms. Then chalcone isomerase (CHI) catalyzes the intramolecular cyclization of chalcone, closing the C ring and generating (2S)-naringenin, a flavanone considered a limiting substrate of the biosynthetic pathway (Sunil and Shetty, 2022). When present in adequate concentrations, naringenin favors the metabolic flow toward anthocyanins (Jiang *et al.*, 2023). Furthermore, the CHI enzyme has four functional types: those capable of producing naringenin (type I) and isoflavonoids (type II), and those optimizing performance (type III and IV, regulators) (Liu *et al.*, 2021; Zhu *et al.*, 2021). These enzymes tend to organize into complexes (metabolons), which favor an efficient and controlled metabolic flow (Wang J. *et al.*, 2022).

2.2.3 Modification to Anthocyanidins (F3H, F3′H/F3′5′H, DFR, and ANS)

In this step, naringenin (flavanone) is hydroxylated at position 3 by the enzyme flavanone 3-hydroxylase (F3H), resulting in dihydrokaempferol (dihydroflavonol). Subsequently, the enzymes flavonoid 3′-hydroxylase (F3′H) and flavonoid 3′,5′-hydroxylase (F3′5′H) act mainly on the dihydrokaempferol, by hydroxylation of the B ring, using it as a substrate to generate dihydroquercetin and dihydromyricetin, respectively (Kumar *et al.*, 2023; Sunil and Shetty, 2022). Here, flavonol synthase (FLS) can direct flavonol synthesis by introducing a double bond between carbons 2 and 3 of dihydroflavonols, resulting in compounds such as kaempferol, quercetin, and myricetin (Deng and Lu, 2017). Then, dihydroflavonols are subjected to the action of dihydroflavonol 4-reductase (DFR), which transforms them into 3,4-*cis*-leucoanthocyanidins, colorless metabolites that act as immediate precursors of pigments. Subsequently, the enzyme anthocyanidin synthase (ANS) catalyzes the oxidation of colorless leucoanthocyanidins, generating colored anthocyanidins cyanidin, pelargonidin, and delphinidin (most commonly explored), which are responsible for the color variation in plant tissues (Kumar *et al.*, 2023; Sunil and Shetty, 2022). Additionally, the product of DFR and ANS can also be catalyzed by leucoanthocyanidin reductase (LAR) and anthocyanin reductase (ANR), respectively, to produce flavan-3-ols, substrates for the biosynthesis of proanthocyanidin, also called condensed tannins, such as (2R, 3R)-*cis*-flavan-3-ols (He *et al.*, 2010; Sunil and Shetty, 2022). In grapes, there are mainly six types of anthocyanidins responsible for their coloration: pelargonidin, cyanidin, delphinidin, peonidin, petunidin, and malvidin. Among these, malvidin is usually the predominant form in most red grape varieties, giving them intense and stable colors. Likewise, some of these molecules, such as delphinidin (3,5,7,3′,4′,5′-hexahydroxyflavyl), have a higher degree of hydroxylation in ring B, which directly influences the intensity and variability of the coloration (He *et al.*, 2010).

2.2.4 Glycosylation (UFGT → Anthocyanin-3-*O*-Glucosides)

Following the biosynthesis of anthocyanidins, unstable compounds responsible for coloration, a series of essential modifications occur in the cytosol, such as glycosylation, methylation, and acylation, which allow for their stabilization and subsequent accumulation in the vacuole. These transformations not only increase the solubility and stability of anthocyanins but also contribute significantly to their structural and functional diversity. In particular, glycosylation plays a central role, as it facilitates the overproduction of stable forms of anthocyanins by converting reactive aglycones into glycosylated derivatives that are more resistant to pH and degradation (Sunil and Shetty, 2022).

Glycosylation is catalyzed by family 1 glycosyltransferases (UGTs), which use activated sugars such as UDP-glucose or UDP-glucuronic acid as glycosyl group donors. In particular, the enzyme 3-*O*-glycosyltransferase (A3GT), a type of UDP-glucose:flavonoid 3-*O*-glucosyltransferase (UFGT), participates in the transfer of a glycosyl group to the 3-OH position of anthocyanidins, forming stable and soluble anthocyanins, such as cyanidin-3-*O*-glucoside (C3G), delphinidin-3-*O*-glucoside (D3G), and pelargonidin-3-*O*-glucoside (P3G) (Sasaki *et al.*, 2014; Yan *et al.*, 2021; Sunil and Shetty, 2022). After the glycosylation process, anthocyanins are directed to the vacuole by transport mechanisms mediated by specialized proteins, members of the glutathione-S-transferase (GST) families and transporters associated with the expulsion of multiple compounds (Multidrug and Toxic Compound Extrusion, MATE), which act as key elements in the recognition and translocation of these molecules, ensuring their proper sequestration and accumulation in the vacuolar compartment (Zhao *et al.*, 2022).

2.3 STRUCTURAL DIVERSITY AND CLASSIFICATION OF ACYLATED ANTHOCYANINS

Acylated anthocyanins are a structurally modified subclass of glycosylated anthocyanins, in which one or more hydroxyl groups of the sugar residues have been esterified with organic acids. This complex natural modification is part of the natural metabolism of numerous plant species and occurs during the biosynthesis of these bioactive compounds. Acylated anthocyanins accumulate in the roots, stems, flowers, and fruits, where they play key roles in pigmentation, defense, and adaptation to the environment (Luo *et al.*, 2022). Chemical and/or enzymatic acylation *in vitro* is also feasible and, like natural acylation, confers greater chemical and thermal stability, solubility in organic solvents, and better bioactivity to acylated anthocyanins compared to nonacylated anthocyanins (Luo *et al.*, 2022; Wang *et al.*, 2022; Yañez-Apam *et al.*, 2023).

The classification of acylated anthocyanins is based on the number of acyl groups attached to their structure, and they are distinguished into four categories: monoacylated, diacylated, triacylated, and tetraacylated. The latter three are commonly grouped under the term polyacylated anthocyanins due to the presence of multiple acyl substituents. Suitable sources of anthocyanins that mainly contain monoacylated anthocyanins include red cabbage, purple corn, black carrots, and other vegetables and berries, while flowers and leaves mainly contain tri- and tetraacylated anthocyanins (Table 2.1). Some foods even contain a mixture of monoacylated and diacylated anthocyanins, or triacylated and tetraacylated anthocyanins, such as red radishes and red-purple flowers, respectively (Vidana Gamage *et al.*, 2022).

Several studies maintain that acylation of anthocyanins is a determining factor that improves the color and stability of these soluble pigments responsible for the red, purple, and blue hues in numerous foods. Acylation is a process that occurs in the flavylium ring (the chromophoric part that gives the red–purple–blue color) of anthocyanins, which is highly sensitive to variations in critical environmental parameters such as pH, temperature, and light exposure (Giusti and Wrolstad, 2003; Cai *et al.*, 2022). In this ring, aromatic acyl groups (e.g., coumaroyl, feruloyl, and caffeoyl) are stacked, generating the different types of polyacylated anthocyanins (di-, tri-, or multiple acyl groups). These rings are arranged

TABLE 2.1 Examples of Acylated Anthocyanins in Different Plants

TYPE	*ACYLATED ANTHOCYANIN*	*FOOD SOURCE*	*REFERENCES*
Monoacylated	Cyanidin-3-(6′-malonylglucoside), pelargonidin-3-(6′-malonylglucoside)	Purple corn	Chatham *et al.* (2020)
Monoacylated	Petunidin-coumaryl-rutinoside-glucoside, peonidin-coumaryl-rutinoside-glucoside, and petunidin-caffeoyl-rutinoside glucoside	Purple potatoes (*Solanum tuberosum* 'Synkeä Sakari')	Chen *et al.* (2020)
Monoacylated and diacylated	Peonidin-3- (6′-hydroxybenzoyl)-sophoroside-5-glucoside and peonidin-3- (6′-hydroxybenzoyl-6″-caffeoyl)-sophoroside-5-glucoside	Purple-fleshed sweet potato	Oliviera *et al.* (2019)
Diacylated	Pelargonidin-3-(*p*-coumaroyl)diglucoside-5-(-malonyl)glucoside and pelargonidin-3- (feruloyl) diglucoside-5- (malonyl)glucoside	Red radish	Li *et al.* (2020)
Diacylated	Cyanidin-derivatives glucosylated to varying degrees and acylated with *p*-coumaryl, caffeoyl, feruloyl, or sinapoyl groups	Red cabbage	Strauch *et al.* (2019)
Triacylated	Malonylated delphinidin 3,3′,5′-triglucosides having 3′,5′-side chains with alternative d-glucose and *p*-coumaric acid units	*Clitoria ternatea* petals	Jeyaraj *et al.* (2021)

around the central flavylium ring, generating a compact sandwich-like structure capable of changing the distribution of electrons, altering the interaction of the molecule with light-promoting color intensification (Trouillas *et al.*, 2016; Luo *et al.*, 2022; Cheng *et al.*, 2023). This generates hydrophobicity and light and heat resistance of the pigments, increasing their intensity and stability, and prolonging their color half-life.

The conjugated double bonds and –OH groups in the aromatic acyl groups of anthocyanins are capable of chelating metals. Even greater acylations will increase the number of sites available for chelation, meaning these acylated pigments will have greater antioxidant capacity and stability against metallic pro-oxidants such as Fe^{2+} and Cu^{2+} (Fenger *et al.*, 2019).

These supramolecular modifications occur in the tri- and tetra-acylated anthocyanins of purple potatoes and red cabbage, expressed in highly stable and resistant purple-blue pigments, which are used as high-performance natural colorants in products for the food industry (Fenger *et al.*, 2021). Experimental evidence obtained from plant matrices such as purple sweet potato, purple carrot, red radish, and *Clitoria ternatea* flowers has shown that extracts rich in acylated anthocyanins, especially the diacylated and polyacylated forms, retain their color intensity for extended periods under storage, heat, or lighting conditions. Likewise, comparative studies with nonacylated sources, such as red grapes, have confirmed a greater susceptibility to color loss in polyacylated forms, which positions acylated anthocyanins as superior compounds for application as natural colorants with high resistance to degradation during storage and processing (Yañez-Apam *et al.*, 2023).

2.4 ACYLATION OF ANTHOCYANINS: MECHANISMS, ENZYMES, AND REGULATION

2.4.1 Types of Acyl Groups

Anthocyanin glycosyl acylation refers to the esterification of hydroxyl groups on the sugar moieties of anthocyanins with aliphatic or aromatic acids (Zhao *et al.*, 2017). Common aromatic acyl donors include hydroxycinnamic acids such as *p*-coumaric, caffeic, ferulic, and sinapic acids, while aliphatic donors include dicarboxylic acids such as malic or acetic acid. Plant species differ in the degree and type of acylation, and recently, Vidana Gamage and colleagues classified major anthocyanin sources by acylation level: fruits like elderberry, blackberry, and blackcurrant mainly contain nonacylated anthocyanins, whereas vegetables such as red radish, purple corn, and black carrot yield primarily monoacylated pigments (Vidana Gamage *et al.*, 2022). Some vegetables (e.g., red cabbage and purple sweet potato) accumulate both mono- and di-acylated forms. Highly acylated (tri- or tetra-) anthocyanins are mostly found in flowers and are associated with very high stability (Vidana Gamage *et al.*, 2022).

2.4.2 Key Enzymes Involved: BAHD and SCPL Acyltransferases

Anthocyanin acylation in plants is catalyzed by two distinct families of anthocyanin acyltransferases (AATs). The first group comprises cytosolic BAHD (named after benzoic acid, anthocyanin, HCAA synthase, and deacetylvindoline 4-*O*-acetyltransferase) AATs, which utilize acyl-CoA thioesters – such as malonyl-CoA, acetyl-CoA, and cinnamoyl-CoAs – as activated acyl donors. These BAHD AATs share a conserved His–X–X–X–Asp–Gly catalytic motif and exhibit high substrate specificity toward both acyl donors and anthocyanin acceptors (Nakayama *et al.*, 2003). Complementing this cytosolic pathway, a second class of AATs – vacuolar serine carboxypeptidase-like (SCPL) acyltransferases – uses acyl-glucose esters (e.g., sinapoylglucose, feruloylglucose, and malylglucose) as acyl donors. These enzymes function within the vacuole and represent a distinct mechanism for anthocyanin modification (Sasaki *et al.*, 2014).

2.4.3 Acyl Donors and Regiospecificity

BAHD AATs typically accept acyl-CoA thioesters: "aliphatic" AATs use donors like malonyl-CoA or acetyl-CoA, whereas "aromatic" AATs use hydroxycinnamoyl-CoAs (*p*-coumaroyl-, caffeoyl-, and cinnamoyl-CoA) (Sasaki *et al.*, 2014; Zhao *et al.*, 2017). In contrast, SCPL AATs use sugar-esters such as sinapoylglucose or feruloylglucose as activated donors. The acyl group from either class of enzyme is transferred to the hydroxyl group on the anthocyanin sugar moiety. In nearly all cases studied, this is the 6-OH of the C3-linked glucose or galactose (Teng *et al.*, 2022; Yan *et al.*, 2016). For example, acylation of bilberry anthocyanins with cinnamic acids occurred at the 6-OH of glucoside/galactoside and the 5-OH of arabinose (Cai *et al.*, 2020).

2.4.4 Transcriptional Control

AAT genes are transcriptionally regulated as part of the late flavonoid pathway. The MYB–bHLH–WD40 (MBW) transcriptional complex plays a central role, activating late biosynthetic genes including those encoding UGTs and AATs. In *Arabidopsis*, MBW complexes directly upregulate genes like *UFGT* and *AAT* under developmental and stress cues (Sun *et al.*, 2022). In purple carrot roots, the R2R3-MYB factor DcMYB7 activates both UGTs and the SCPL-type AAT gene *DcSAT1/DcSCPL1*, coordinating glycosylation and acylation of cyanidin derivatives (Curaba *et al.*, 2020).

2.4.5 Environmental and Hormonal Influences

Environmental cues modulate anthocyanin acylation through MBW regulation and acyltransferase activity. Light and temperature shifts often induce MBW activity: in apple, cold triggers the phosphorylation of bHLH (MdbHLH3) and its binding to MYB1, activating downstream genes. Jasmonic acid (JA) signaling also enhances acylation *via* degradation of Jasmonate ZIM-domain (JAZ) repressors and subsequent activation of MYB–bHLH complexes. External application of JA frequently increases acylated anthocyanins in fruits and leaves (Cappellini *et al.*, 2021).

2.4.6 Epigenetic and Posttranslational Regulation

Anthocyanin acylation is modulated by chromatin changes and protein modifications. Stress-induced promoter demethylation enhances anthocyanin biosynthetic gene expression, including AATs. For example, cold can demethylate promoters of anthocyanin genes in peach, facilitating transcription. Histone modifications (e.g., H3K9 acetylation or H3K4 methylation) and histone variant dynamics (e.g., H2A.Z eviction) also regulate gene accessibility. Posttranslationally, phosphorylation of transcription factors like PAP1 (MYB75) and MdbHLH3 enhances their activity under light or cold stress (Cappellini *et al.*, 2021).

In Figure 2.2, a schematic of the general effects of both environmental and epigenetic regulation is presented.

2.4.7 Example of Reaction Mechanism

A typical acylation mechanism involves an anthocyanin-3-*O*-glycoside (e.g., Cyn-3-*O*-Glc) undergoing esterification *via* BAHD or SCPL AATs. The acyl donor – an acyl-CoA (e.g., sinapoyl-CoA) or acylglucose (e.g., sinapoylglucose) – is presented to the enzyme, which catalyzes the nucleophilic attack by the sugar hydroxyl (usually at 6-OH) to yield an ester-linked anthocyanin (Zhao *et al.*, 2017; Nakayama *et al.*, 2003).

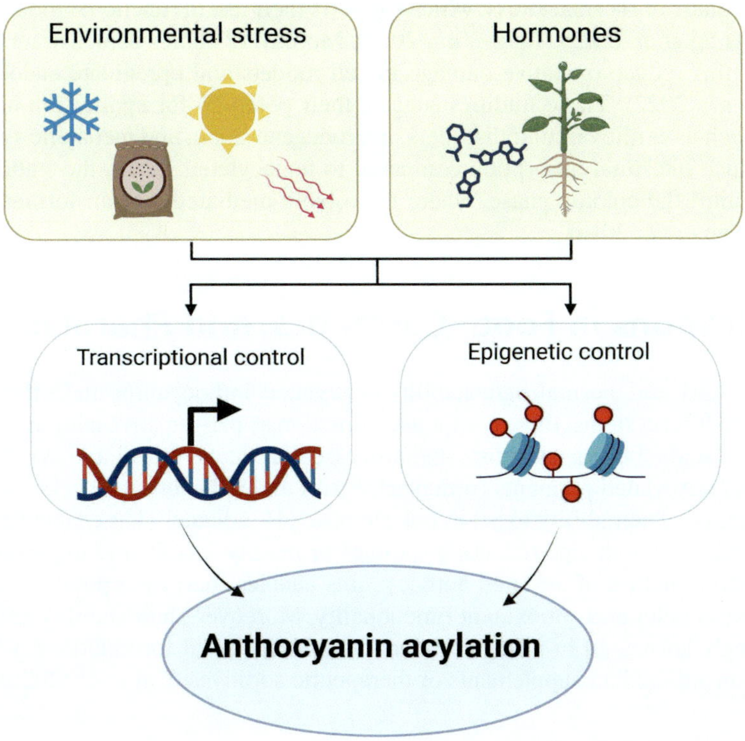

FIGURE 2.2 Regulatory framework linking environmental and hormonal signals to the transcriptional and epigenetic control of anthocyanin acylation (conceptual diagram).

2.5 BIOLOGICAL AND INDUSTRIAL SIGNIFICANCE

2.5.1 Role in Plant Defense and Pollinator Attraction

Acylated anthocyanins enhance plant fitness by improving pigment stability and shifting the hue toward blue, which can aid in pollinator attraction and photoprotection. Their increased resistance to pH, heat, and light allows acylated pigments to persist longer in vacuoles, especially under environmental stresses such as high solar irradiation or cold (Giusti and Wrolstad, 2003). The intramolecular stacking of acyl groups protects the chromophore, stabilizing the pigment and preserving its function in photoprotective and antioxidative roles (Zhao *et al.*, 2017; Vidana Gamage *et al.*, 2022). These stabilized pigments absorb UV light and scavenge reactive oxygen species, protecting plant tissues from radiation damage and oxidative stress (Teng *et al.*, 2022). Additionally, acylation-induced color modulation enhances the visual appeal of fruits and flowers, contributing to successful pollination and seed dispersal (Giusti and Wrolstad, 2003).

2.5.2 Health Benefits

Anthocyanins are recognized for their antioxidant and anti-inflammatory effects, and acylated derivatives, due to their enhanced structural stability, may better retain bioactivity through digestion and storage. Acylated anthocyanins from bilberry, raspberry, and black rice have been shown to maintain antioxidant capacity after enzymatic or thermal stress, which supports their use in functional foods and nutraceuticals (Fei *et al.*, 2021; Kong *et al.*, 2023; Teng *et al.*, 2022). Moreover, studies demonstrate that enzymatically acylated anthocyanins reduce oxidative damage in cell models and upregulate endogenous antioxidant enzymes (Teng *et al.*, 2022). These findings suggest their potential for application in the prevention of chronic diseases such as cardiovascular disorders, neurodegeneration, and metabolic syndrome. Although acylation may reduce intestinal absorption compared to nonacylated forms, the stability advantage can preserve efficacy until the colonic phase, where microbiota-mediated biotransformation could enhance systemic effects (Yan *et al.*, 2016).

2.5.3 Applications in Food, Cosmetics, and Pharmaceuticals

The increased solubility and thermal/light stability of acylated anthocyanins make them highly attractive for industrial use. In food systems, these compounds function as pH-sensitive natural colorants with superior performance in acidic beverages, dairy, and processed products (Giusti and Wrolstad, 2003; Vidana Gamage *et al.*, 2022). Acylated pigments from black carrot and red cabbage are commonly used for their superior color retention. Their ability to act as colorimetric pH indicators has enabled application in intelligent packaging, with color change reflecting spoilage or quality loss (Fei *et al.*, 2021). In the cosmetic industry, the lipophilic nature of acylated anthocyanins enables their incorporation into emulsions and oils, providing natural color and antioxidant functionality. Moreover, their stability and functional versatility are increasingly harnessed in pharmaceutical and nutraceutical formulations, where they serve as active ingredients in antioxidant supplements or therapeutic additives (Cai *et al.*, 2020).

2.5.4 Metabolic Engineering to Enhance Acylation in Crops

Genetic engineering of anthocyanin pathways in crops targets the overexpression or introduction of specific acyltransferases to elevate acylated pigment production. For example, heterologous expression of AATs in tobacco has yielded highly acylated anthocyanins with improved stability (Nakayama *et al.*, 2003). Likewise, manipulation of transcriptional regulators such as MYB and bHLH partners can drive acylation through the co-expression of structural genes (Curaba *et al.*, 2020). The integration of BAHD or SCPL-type acyltransferases under tissue-specific or inducible promoters in tomato, maize, or potato could channel metabolic flux toward acylated pigments, enhancing stress resilience and color quality. Recent advances also suggest that modifying acyl-CoA donor availability or flux through precursor pathways may further increase acylation efficiency (Kong *et al.*, 2023). Clustered Regularly Interspaced Short Palindromic Repeats (CRISPR)/Cas approaches could be applied to downregulate competing pathways or repressors, facilitating the accumulation of desired acylated derivatives.

2.5.5 Biotechnological Approaches

Beyond plant systems, microbial and cell culture-based production platforms offer scalable routes to acylated anthocyanins. Plant cell suspension cultures can be fed specific acyl-donors to produce targeted acylated derivatives, which has been demonstrated in carrot and sweet potato systems (Giusti and Wrolstad, 2003). More recently, enzymatic acylation strategies *in vitro* – using lipases or esterases – have produced lipid-soluble anthocyanin derivatives with enhanced antioxidant properties (Teng *et al.*, 2022; Cai *et al.*, 2020). For example, acylation of bilberry anthocyanins with cinnamic acids increased lipid solubility and thermal stability, broadening industrial applicability (Cai *et al.*, 2020). In microbial platforms, synthetic biology efforts aim to reconstruct the entire anthocyanin biosynthetic pathway in yeast or *Escherichia coli*, including genes encoding UGTs and AATs (Nakayama *et al.*, 2003). While full microbial production of acylated anthocyanins remains challenging due to complex co-factor and compartmentalization requirements, continued refinement could eventually yield a sustainable, fermentation-based production system.

2.5.6 Unresolved Questions and Future Research Directions

Despite extensive progress, several gaps remain in our understanding of anthocyanin acylation. Many plant species lack full characterization of their acyltransferase genes, and structure–function relationships among AATs remain to be elucidated. Substrate promiscuity versus selectivity in BAHD and SCPL enzymes is not fully understood, limiting rational design for metabolic engineering (Nakayama *et al.*, 2003; Sasaki *et al.*, 2014). On the regulatory side, the interaction of MBW factors with epigenetic mechanisms and hormonal networks that govern AAT gene expression needs further investigation. Additionally, the transport and vacuolar storage of acylated anthocyanins, and how they differ from nonacylated forms, remain poorly characterized. From an industrial perspective, optimizing extraction, purification, and formulation of acylated anthocyanins with high yield and activity is a continuing challenge. Future research should prioritize (1) the discovery and functional analysis of novel AATs in underexplored crops, (2) decoding environmental and epigenetic regulation of acylation, and (3) advancing bioengineered and microbial platforms for sustainable production of stable anthocyanins with high added value.

REFERENCES

Anwar, M., Chen, L., Xiao, Y., Wu, J., Zeng, L., Li, H., Wu, Q., & Hu, Z. (2021). Recent advanced metabolic and genetic engineering of phenylpropanoid biosynthetic pathways. *International Journal of Molecular Sciences*, 22(17), Article 17. https://doi.org/10.3390/ijms22179544

Cai, D., Li, X., Chen, J., Jiang, X., Ma, X., Sun, J., Tian, L., Vidyarthi, S. K., Xu, J., Pan, Z., & Bai, W. (2022). A comprehensive review on innovative and advanced stabilization approaches of anthocyanin by modifying structure and controlling environmental factors. *Food Chemistry*, 366, 130611. https://doi.org/10.1016/j.foodchem.2021.130611

Cai, J., Zeng, F., Zheng, S., Huang, X., Zhang, J., Zhang, P., & Fei, P. (2020). Preparation of lipid-soluble bilberry anthocyanins through acylation with cinnamic acids and their antioxidation activities. *Journal of Agricultural and Food Chemistry*, 68(27), 7467–7473. https://doi.org/10.1021/acs.jafc.0c01912

Cappellini, F., Marinelli, A., Toccaceli, M., Tonelli, C., & Petroni, K. (2021). Anthocyanins: From mechanisms of regulation in plants to health benefits in foods. *Frontiers in Plant Science*, 12, 748049. https://doi.org/10.3389/fpls.2021.748049

Chatham, L. A., Howard, J. E., & Juvik, J. A. (2020). A natural colorant system from corn: Flavone-anthocyanin copigmentation for altered hues and improved shelf life. *Food Chemistry*, 310, 125734. https://doi.org/10.1016/j.foodchem.2019.125734

Chen, K., Wei, X., Zhang, J., Pariyani, R., Jokioja, J., Kortesniemi, M., Linderborg, K. M., Heinonen, J., Sainio, T., Zhang, Y., & Yang, B. (2020). Effects of anthocyanin extracts from bilberry (*Vaccinium myrtillus* L.) and purple potato (*Solanum tuberosum* L. var. 'Synkeä Sakari') on the plasma metabolomic profile of Zucker diabetic fatty rats. *Journal of Agricultural and Food Chemistry*, 68(35), 9436–9450. https://doi.org/10.1021/acs.jafc.0c04125

Cheng, S., Wu, T., Gao, J., Han, X., Huang, W., You, Y., Zhan, J., Cheng, S., Wu, T., Gao, J., Han, X., Huang, W., You, Y., & Zhan, J. (2023). Color myth: Anthocyanins reactions and enological approaches achieving their stabilization in the aging process of red wine. *Food Innovation and Advances*, 2(4), 255–271. https://doi.org/10.48130/FIA-2023-0027

Curaba, J., Bostan, H., Cavagnaro, P. F., Senalik, D., Mengist, M. F., Zhao, Y., Simon, P. W., & Iorizzo, M. (2020). Identification of an SCPL gene controlling anthocyanin acylation in carrot (*Daucus carota* L.) root. *Frontiers in Plant Science*, 10, 1769. https://doi.org/10.3389/fpls.2019.01770

Dangles, O., & Fenger, J.-A. (2018). The chemical reactivity of anthocyanins and its consequences in food science and nutrition. *Molecules*, 23(8), 1970. https://doi.org/10.3390/molecules23081970

Deng, Y., & Lu, S. (2017). Biosynthesis and regulation of phenylpropanoids in plants. *Critical Reviews in Plant Sciences*, 36(4), 257–290. https://doi.org/10.1080/07352689.2017.1402852

Fei, P., Zeng, F., Zheng, S., Chen, Q., Hu, Y., & Cai, J. (2021). Acylation of blueberry anthocyanins with maleic acid: Improvement of the stability and its application potential in intelligent color indicator packing materials. *Dyes and Pigments*, 184, 108852. https://doi.org/10.1016/j.dyepig.2020.108852

Fenger, J.-A., Moloney, M., Robbins, R. J., Collins, T. M., & Dangles, O. (2019). The influence of acylation, metal binding and natural antioxidants on the thermal stability of red cabbage anthocyanins in neutral solution. *Food & Function*, 10(10), 6740–6751. https://doi.org/10.1039/C9FO01884K

Fenger, J.-A., Sigurdson, G. T., Robbins, R. J., Collins, T. M., Giusti, M. M., & Dangles, O. (2021). Acylated anthocyanins from red cabbage and purple sweet potato can bind metal ions and produce stable blue colors. *International Journal of Molecular Sciences*, 22(9), 4551. https://doi.org/10.3390/ijms22094551

Ghareaghajlou, N., Hallaj-Nezhadi, S., & Ghasempour, Z. (2021). Red cabbage anthocyanins: Stability, extraction, biological activities and applications in food systems. *Food Chemistry*, 365, 130482. https://doi.org/10.1016/j.foodchem.2021.130482

Giusti, M. M., & Wrolstad, R. E. (2003). Acylated anthocyanins from edible sources and their applications in food systems. *Biochemical Engineering Journal*, 14(3), 217–225. https://doi.org/10.1016/S1369-703X(02)00221-8

He, F., Mu, L., Yan, G.-L., Liang, N.-N., Pan, Q.-H., Wang, J., Reeves, M. J., & Duan, C.-Q. (2010). Biosynthesis of anthocyanins and their regulation in colored grapes. *Molecules*, 15(12), Article 12. https://doi.org/10.3390/molecules15129057

Jeyaraj, E. J., Lim, Y. Y., & Choo, W. S. (2021). Extraction methods of butterfly pea (*Clitoria ternatea*) flower and biological activities of its phytochemicals. *Journal of Food Science and Technology*, 58(6), 2054–2067. https://doi.org/10.1007/s13197-020-04745-3

Jiang, L., Gao, Y., Han, L., Zhang, W., & Fan, P. (2023). Designing plant flavonoids: Harnessing transcriptional regulation and enzyme variation to enhance yield and diversity. *Frontiers in Plant Science*, 14, 1220062. https://doi.org/10.3389/fpls.2023.1220062

Jokioja, J., Yang, B., & Linderborg, K. M. (2021). Acylated anthocyanins: A review on their bioavailability and effects on postprandial carbohydrate metabolism and inflammation. *Comprehensive Reviews in Food Science and Food Safety*, 20(6), 5570–5615. https://doi.org/10.1111/1541-4337.12836

Kaur, A., Sharma, K., Pawar, S. V., & Sembi, J. K. (2025). Genome-wide characterization of *PAL*, *C4H*, and *4CL* genes regulating the phenylpropanoid pathway in *Vanilla planifolia*. *Scientific Reports*, 15(1), 10714. https://doi.org/10.1038/s41598-024-81968-w

Kong, Y., Wang, X., Wu, Z., Li, Y., Xu, F., & Xie, F. (2023). Enzymatic acylation of black rice anthocyanins and evaluation of antioxidant capacity and stability of their derivatives. *Foods*, 12(24), 4505. https://doi.org/10.3390/foods12244505

Kumar, K., Debnath, P., Singh, S., & Kumar, N. (2023). An overview of plant phenolics and their involvement in abiotic stress tolerance. *Stresses*, 3(3), Article 3. https://doi.org/10.3390/stresses3030040

Kumari, A., Kumar, V., Ovadia, R., & Oren-Shamir, M. (2023). Phenylalanine in motion: A tale of an essential molecule with many faces. *Biotechnology Advances*, 68, 108246. https://doi.org/10.1016/j.biotechadv.2023.108246

Li, A., Xiao, R., He, S., An, X., He, Y., Wang, C., Yin, S., Wang, B., Shi, X., & He, J. (2019). Research advances of purple sweet potato anthocyanins: Extraction, identification, stability, bioactivity, application, and biotransformation. *Molecules*, 24(21), Article 21. https://doi.org/10.3390/molecules24213816

Li, W., Pang, X., Xiao, J., Wang, X., He, R., & Zhao, X. (2020). Degradation kinetics of pelargonidin-3-(p-coumaroyl)diglucoside-5-(malonyl)glucoside and pelargonidin-3- (feruloyl)diglucoside-5-(malonyl)glucoside in red radish during air-impingement jet drying. *LWT*, 127, 109390. https://doi.org/10.1016/j.lwt.2020.109390

Liu, Y., Wu, L., Deng, Z., & Yu, Y. (2021). Two putative parallel pathways for Naringenin biosynthesis in *Epimedium wushanense*. *RSC Advances*, 11(23), 13919–13927. https://doi.org/10.1039/D1RA00866H

Luo, X., Wang, R., Wang, J., Li, Y., Luo, H., Chen, S., Zeng, X., & Han, Z. (2022). Acylation of anthocyanins and their applications in the food industry: Mechanisms and recent research advances. *Foods*, 11(14), 2166. https://doi.org/10.3390/foods11142166

Montilla, E. C., Arzaba, M. R., Hillebrand, S., & Winterhalter, P. (2011). Anthocyanin composition of black carrot (*Daucus carota* ssp. *sativus* var. *atrorubens*) and purple sweet potato (*Ipomoea batatas*) extracts and their use as food colorants. *Food Chemistry*, 125(2), 451–457. https://doi.org/10.1016/j.foodchem.2010.08.030

Mozos, I., Flangea, C., Vlad, D. C., Gug, C., Mozos, C., Stoian, D., Luca, C. T., Horbańczuk, J. O., Horbańczuk, O. K., & Atanasov, A. G. (2021). Effects of anthocyanins on vascular health. *Biomolecules*, 11(6), Article 6. https://doi.org/10.3390/biom11060811

Nakayama, T., Suzuki, H., & Nishino, T. (2003). Anthocyanin acyltransferases: Specificities, mechanism, phylogenetics, and applications. *Journal of Molecular Catalysis B: Enzymatic*, 23(2–6), 117–132. https://doi.org/10.1016/S1381-1177(03)00078-X

Oliveira, H., Basílio, N., Pina, F., Fernandes, I., de Freitas, V., & Mateus, N. (2019). Purple-fleshed sweet potato acylated anthocyanins: Equilibrium network and photophysical properties. *Food Chemistry*, 288, 386–394. https://doi.org/10.1016/j.foodchem.2019.02.132

Sasaki, N., Nishizaki, Y., Ozeki, Y., & Miyahara, T. (2014). The role of acyl-glucose in anthocyanin modifications. *Molecules*, 19(11), 18747–18766. https://doi.org/10.3390/molecules191118747

Sigurdson, G. T., Tang, P., & Giusti, M. M. (2017). Natural colorants: Food colorants from natural sources. *Annual Review of Food Science and Technology*, 8, 261–280. https://doi.org/10.1146/annurev-food-030216-025923

Strauch, R. C., Mengist, M. F., Pan, K., Yousef, G. G., Iorizzo, M., Brown, A. F., & Lila, M. A. (2019). Variation in anthocyanin profiles of 27 genotypes of red cabbage over two growing seasons. *Food Chemistry*, 301, 125289. https://doi.org/10.1016/j.foodchem.2019.125289

Sun, X., Zhang, Q., & Zhou, H. (2021). Anthocyanins: from biosynthesis regulation to crop improvement. *Botany Letters*, 168(4), 546–557. https://doi.org/10.1080/23818107.2021.1909498

Sun, X., Zhang, Z., Li, J., Zhang, H., Peng, Y., & Li, Z. (2022). Uncovering hierarchical regulation among MYB-bHLH-WD40 proteins and manipulating anthocyanin pigmentation in rice. *International Journal of Molecular Sciences*, 23(15), 8203. https://doi.org/10.3390/ijms23158203

Sunil, L., & Shetty, N. P. (2022). Biosynthesis and regulation of anthocyanin pathway genes. *Applied Microbiology and Biotechnology*, 106(5), 1783–1798. https://doi.org/10.1007/s00253-022-11835-z

Teng, H., Mi, Y., Cao, H., & Chen, L. (2022). Enzymatic acylation of raspberry anthocyanin: Evaluations on its stability and oxidative stress prevention. *Food Chemistry*, 372, 130766. https://doi.org/10.1016/j.foodchem.2021.130766

Trouillas, P., Sancho-García, J. C., De Freitas, V., Gierschner, J., Otyepka, M., & Dangles, O. (2016). Stabilizing and modulating color by copigmentation: Insights from theory and experiment. *Chemical Reviews*, 116(9), 4937–4982. https://doi.org/10.1021/acs.chemrev.5b00507

Truong, V. D., Deighton, N., Thompson, R. T., McFeeters, R. F., Dean, L. O., Pecota, K. V., & Yencho, G. C. (2010). Characterization of anthocyanins and anthocyanidins in purple-fleshed sweet potatoes by HPLC-DAD/ESI-MS/MS. *Journal of Agricultural and Food Chemistry*, 58(1), 404–410. https://doi.org/10.1021/jf902799a

Vidana Gamage, G. C., Lim, Y. Y., & Choo, W. S. (2022). Sources and relative stabilities of acylated and nonacylated anthocyanins in beverage systems. *Journal of Food Science and Technology*, 59(3), 831–845. https://doi.org/10.1007/s13197-021-05054-z

Wang, J., Jiang, Y., Sun, T., Zhang, C., Liu, X., & Li, Y. (2022). Genome-wide classification and evolutionary analysis reveal diverged patterns of chalcone isomerase in plants. *Biomolecules*, 12(7), 961. https://doi.org/10.3390/biom12070961

Wang, P., Liu, J., Zhuang, Y., & Fei, P. (2022). Acylating blueberry anthocyanins with fatty acids: Improvement of their lipid solubility and antioxidant activities. *Food Chemistry: X*, 15, 100420. https://doi.org/10.1016/j.fochx.2022.100420

Xu, J., Su, X., Lim, S., Griffin, J., Carey, E., Katz, B., Tomich, J., Smith, J. S., & Wang, W. (2015). Characterisation and stability of anthocyanins in purple-fleshed sweet potato P40. *Food Chemistry*, 90–96. https://doi.org/10.1016/j.foodchem.2014.08.123

Xu, M., Fang, D., Kimatu, B. M., Lyu, L., Wu, W., Cao, F., & Li, W. (2024). Recent advances in anthocyanin-based films and its application in sustainable intelligent food packaging: A review. *Food Control*, 162, 110431. https://doi.org/10.1016/j.foodcont.2024.110431

Yan, H., Pei, X., Zhang, H., Li, X., Zhang, X., Zhao, M., Chiang, V. L., Sederoff, R. R., & Zhao, X. (2021). MYB-mediated regulation of anthocyanin biosynthesis. *International Journal of Molecular Sciences*, 22(6), 3103. https://doi.org/10.3390/ijms22063103

Yan, Z., Li, C., Zhang, L., Liu, Q., Ou, S., & Zeng, X. (2016). Enzymatic acylation of anthocyanin isolated from black rice with methyl aromatic acid ester as donor: Stability of the acylated derivatives. *Journal of Agricultural and Food Chemistry*, 64(5), 1137–1143. https://doi.org/10.1021/acs.jafc.5b05031

Yañez-Apam, J., Domínguez-Uscanga, A., Herrera-González, A., Contreras, J., Mojica, L., Mahady, G., & Luna-Vital, D. A. (2023). Pharmacological activities and chemical stability of natural and enzymatically acylated anthocyanins: A comparative review. *Pharmaceuticals*, 16(5), 638. https://doi.org/10.3390/ph16050638

Zhao, C.-L., Yu, Y.-Q., Chen, Z.-J., Wen, G.-S., Wei, F.-G., Zheng, Q., Wang, C.-D., & Xiao, X.-L. (2017). Stability-increasing effects of anthocyanin glycosyl acylation. *Food Chemistry*, 214, 119–128. https://doi.org/10.1016/j.foodchem.2016.07.073

Zhao, X., Zhang, Y., Long, T., Wang, S., & Yang, J. (2022). Regulation mechanism of plant pigments biosynthesis: Anthocyanins, carotenoids, and betalains. *Metabolites*, 12(9), 871. https://doi.org/10.3390/metabo12090871

Zhu, J., Zhao, W., Li, R., Guo, D., Li, H., Wang, Y., Mei, W., & Peng, S. (2021). Identification and characterization of chalcone isomerase genes involved in flavonoid production in *Dracaena cambodiana*. *Frontiers in Plant Science*, 12, 616396. https://doi.org/10.3389/fpls.2021.616396

Occurrence of Acylated Anthocyanins in Edible and Non-Edible Sources

3

Wen Tao, Muci Wu, Jingren He, and Rui Zhang

3.1 OVERVIEW OF STRUCTURAL CHARACTERISTICS OF ACYLATED ANTHOCYANINS

Anthocyanins (ACNs) are hydrophilic flavonoids found in the vacuoles of cells and are responsible for the red, blue, and purple colors of many plants present in nature. These compounds present a basic flavonoid structure, usually represented as a typical C6–C3–C6 carbon chain of two benzene rings (A- and B-rings) linked by a heterocyclic ring (C-ring), forming a π-conjugated 2-phenylbenzopyrylium cation (flavylium cation) with chromophore properties (Jokioja et al., 2021). The general structure of ACN aglycones is formed by hydroxylation at the 4′, 3, 5, and 7 carbon positions (Jokioja et al., 2021). To date, more than 700 different ACNs have been identified in nature (Santos-Buelga et al., 2014). Interestingly, only six common ACN aglycones (anthocyanidins), including delphinidin (Dp), cyanidin (Cy), petunidin (Pt), peonidin (Pn), pelargonidin (Pg), and malvidin (Mv), are found in edible plants, and they differ in the hydroxy and methoxy substituents at carbons 3′ and 5′ of ring B, giving rise to different ACN structures (Figure 3.1) (Tsuda, 2012). Due to the instability of aglycones, sugar moieties (*e.g.,* glucose, galactose, rhamnose, xylose, or arabinose) are almost conjugated as mono-, di-, or triglycosides to the aglycone backbone *via* an O-glycosidic bond of the hydroxy group at position C3 (*i.e.,* C3–OH group in ring C), which plays a critical role in the stabilization of ACNs (L.-S. Wang & Stoner, 2008). However, glycosylation patterns have also been confirmed at positions C5 and C7, and less frequently at positions C3′ and C5′ (de Pascual-Teresa & Sanchez-Ballesta, 2008). Monoglycosylation is the simplest pattern among all the glycosylated ACNs. In fact, the majority naturally occurring ACNs are polyglycosylated with sequential sugars attached to the same position, or with single/sequential sugars attached to different positions of the same core skeleton and usually attached to the hydroxyl group at C3 and C5 of the aglycones (Teixeira et al., 2023). The sugar moieties attached to the aglycones of ACNs can then be further acylated with aromatic acids (*e.g., p*-coumaric, caffeic, ferulic, sinapic, gallic, or *p*-hydroxybenzoic acid) or aliphatic acids (*e.g.,* malonic, acetic, malic, succinic, or oxalic acid) by covalent ester linkage *via* cytoplasmic or vacuolar acyltansferases in the post-biosynthetic pathway (Sasaki et al., 2014). In addition, some anthocyanins (*e.g.,* lobelinin A isolated from the bluish violet petals of *Lobelia erinus* flowers, which has one *p*-coumaroyl,

DOI: 10.1201/9781003519591-4

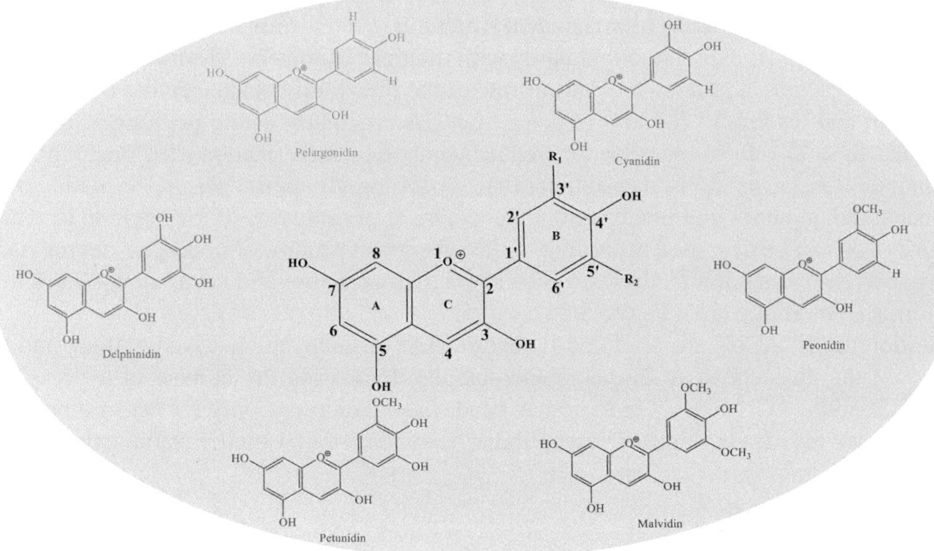

FIGURE 3.1 General structure of ACN aglycones in their flavylium cation form and six common anthocyanidins in nature: R_1 – H, OH, OCH_3; R_2 – H, OH, OCH_3.

one malonyl, and two caffeoyl groups) are even simultaneously acylated by both aromatic and aliphatic acids (Kondo et al., 1989). These acyl substituents are commonly bound to the sugar moieties at the C3 or C5 position of the flavonoid skeleton and esterified at the 6-OH or less frequently at the 4-OH of the sugar residues. Furthermore, some complex acylated anthocyanins attached with different types of sugar moieties or with different positions of sugar moieties have been found in previous studies (Giusti & Wrolstad, 2003). Acylated ACNs are categorized mainly into four categories based on the number of acyl groups: monoacylated, diacylated, triacylated, and tetraacylated. All di-, tri-, and tetraacylated ACNs are referred to as polyacylated anthocyanins. Acylation is one of the crucial factors for creating diversity within ACN molecules, and the types, numbers, and acylated positions of the acyl groups significantly increase the types of ACNs (Andersen & Jordheim, 2010).

3.2 COLORS AND CHEMICAL STABILITY

In terms of chemical structure, ACNs are water-soluble glycosides and acylglycosides of anthocyanidins in the form of polyhydroxylated and polymethoxylated heterosides derived from flavylium or 2-phenylbenzopyrilium ions. The colors and stability of ACNs are highly dependent on several factors, including pH conditions, structural features (amount and position of hydroxylation and methylation in the B-ring, as well as different sugars and/or acids attached to their aromatic rings), complexation with metal ions, and the presence of other plant co-pigments (*e.g.*, flavonoids, alkaloids, amino acids, organic acids, nucleotides, or polysaccharides) (Alappat & Alappat, 2020; Teixeira et al., 2023). Due to the ionic nature of the molecular structure, the most interesting factor affecting the color and stability of ACNs is pH value. It is known that ACNs exist in a dynamic equilibrium of different chemical forms, depending on the pH value of their surrounding environments. Thus, in an acidic environment (pH ≈ 1), ACNs are found in the most stable form of the flavylium cation (red color), which has a positively charged pyrylium ring. As the pH increases, ACNs undergo two different types of reactions (proton-transfer reactions and hydration reactions) and continue to generate new chemical structures. Proton-transfer reactions usually occur relatively quickly, resulting in blue quinonoid bases with multiple tautomers. Meanwhile, the pH-dependent hydration reaction occurs at a relatively slower rate in the flavylium nucleus, giving rise to the colorless hemiketal form and leading to the loss of color in ACNs at slightly acidic pH ranges. In addition, the hemiketal structures also form the yellow *cis*-chalcone upon tautomerization (ring opening), and finally, *trans*-chalcone is formed by the isomerization of the *cis*-analog. At neutral pH, ACNs mainly occur in the form of uncharged quinones and in a blue to violet color. At pH above 7, ACNs begin to lose their structural integrity and can be degraded according to the substituent groups, or undergo a deprotonation reaction to form corresponding anionic bases, which alters their structure and keeps an equilibrium between different forms (Pina et al., 2015; Figure 3.2).

In addition to pH, ACNs are modified by hydroxylation, methylation, glycosylation, and acylation, which increase the diversity of ACN colors and stability. Increasing the number of hydroxyl groups in the B-ring makes the ACN bluer. On the other hand, methylation not only results in a red shift in the color of ACNs but also leads to a low susceptibility to oxidation and higher stabilization of the ACNs.

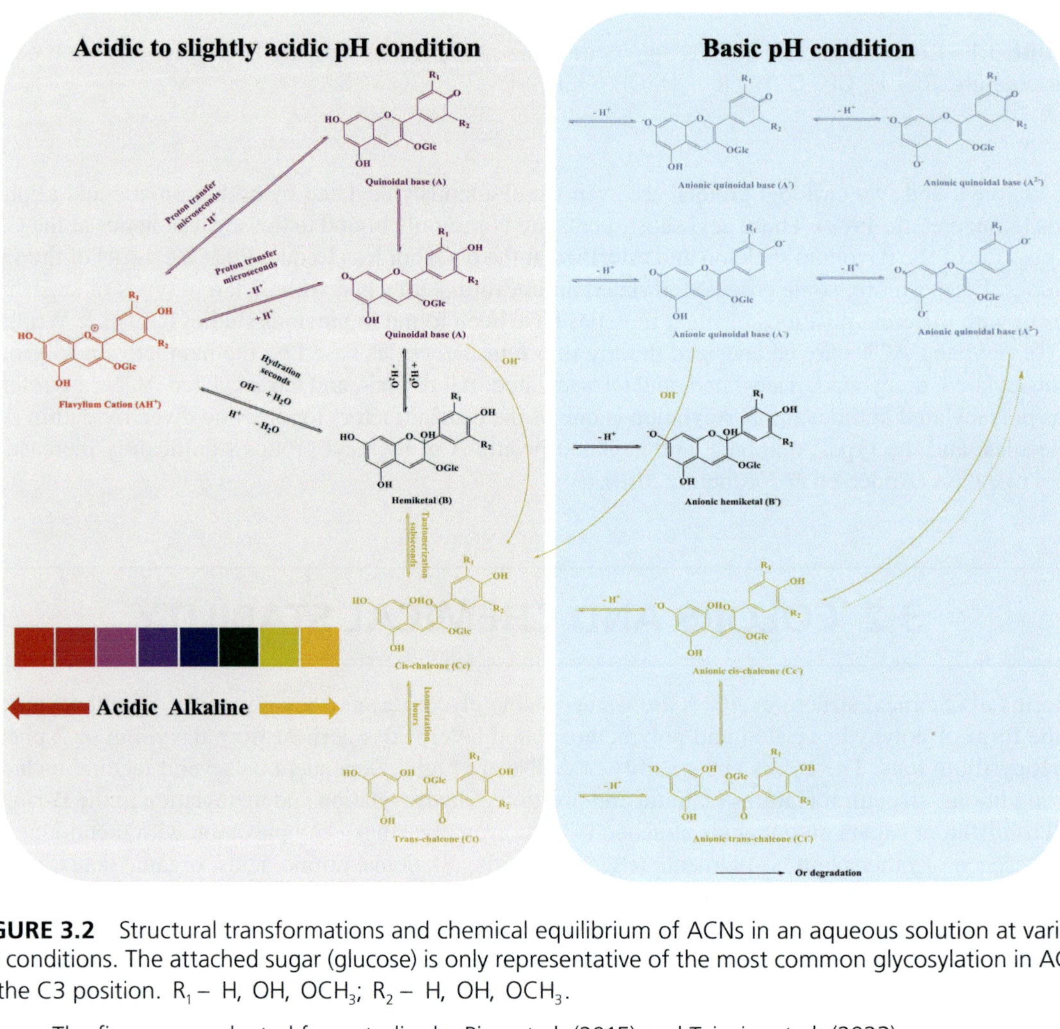

FIGURE 3.2 Structural transformations and chemical equilibrium of ACNs in an aqueous solution at various pH conditions. The attached sugar (glucose) is only representative of the most common glycosylation in ACNs at the C3 position. R_1 – H, OH, OCH_3; R_2 – H, OH, OCH_3.

Source: The figure was adapted from studies by Pina et al. (2015) and Teixeira et al. (2023).

Glycosylated ACNs show a hypsochromic shift in the absorption maxima of the spectra and increase in their stability, which is beneficial for storage in the vacuoles. Furthermore, the glycosyl moieties of ACNs can be further modified by aromatic and/or aliphatic acyl moieties. While aliphatic acylation does not lead to a change in color, aromatic acylation leads to a shift toward blue (Alappat & Alappat, 2020). In addition, the use of metals to stabilize the color of ACNs could be a successful method. For example, the interaction between ACNs and molybdenum (Mo) is thought to be responsible for the stabilization of the blue color in *Hindu* cabbage tissues (Riaz et al., 2016). ACNs are also sensitive to temperature, light, oxygen exposure, solvents, the presence of enzymes, other polyphenols, metal ions, and other factors. Therefore, plants have developed different strategies to increase the stability of their secondary metabolites. Acylating with different aromatic and aliphatic units also appears to be a valid strategy that results in an overall increase in the resistance of ACNs to different degradation factors.

Compared to nonacylated ACNs, acylation increases the stability of ACNs depending on the type of acyl groups (aromatic or aliphatic acid and their structure and size), their number, and their attachment sites (Jokioja et al., 2021). Physically, acylation typically changes the polarity depending on the acyl groups, increases the structural size of the molecules, and changes the spatial structures. In particular, acylation with aromatic acyl groups increases the molecular size of ACNs, which can confer steric hindrance effects on the ACNs. The aforementioned physical effects can further influence the chemical reactivity of ACNs, leading to several interactions and transformations: the increased ability of ACNs to partition into nonpolar or low-polar domains of media *in vitro* or *in vivo* (Jang et al., 2005); the ability of the hydroxyl group of ACNs to donate hydrogen atoms to free radicals (Kay et al., 2004); or the conversion of ACN flavylium cations into zwitterions, which is specifically induced in acylated ACNs by some aliphatic dicarboxylic acids (*e.g.,* malic, malonic, and succinic acids) or by some aromatic acids (*e.g., p*-coumaric and sinapic acids) (Harborne & Grayer, 1988). Stereochemically, efficient enhancement of the stability of ACNs may be due to the stacking effect of the acyl groups, especially the aromatic ones, with the pyrylium ring of the anthocyanidin flavylium cations. A typical example of cyanidin-3-*O*-(3″-*O-trans* (E)-caffeoyl-6″-*O*-E-caffeoyl-β-D-glucopyranoside) (Figure 3.3) was reported to easily form a "sandwich"-type structure in which the flexible and collapsible sugar chain acts like a ribbon, placing the two caffeoyl groups on either side of the plane of the 2-phenylbenzopyran scaffold. The intramolecular hydrophobic interactions between the benzopyrylium ring and the aromatic acyl groups can provide an effective physical barrier to prevent the nucleophilic attack of the OH- in media at the C2 and/or C4 position of anthocyanidins at acidic to neutral pH ranges to produce a colorless pseudobase or chalcone,

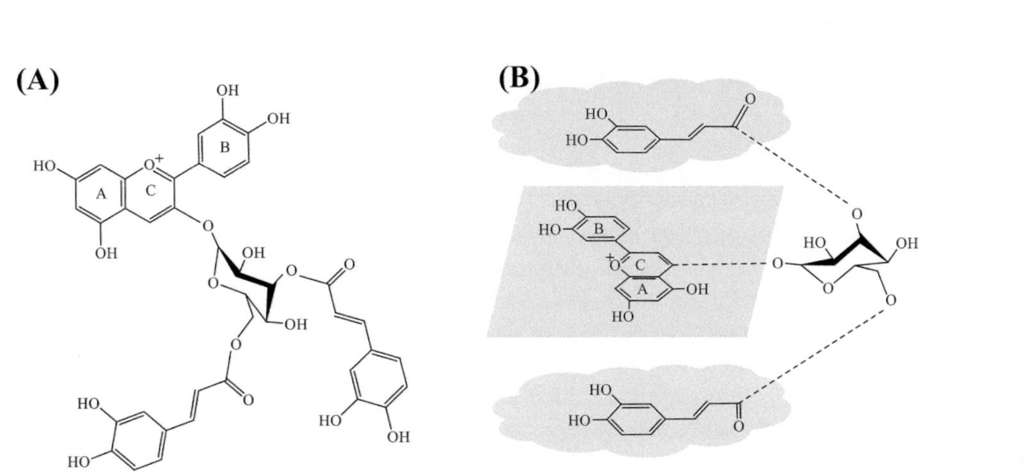

FIGURE 3.3 Chemical structure (A) and suggested intramolecular co-pigmentation complex (B) of cyanidin-3-*O*-(3″-*O*-E-caffeoyl-6″-*O*-E-caffeoyl-β-D-glucopyranoside).

Source: The figure was adapted and modified slightly from study by Zhao et al. (2017).

greatly improving the stability of the ACNs (Zhao et al., 2017). Besides, the mechanisms by which acylation increases the chemical stability of acylated ACNs *in vitro* and *in vivo* primarily also involve the photochemical, biochemical, and environmental aspects under specific conditions. The effect of acylation on the chemistry of ACNs will be extensively reviewed in other chapters.

3.3 EDIBLE SOURCES OF ACYLATED ANTHOCYANINS

Acylation serves as an effective strategy to improve the chemical and color stability of ACNs through intramolecular and/or intermolecular co-pigmentation and self-association reactions. Therefore, sources of acylated ACNs may provide the desired stability for food applications (Giusti & Wrolstad, 2003). A large number of structurally different acylated ACNs have already been found in common fruits, vegetables, tubers, grains, and legumes, including grapes, berries, purple sweet potato, black carrot, red cabbage, and black rice, and some less-known edible sources such as pansy, carnation, roses, and other edible flowers. Depending on the different plants, the proportion of acylated ACNs in total ACNs ranges from less than 1% to almost 100% (Jokioja et al., 2021).

3.3.1 Fruits

Common ACN sources from fruits contain mainly nonacylated polyglycoslated ACNs, characterized by different sugar substituents (glucoside, rutinoside, or sambubioside) attached at the same or different sites at the C3 and/or C5 position. Cyanindin-3-*O*-glucoside, cyanidin-3-*O*-rutinoside, delphinidin-3-*O*-glucoside, and delphinidin-3-*O*-rutinoside are among the most abundant nonacylated ACNs found in fruits (Wu et al., 2006). Nevertheless, there are still many ACNs exited in some fruits that are acylated with one or more aromatic acids (*e.g.,* coumaric acid, caffeic acid, and ferulic acid) and aliphatic acids (*e.g.,* malonic acid, acetic acid, and succinic acid). For example, Slimestad et al. detected two acylated ACNs (*i.e.,* delphinidin 3-*O*-(6″-coumaroyl)-glucoside and cyanidin 3-*O*-(6″-coumaroyl)-glucoside) from the ethyl acetate phase of the black currant extract, although the amounts of these compounds were negligible (Slimestad & Solheim, 2002). Five cyanidin derivatives acylated with coumaric acid, acetic acid, and malonic acid (*i.e.,* cyanidin-sambubioside-malonyl-glucoside, cyanidin-sambubioside-acetyl-glucoside, cyanidin-3-(Z)-*p*-coumaroyl-sambubioside-5-glucoside, cyanidin-3-(E)-*p*-coumaroyl-sambubioside-5-glucoside, and cyanidin-3-*p*-coumaroyl-sambubioside) were detected at low levels in different elderberry interspecific hybrids (Mikulic-Petkovsek et al., 2014). Strawberry ACNs derive from pelargonidin and cyanidin aglycones, with pelargonidin-3-glucoside being the major ACN in the fruits, and a small proportion of cyanidin-3-glucoside and pelargonidin-3-rutinoside also appears to be identified in all varieties. These three compounds represented more than 95% of total ACNs in strawberries. Besides, pelargonidin-3-arabinoside and cyanidin-3-rutinoside have been reported in some strawberry cultivars, as well as various acylated ACNs formed by acylation with different aliphatic acids (*e.g.,* malic, malonic, succinic, and acetic acids). In particular, pelargonidin-3–(6-malonyl)-glucoside was one of the main pigments in several Japanese cultivars, accounting for 5%–30% of the total ACNs content (Yoshida et al., 2002). Other acylated ACNs such as pelargonidin-3-acetyl-glucoside (Hong & Wrolstad, 1990) and pelargonidin-succinyl-glucoside (Bakker et al., 1994) were also identified in some rare strawberry varieties. As another edible source, ACNs in grape skins are mainly represented by five aglycones: delphinidin, cyanidin, petunidin, peonidin, and malvidin, and trace levels of pelargonidin, which can be either 3-OH monoglucosylated or 3,5-OH diglucosylated (Andreeva, 2020). Glucosylated derivatives of these ACNs, esterified at the C6 position of glucose with acetyl or *p*-coumaroyl groups, have also been detected at low concentrations. Kőrösi et al. (2022) identified 21 different anthocyanins in the grape skin or juice of 15 various teinturier varieties collected from the gene bank located at Pécs in the southwest of Hungary. Among all ACNs, nine were glucosylated in the C3 or/and C5 positions, and 12 ACNs were acylated with acetyl or *p*-coumaroyl groups, namely, petunidin-3-acetyl-glucoside, peonidin-3-acetyl-glucoside, malvidin-3-acetyl-glucoside, peonidin-3-*p*-coumaroyl-glucoside, malvidin-3-*p*-coumaroyl-glucoside, delphinidin-3-acetyl-glucoside, malvidin-3-*O*-(6-*O*-acetyl)-glucoside-5-*O*-glucoside,delphinidin-3-*O*-(6-*O*-*p*-coumaroyl)-glucoside,cyanidin-3-*O*-(6-*O*-*p*-coumaroyl)-glucoside, malvidin-3-*O*-(6-*O*-coumaroyl)-glucoside-5-*O*-glucoside, petunidin-3-*O*-(6-*O*-*trans*-*p*-coumaroyl)-glucoside, and peonidin-3-*O*-(6-*O*-*trans*-*p*-coumaroyl)-glucoside (Kőrösi et al., 2022). In addition to the above acylated ACNs, cyanidin-3-acetyl-glucoside was also identified from grape varieties cultivated in two Portuguese wine regions (Dão and Douro). Regardless of the grape variety, malvidin derivatives are the most predominant in the grape skin with more than 50% content, providing the characteristic red color and stability of young red wines, due to its resistance to oxidation (Costa et al., 2014). Additionally, less than 2% of malvidin-acetyl-glucoside and cyanidin-3-acyl-glucoside were detected in black plum (Wu et al., 2006; X. Zhang et al., 2019).

However, there are some special cases in fruits such as *Lycium* that are excellent edible source of acylated ACNs mainly composed of delphinidin and petunidin derivatives acylated with one coumaric acid or caffeic acid at the sugar moieties at the C3 position, including delphinidin-3-*O*-[6-*O*-(4-*O*-*p*-coumaroyl-α-L-rhamnopyranosyl)-β-D-glucopyranoside]-5-*O*-[β-D-glucopyranoside], petunidin-3-*O*-[6-*O*-(4-*O*-*trans*-*p*-coumaroyl-α-L-rhamnopyranosyl)-β-D-glucopyranoside]-5-*O*-[β-D-glucopyranoside], malvidin-3-*O*-[6-*O*-(4-*O*-*p*-coumaroyl-α-L-rhamnosyl)-β-D-glucopyranoside]-5-*O*-β-D-glucopyranoside, petunidin-3-*O*-(glucosyl-*p*-coumaroyl)-rutinoside-5-*O*-glucoside, petunidin-3-*O*-(caffeoyl)-rutinoside-5-*O*-glucoside, petunidin-3-*O*-(*p*-coumaroyl)-rutinoside, petunidin-3-*O*-[rhamnopyranosyl-(*p*-coumaroyl)]-5-*O*-glucoside, and delphinidin-3-*O*-(6″-*p*-coumaryl)-glucoside (Liu et al., 2020; Tian et al., 2021). Besides, *Nitraria tangutorun* has also been reported as another source of acylated ACNs, and cyanidin-3-*O*-(trans-*p*-coumaroyl)-diglucoside has been identified as the main ACN (Sang et al., 2019). In addition to the berries described, some red-fleshed fruits (*e.g.,* peach or apple) have also been found to contain a certain quantity of ACNs monoacylated with coumaric acid. For example, malvidin-3-coumaroylglucoside-acetaldehyde as the only acylated ACN with contents ranging from 0.11 to 0.27 mg/100 g fresh weight was detected among all analyzed European commercial blood-flesh peach cultivars (Monsan 1 (cov), Monsantrois (cov), Monlori (cov), and Monid (cov)). This pyranoanthocyanin is a well-known ACNs-derived pigment found in red or rosé wines and was reported in *Prunus persica* for the first time (Aubert & Chalot, 2020). Besides, Yang et al. detected a mononacylated ACN (delphindin-3-*O*-(6″-*O*-*p*-coumaryl)-glucoside) in both the fruit and leaves of red-fleshed apple (Yang et al., 2024). Furthermore, some fruits have been reported to contain a certain amount of acylated ACNs although they do not present bright colors. For instance, Cebadera-Miranda et al. detected six different mono- or diacylated ACNs in three varieties (Sanguinelli, Tarocco Rosso, and Tarocco Ippolito) of blood oranges, and cyanidin-3-(6″-malonyl)-glucoside showed the highest content in total ACNs, ranging from 2.75 to 10.85 μg/100 g of the edible portion (Cebadera-Miranda et al., 2019). In addition to 97% of cyanidin-3-glucoside, a small amount of cyanidin-3-(6″-malonylglucoside) (2%) was also detected in the rind of the passion fruit (Kidøy et al., 1997).

3.3.2 Vegetables and Tubers

Vegetables and tubers including purple-fleshed sweet potato, red cabbage, red radish, and purple carrot represent the richest sources of acylated ACNs, presenting an extensive diversity of these compounds. Varieties of purple sweet potato (*Ipomoea batatas* L.) are rich in various acylated ACNs, with *p*-hydroxy-benzoic acid, ferulic acid, and caffeic acid in their structure, which is characteristic of the tubers. For instance, in the variety of purple sweet potato "Jihei No. 1," a total of six monoacylated ACNs and four diacylated ACNs were identified, and cyanidin-3-(6″-caffeoyl)-sophoroside-5-glucoside, peonidin-3-caffeoyl-sophoroside-5-glucoside, peonidin-3-caffeoyl-*p*-hydroxybenzoyl-sophoroside-5-glucoside, and peonidin-3-(6″-caffeoyl)(6‴-feruloyl)-sophoroside-5-glucoside were the four major ACNs, accounting for half of the total ACNs content (Li et al., 2013). In addition to peonidin and cyanidin derivatives, Lee et al. (2013) six pelargonidin-based ACNs were separated from four purple-fleshed sweet potato cultivars from

Korea. Besides, purple sweet potato leaves have similar ACN profiles as the tubers and are also can be regarded as an excellent acylated ACNs source. Interestingly, the leaves exhibited a significantly higher content of cyanidin derivatives than the tubers (Li et al., 2019). On the other hand, *Brassica* vegetables mainly comprise red cabbage (*Brassica oleracea* L. var *Capitata* f. *rubra*) (Fierri et al., 2023; Strauch et al., 2019), broccoli sprouts (*Brassica oleracea* 'Viola') (Moreno et al., 2010), red and purple mustards (*Brassica juncea*) (Sun et al., 2013), purple kohlrabi (*Brassica oleracea* var. *gongylodes*) (Sun et al., 2013), red mizuna (*Brassica rapa* var. *nipposinica*) (Park et al., 2020), red kale (*Brassica oleracea* var. *acephala*) (Olsen et al., 2010), and purple cauliflower (*Brassica oleracea* L. var. *botrytis*) (Scalzo et al., 2008) and are also excellent acylated ACNs-rich sources and have unusually complex structures. In these vegetables, the major acylated ACNs are cyanidin-3-*O*-diglucoside-5-*O*-glucoside derivatives with one or two acyl groups (sinapoyl, *p*-coumaroyl, caffeoyl, or feruloyl) connected to the diglucoside at the C3 position (Figure 3.4). Fierri et al. (2023) and Strauch et al. (2019) detected 23 different (12 diacylated and 11 monoacylated) structurally acylated ACNs at the base of the cyanidin-3-*O*-diglucoside-5-*O*-glucoside structure, accounting for more than 80% of the total ACNs. Peonidin-3-(*p*-coumaroyl)-diglucoside-5-glucoside and peonidin-3-(*p*-coumaroyl)(sinapoyl)-diglucose-5-glucoside were also detected as special acylated ACNs in new varieties of red cabbage, although their content was very low (Strauch et al., 2019). In the case of broccoli sprouts, purple kohlrabi, red and purple mustards, the produced fragments are basically consistent with those of other common *Brassica* vegetables, with one or two same/different aromatic groups (*p*-coumaric, sinapic, or ferulic acid) attached to the C3 glycosidic substituent. Some of ACNs are further monoacylated with one aliphatic group (malonic acid) on the sugar moiety at C5 (Figure 3.4) (Moreno et al., 2010; Sun et al., 2013). Additionally, in the study by Scalzo et al. (2008), some diacylated ACNs (*e.g.*, cyanidin-3-(6-*p*-coumaryl)-sophoroside-5–(6-sinapoyl)-glucoside and cyanidin-3-(6-feruloyl)-sophoroside-5-(6-sinapoyl)-glucoside) with two different aromatic groups attached to the sugar moieties at C3 and C5, respectively, were also found in purple cauliflower from Sicily. On the other hand, other vegetables such as black carrots, radish sprouts, and purple and red radish also belong to family Brassicaceae and are reported as rich sources of acylated ACNs. For example, Koley et al. (2020) putatively identified 28 ACNs, including 22 cyanidin glycosides, 5 pelargonidin glycosides, and 1 delphinidin glycoside in purple radish. Among these, 22 were acylated glycosides, and the predominant structural backbone was anthocyanin-3-*O*-diglucoside-5-*O*-glucoside, anthocyanin-3-*O*-triglucoside-5-*O*-glucoside, or anthocyanin-3-*O*-diglucoside-5-*O*-diglucoside (Figure 3.4). Throughout the samples, of the total ACNs, the aforementioned acylated ACNs accounted for 74% in the leaf lamina, 90% in the leaf petiole, 97% in the root periderm, and 100% in the root xylem. Whereas diacylated ACNs were dominant in the root xylem, root periderm, leaf lamina, and leaf petiole accounted for 88%, 52%, and 67%, respectively. Besides, Jing et al. (2012) detected 17 different pelargonidin-3-diglucoside-5-glucoside derivatives acylated with caffeic acid, *p*-coumaric acid, ferulic acid, and malonic acid in three Chinese radish cultivars.

In addition to vegetables from the family Brassicaceae, the Solanaceae family contains many horticultural species of economic importance that are also considered to be abundant sources of acylated ACNs, including tomato (*Solanum lycopersicum*), pepper (*Capsicum* spp.), eggplant (*Solanum melongena*), and potato (*Solanum tuberosum*). For instance, the most common ACN structure in pepper and eggplant fruits is delphinidin-3-(*p*-coumaroyl)-rutinoside-5-glucoside, and the *p*-coumaroyl moiety could be substituted by either a feruloyl or a caffeoyl acyl moiety (Azuma et al., 2008; Sadilova et al., 2006). Besides, more structural variations of ACNs can be found in colored potato tubers. In red potato tubers, pelargonidin-3-(*p*-coumaroyl)-rutinoside-5-glucoside is the major ACN, with lower levels of peonidin-3-(*p*-coumaroyl)-rutinoside-5-glucoside and pelargonidin-3-(trans-feruloyl)-rutinoside-5-glucoside (Sampaio et al., 2021). Nevertheless, petunidin-3-(*p*-coumaroyl)-rutinoside-5-glucoside is the predominant ACN in purple potato tubers. In addition, 3-(*p*-coumaroyl)-rutinoside-5-glucosides of malvidin and delphinidin have been found in different purple cultivars (Sampaio et al., 2021). It is worth noting that the color deepening is strongly associated with increased levels of malvidin glycosides. Lachman et al. (2009) found that potato tubers of cv. British Columbia Blue contained almost exclusively cyanidin derivatives. Similar to peppers and eggplants, delphinidin and petunidin derivatives have been mainly detected in purple tomato fruits. In fruits of an *Aft/Aft atv/atv* tomato genotype,

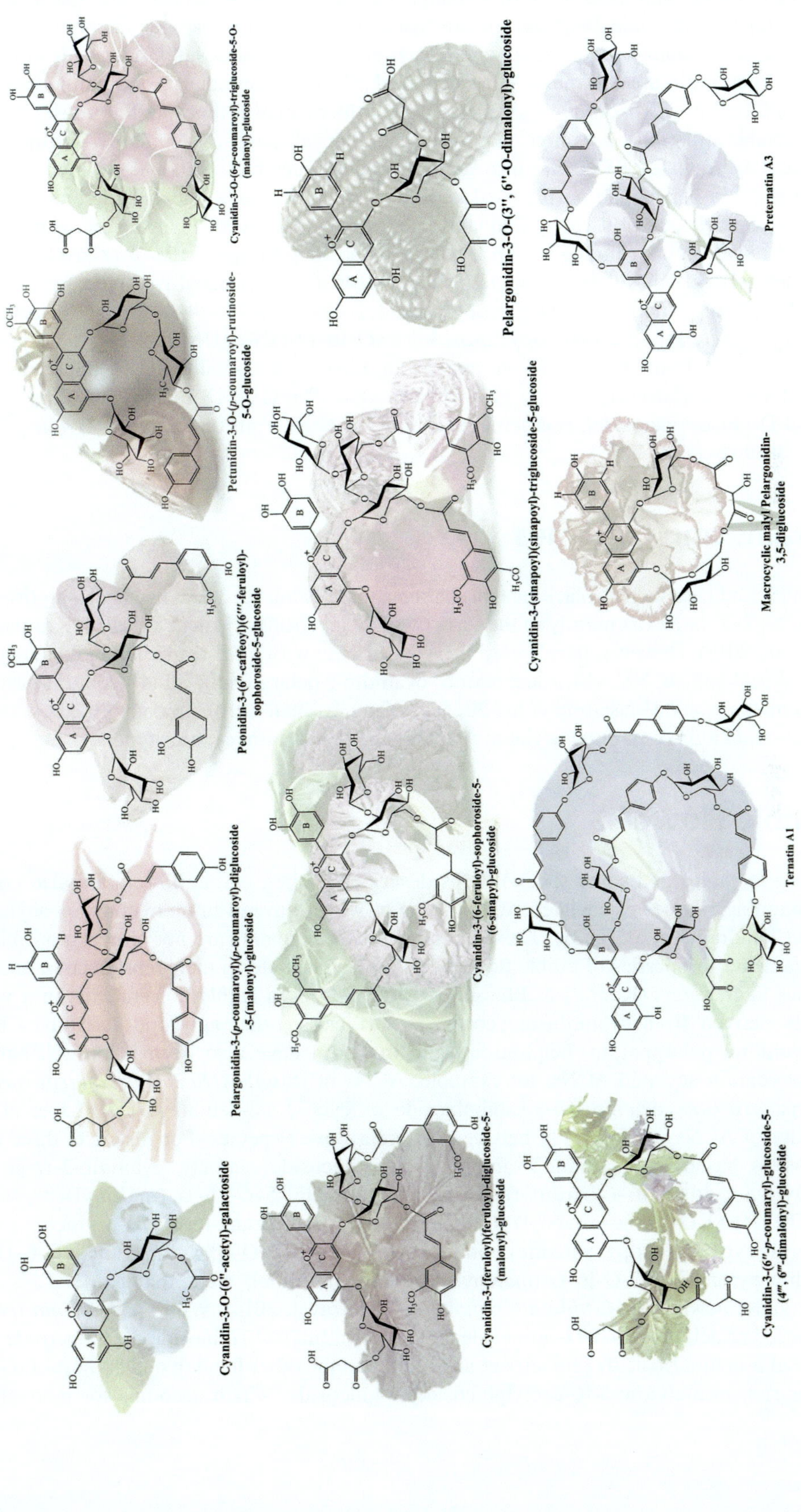

FIGURE 3.4 Chemical structures of acylated ACNs derived from some edible sources.

delphinidin-3-rutinoside and petunidin-3-(*p*-coumaroyl)-rutinoside-5-glucoside are the major ACNs (Mes et al., 2008). In transgenic *Del/Ros1* tomato fruits, where two transcription factors that control the ACN pathway in *Antirrhinum* are overexpressed, 3-(*p*-coumaroyl)-rutinoside-5-glucoside conjugates of delphinidin and petunidin are the main ACNs (Su et al., 2016). In summary, six common anthocyanidins have been found in solanaceous vegetables. Despite the diverse anthocyanidin profiles observed in these four vegetables, the most common ACN form is anthocyanidin-3-(*p*-coumaroyl)-rutinoside-5-glucoside (Figure 3.4). Acylated ACNs are also found in other vegetables such as red onion, red chicory, purple shamrock, and perilla and purple basil leaves, where mono- and diacylated anthocyanidins predominate and are acylated with hydroxycinnamic acid or malonic acid. It is worthy to note that two novel polyacylated ACNs (cyanidin-3-*O*-[β-D-glucopyranoside]-7-*O*-[6-*O*-(E)-(4-*O*-(6-*O*-(E)-caffeoyl-β-D-glucopyranosyl)-caffeoyl)-β-D-glucopyranoside]-3′-*O*-[6-*O*-(E)-caffeoyl-β-D-glucopyranoside] and cyanidin-3-*O*-[4, 6-di-*O*-malonyl-β-D-glucopyranoside]-7-*O*-[6-*O*-(E)-(4-*O*-(6-*O*-(E)-caffeoyl-β-D-glucopyranosyl)-caffeoyl)-β-D-glucopyranoside]-3′-*O*-[6-*O*-(E)-caffeoyl-β-D-glucopyranoside], bicolmalonin and bicolnin) and a known one (cyanidin-3-*O*-[6-*O*-malonyl-β-D-glucopyranoside]-7-*O*-[6-*O*-(E)-(4-*O*-(6-*O*-(E)-caffeoyl-β-D-glucopyranosyl)-caffeoyl)-β-D-glucopyranoside]-3′-*O*-[6-*O*-(E)-caffeoyl-β-D-glucopyranoside], rubrocinerarin) were isolated from the leaves of *Gynura bicolor* as major ACNs (Shimizu et al., 2010).

3.3.3 Grains and Legumes

Grains, legumes, and nuts such as black rice, black bean, and peanut seed coat have been reported be fine sources of ACNs, but they exist mainly in the form of ACN glycosides (Han et al., 2015; Kraithong et al., 2025; Peng et al., 2019). However, purple corn is a special case with a relatively higher content of mono-acylated and diacylated ACNs, which are mainly cyanidin-, pelargonidin-, and peonidin-based ACNs acylated with malonic acid (Massironi et al., 2025; Z. Yang & Zhai, 2010). To date, little is known about grains, legumes, and nuts containing acylated ACNs as the major anthocyanins type.

3.3.4 Edible Flowers

In addition to vegetables and some special crops described earlier, edible flowers are also considered to be an outstanding sources of acylated ACNs, some of which have even more complex acylation patterns (Gayan Chandrajith et al., 2021). These abundant ACN pigments are not only responsible for the attractive ornamental properties of edible flowers but also contribute to their reproduction and defense against various aggressions (Kumari & Bhargava, 2021). ACNs in edible flowers exist in a variety of forms, mainly derived from cyanidin aglycones. Nevertheless, other aglycone derivatives including peonidin, petunidin, pelargonidin, delphinidin, and malvidin have also been identified. Not all the edible flowers contain acylated ACNs, for example, ACNs in Roselle (*Hibiscus sabdariffa*) exist in a nonacylated pattern (*i.e.,* delphinidin-3-sambubioside or cyanidin-3-sambubioside) (Li et al., 2024). The complexity of ACNs varies highly between different flower species. For instance, three different triacylated ACNs based on cyanidin-3-sambubioside-5-glucoside, namely, cyanidin-3-*O*-[2-*O*-(2-*O*-(trans-feruloyl)-β-xylosyl)-6-*O*-(trans-feruloyl)-β-glucoside]-5-*O*-[6-*O*-(malonyl)-β-glucoside], cyanidin-3-*O*-[2-*O*-(2-*O*-(trans-sinapoyl)-β-xylopyranosyl)-6-*O*-(trans-feruloyl)-β-glucopyranoside]-5-*O*-[6-*O*-(malonyl)-β-glucopyranoside], and cyanidin-3-*O*-[2-*O*-(β-xylopyranosyl)-6-*O*-(trans-feruloyl)-β-glucopyranoside]-5-*O*-[6-*O*-(malonyl)-β-glucopyranoside], were the main ACNs detected in the purple-violet flowers of *Matthiola longipetala* (Tatsuzawa, 2014). Wild pansy (*Viola tricolor*) is mainly composed of delphinidin-3-*O*-rutinoside and delphinidin-3-*O*-rutinoside-5-*O*-glucoside acylated with caffeic and *p*-coumaric acids (Teixeira et al., 2024). On the other hand, wood cranesbill (*Geranium sylvaticum*) is rich in malvidin-3-(6-acetylglucoside)-5-glucoside, which accounts for more than 85%

of the total ACNs (Andersen et al., 1995). Infrequently, one acyl group may simultaneously link the two glycosyls located at different positions on the anthocyanidins and further lead to a special and rare rearrangement. For the anthocyanidin-3,5-diglucoside in *Dianthus caryophyllus* and *Dianthus caryophyllus* cv petals, a maloyl is concurrently linked the glucosyls at the C3 and C5 positions to form a cyclic structure (Sasaki et al., 2013), and this type of acylated ACN has been found only in Caryophyllaceae (Figure 3.4) (Fukui et al., 2003; Sasaki et al., 2013). In addition to the mono- and diacylated ACNs described earlier, some studies have also identified tri-, tetra-, and penta-acylated ACNs in some special edible flowers. Additionally, the sites are more uncommon when more acyl groups are present. ACNs of butterfly pea flower (*Clitoria ternatea*) known as ternatins appear blue in color and are acylated based on delphinidin. Their structures were characterized as malonylated delphinidin-3, 3′, 5′-triglucosides having 3′, 5′-side chains with alternative D-glucose and *p*-coumaric acid units. A total of 14 structurally different polyacylated delphinidin glucosides were identified in all the blue petal lines, namely, ternatins A1–A3, B1–B4, C1–C4, and D1–D3, while some studies also have identified several other delphinidin derivatives (*e.g.*, delphinidin-3–(6-succinyl)-*O*-glucoside-3′-(6-*p*-coumaroyl)-*O*-glucoside-5′-(6-*p*-coumaroyl)-*O*-diglucoside and delphinidin-3-*O*-glucoside-3′-(6-succinyl)-*O*-glucoside-5′-(6-*p*-coumaroyl)-*O*-glucoside) (Shen et al., 2016; Tao et al., 2025; Terahara et al., 1998; Zakaria et al., 2018). Another example, a new tetra-acylated ACN as one of the major ACNs (cyanidin-3-*O*-[6-*O*-(malonyl)-β-D-glucopyranoside-7-*O*-[6-*O*-(trans-*p*-coumaryl)-β-D-glucopyranoside]-3′-*O*-[6-*O*-[*trans*-4-*O*-[6-*O*-(*trans*-*p*-coumaryl)-β-D-glucopyranosyl]-*p*-coumaryl]-β-D-glucopyranoside]) was isolated from the red-purple flowers of *Laeliocattleya* cv Mini Purple, with three glucosyl groups possessing four acyl groups attached to the C3, C7, and C3′ position of cyanidin (Tatsuzawa et al., 1994). Additionally, some polyacylated ACNs may simultaneously contain two or more different types of acyl groups. Lobelinin A, which was isolated from the bluish violet petals of *Lobelia erinus* flowers, possesses one coumaroyl, one malonyl, and two caffeoyl groups, and Lobelinin B has four different types of acyl groups, namely, coumaroyl, malonyl, caffeoyl, and feruloyl groups (Kondo et al., 1989). Naturally, several ACNs possessing different numbers of acyl groups can be found in one plant organ. Terahara et al. (2001) detected one diacylated (cyanidin-3-(di-*p*-coumaroyl)-sophoroside-5-glucoside) and two triacylated anthocyanins (delphinidin-3-(*p*-coumaroyl-feruloyl)-sophoroside-5-(malonyl)-glucoside and delphinidin-3-(diferuloyl)-sophoroside-5-(malonyl)-glucoside) from *Ajuga reptans* L. flowers. Pale et al. (2003, 1998) identified cyanidin-3-*O*-[2-*O*-(6-*O*-E-caffeoyl-β-D-glucopyranosyl)]-{6-*O*-E-caffeoyl]-β-D-glucopyranosyl}-5-*O*-β-D-glucopyranoside and cyanidin-3-*O*-[2-*O*-(6-*O*-E-*p*-coumaroyl-β-D-glucopyranosyl)]-{6-*O*-E-caffeoyl]-β-D-glucopyranosyl}-5-*O*-β-D-glucopyranoside, cyanidin-3-*O*-[2-*O*-(6-*O*-E-*p*-coumaroyl-β-D-glucopyranosyl)]-{6-*O*-[4-*O*-(6-*O*-E-*p*-coumaroyl-β-D-glucopyranosyl)-E-caffeoyl]-β-D-glucopyranosyl}-5-*O*-β-D-glucopyranoside and cyanidin-3-*O*-[2-*O*-(6-*O*-E-caffeoyl-β-D-glucopyranosyl)]-{6-*O*-[4-*O*-(6-*O*-E-3,5-dihydroxycinnamoyl-β-D-glucopyranosyl)-E-caffeoyl]-β-D-glucopyranosyl}-5-*O*-β-D-glucopyranoside from *Ipomoea asarifolia* (Desv.) Roem. and Schult. flowers, respectively, which were diacylated, triacylated, and tetra-acylated with hydroxycinnamic acids.

Furthermore, some uncommon acylated ACNs possessing more than two acyl moieties have been found in some edible sources (Steingass et al., 2023). Among these polyacylated ACNs, a rare group of acylated B-ring-substituted compounds has been reported. An example for multiple B-ring acylated ACNs is the main coloring compound in *Tradescantia pallida* leaves – the tetra-acylated ACN tradescantin, which is the most abundant compound assigned to a cyanidin derivative carrying four feruloyl, one pentose, and three hexose moieties, *i.e.*, cyanidin-tetra(feruloyl)-pentoside-trihexoside. Other remaining ACNs may represent tradescantin-like structures merely carrying three feruloyl moieties, or three feruloyl moieties and one additional caffeoyl and one coumaroyl residue, respectively (Steingass et al., 2023). Additionally, acylated ACNs also can be detected in some herbal plants. A triacylated and B-ring substituted ACN called rhoeonin as the primary ACN had previously been reported in the fresh leaves from boat lily (*Rhoeo spathacea*) (Tan et al., 2014). Table 3.1 presents a list of acylated ACN edible sources and their ACN composition, and Figure 3.4 presents their chemical structures.

TABLE 3.1 Acylated ACN Composition of Common and Lesser-Known Edible Sources

SOURCE	*SCIENTIFIC NAME*	*ACYLATED ACN IDENTIFIED*	*ACYL GROUPS*	*REFERENCES*
Fruits				
Grapes	*Vitis vinifera*	Cyanidin-3-acetyl-glucoside, petunidin-3-acetyl-glucoside, peonidin-3-acetyl-glucoside, malvidin-3-acetyl-glucoside, peonidin-3-*p*-coumaroyl-glucoside, malvidin-3-*p*-coumaroyl-glucoside, delphinidin-3-acetyl-glucoside, malvidin-3-*O*-(6-*O*-acetyl)-glucoside-5-*O*-glucoside, delphinidin-3-*O*-(6-*O*-*p*-coumaroyl)-glucoside, cyanidin-3-*O*-(6-*O*-*p*-coumaroyl)-glucoside, malvidin-3-*O*-(6-*O*-*p*-coumaroyl)-glucoside-5-*O*-glucoside, petunidin-3-*O*-(6-*O*-*trans*-*p*-coumaroyl)-glucoside, peonidin-3-*O*-(6-O-*trans*-*p*-coumaroyl)-glucoside	Acetyl, *p*-coumaroyl	Costa et al. (2014), Kőrösi et al. (2022)
Blackcurrant	*Ribes nigrum* L.	Delphinidin 3-*O*-(6″-coumaroyl)-glucoside, cyanidin 3-*O*-(6″-coumaroyl)-glucoside	Coumaroyl	Slimestad & Solheim (2002)
Gooseberry	*Ribes grossularia* L.	Cyanidin-3-*O*-β-(6″-E-caffeoyl)-glucopyranoside, cyanidin-3-*O*-β-(6″-E-*p*-coumaroyl)-glucopyranoside	*p*-Coumaroyl, caffeoyl	Jordheim et al. (2007)
Jostaberry	*Ribes* x *nidigrolaria*	Cyanidin-3-*O*-β-(6″-E-caffeoyl)-glucopyranoside, cyanidin-3-*O*-β-(6″-E-*p*-coumaroyl)-glucopyranoside	*p*-Coumaroyl, caffeoyl	Jordheim et al. (2007)
Red flowering currant	*Ribes sanguineum*	Cyanidin-3–(6″-coumaroyl)-glucoside	Coumaroyl	Jordheim et al. (2007)
Elderberry	*Sambucus nigra* L.	Cyanidin-sambubioside-malonyl-glucoside, cyanidin-sambubioside-acetyl-glucoside, cyanidin-3-(Z)-*p*-coumaroyl-sambubioside-5-glucoside, cyanidin-3-(E)-*p*-coumaroyl-sambubioside-5-glucoside, cyanidin-3-*p*-coumaroyl-sambubioside	Malonyl, acetyl, *p*-coumaroyl	Mikulic-Petkovsek et al. (2014)
Blackberry	*Rubus* spp.	Cyanidin-3-*O*-malonyl-glucoside, cyanidin-3-*O*-dioxalyl-glucoside	Malonyl, dioxalyl	Zhang et al. (2012)
Blueberry	*Vaccinium* spp.	Cyanidin-3-*O*-(6″-acetyl)-glucoside, cyanidin-3-*O*-(6″-acetyl)-arabinoside, cyanidin-3-*O*-(6″-acetyl)-galactoside, delphinidin-3-*O*-(6″-acetyl)-glucoside, delphinidin-3-*O*-(6″-acetyl)-galactoside, malvidin-3-*O*-(6″-acetyl)-galactoside, malvidin-3-*O*-(6″-acetyl)-glucoside, peonidin-3-*O*-(6″-acetyl)-galactoside, peonidin-3-*O*-(6″-acetyl)-glucoside, petunidin-3-*O*-(6″-acetyl)-galactoside, petunidin-3-*O*-(6″-acetyl)-glucoside	Acetyl	Wang et al. (2024)
Strawberry	*Fragaria* × *ananassa*	Pelargonidin-3-acetyl-glucoside, pelargonidin-succinyl-glucoside, pelargonidin-3-(6-malonyl)-glucoside, cyanidin-3-malonyl-glucoside, pelargonidin-3-malonyl-glucoside, cyanidin-3-malonylglucosyl-5-glucoside	Acetyl, succinyl, malonyl	Bakker et al. (1994), Hong & Wrolstad (1990), Y. Yoshida et al. (2002)

Black plum (peel)	*Prunus salicina* L.	Malvidin-3-acetyl-glucoside, cyanidin-3-acyl-glucoside	Acetyl	Wu et al. (2006), X. Zhang et al. (2019)
Lycium	*Lycium ruthenicum* Murray	Delphinidin-3-*O*-[6-*O*-(4-*O*-*p*-coumaroyl-α-L-rhamnopyranosyl)-β-D-glucopyranoside]-5-*O*-[β-D-glucopyranoside], petunidin-3-*O*-[6-*O*-(4-*O*-*trans*-*p*-coumaroyl-α-L-rhamnopyranosyl)-β-D-glucopyranoside]-5-*O*-[β-D-glucopyranoside], malvidin-3-*O*-[6-*O*-(4-*O*-*p*-coumaroyl-α-L-rhamnosyl)-β-D-glucopyranoside]-5-*O*-β-D-glucopyranoside, petunidin-3-*O*-(glucosyl-*p*-coumaroyl)-rutinoside-5-*O*-glucoside, petunidin-3-*O*-(caffeoyl)-rutinoside-5-*O*-glucoside, petunidin-3-*O*-(*p*-coumaroyl)-rutinoside, petunidin-3-*O*-[rhamnopyranosyl-(*p*-coumaroyl)]-5-*O*-glucoside, delphinidin-3-*O*-(6′-*p*-coumaryl)-glucoside	*P*-coumaroyl, caffeoyl	Liu et al. (2020), Peng et al. (2020), Tian et al. (2021)
Nitraria tangutorum Bobr. fruit	*N. tangutorum*	Cyanidin-3-*O*-(caffeoyl)-diglucoside, delphinidin-3-*O*-(*cis*-*p*-coumaroyl)-glucoside-5-*O*-glucoside, delphinidin-3-*O*-(trans-*p*-coumaroyl)-glucoside-5-*O*-glucoside, cyanidin-3-*O*-(feruloyl)-diglucoside, pelargonidin-3-*O*-(caffeoyl)-diglucoside, cyanidin-3-*O*-(*cis*-*p*-coumaroyl)-diglucoside, cyanidin-3-*O*-(*trans*-*p*-coumaroyl)-diglucoside, pelargonidin-3-*O*-(*p*-coumaroyl)-diglucoside, delphinidin-3-*O*-(*cis*-*p*-coumaroyl)-glucoside, cyanidin-3-*O*-(*p*-coumaroyl)-glucoside, delphinidin-3-*O*-(*trans*-*p*-coumaroyl)-glucoside	Caffeoyl, *p*-coumaroyl, feruloyl	Sang et al. (2019)
Red-fleshed peach	*Prunus persica* L. Batsch 'M1', *Prunus persica* L. Batsch 'M3', *Prunus persica* L. Batsch 'ML', *Prunus persica* L. Batsch 'MN'	Malvidin-3-coumaroyl-glucoside-acetaldehyde	Coumaroyl	Aubert & Chalot (2020)
Red-fleshed apple	*Malus pumila* var. *medzwetzkyana* Dieck	Delphindin-3-*O*-(6″-*O*-*p*-coumaroyl)-glucoside	*p*-Coumaroyl	Yang et al. (2024)
Blood orange "sanguinelli," "Tarocco Rosso," "Tarocco Ippolito"	*Citrus sinensis* L. cv	Delphinidin-3–(6″-malonyl)-glucoside, cyanidin-(malonyl)-diglucoside, cyanidin-3-(6″-malonyl)-glucoside, cyanidin-3-(6″-dioxalyl)-glucoside, peonidin-3-(6″-malonyl)-glucoside, cyanidin-malonyl-(dioxalyl)-hexoside	Malonyl, dioxalyl	Cebadera-Miranda et al. (2019)
Passion fruit (rind)	*Passiflora suberosa*	Cyanidin-3–(6″-malonyl)-glucoside	Malonyl	Kidøy et al. (1997)

(Continued)

TABLE 3.1 (Continued) Acylated ACN Composition of Common and Lesser-Known Edible Sources

SOURCE	*SCIENTIFIC NAME*	*ACYLATED ACN IDENTIFIED*	*ACYL GROUPS*	*REFERENCES*
Vegetables and Tubes				
Purple sweet potato (tubes) "Jihei No. 1" "P40" "NC415" "Stokes purple" "Borami" "Mokpo 62" "Shinzami" "Zami"	Ipomoea batatas L.	Cyanidin-3-(*p*-hydroxybenzoyl)-sophoroside-5-glucoside, peonidin-3-(*p*-hydroxybenzoyl)-sophoroside-5-glucoside, cyanidin-3-(6″-feruloyl)-sophoroside-5-glucoside, peonidin-3-(6″-feruloyl)-sophoroside-5-glucoside, cyanidin-3-(6″-caffeoyl)-sophoroside-5-glucoside, cyanidin-3-(caffeoyl)(*p*-hydroxybenzoyl)-sophoroside-5-glucoside, cyanidin-3-(6″-caffeoyl)(6‴-feruloyl)-sophoroside-5-glucoside, peonidin-3-caffeoyl-sophoroside-5-glucoside, peonidin-3-(caffeoyl)(*p*-hydroxybenzoyl)-sophoroside-5-glucoside, peonidin-3-(6″-caffeoyl-6‴-feruloyl)-sophoroside-5-glucoside, cyanidin-3–(6″,6‴-dicaffeoyl)-sophoroside-5-glucoside, peonidin-3-(6″-caffeoyl-6‴-feruloyl)-sophoroside-5-glucoside, peonidin-dicaffeoyl-sophoroside-5-glucoside, pelargonidin-3-(6‴-caffeoyl)-sophoroside-5-glucoside, pelargonidin-3-(feruloyl)-sophoroside-5-glucoside, pelargonidin-3-(caffeoyl)-sophoroside-5-glucoside, pelargonidin-3-(dicaffeoyl)-sophoroside-5-glucoside, pelargonidin-3-(caffeoyl)(*p*-hydroxybenzyls) sophoroside-5-glucoside, pelargonidin-3-(caffeoyl)(feruloyl)- sophoroside-5-glucoside, pelargonidin-3-(caffeoyl)(*p*-coumaroyl)- sophoroside-5-glucoside	*P*-Hydroxybenzoyl, feruloyl, caffeoyl	Lee et al. (2013), Li et al. (2013), Truong et al. (2010), Xu et al. (2015)
Purple sweet potato (leaves) "Fushu No. 23" "Fushu No. 317"	*Ipomoea batatas* L.	Cyanidin-3-(*p*-hydroxybenzoyl)-sophoroside-5-glucoside, cyanidin-3-(caffeoyl)-sophoroside-5-glucoside, cyanidin-3-(caffeoyl)(*p*-coumaroyl)-sophoroside-5-glucoside, cyanidin-3-(caffeoyl)(feruloyl)-sophoroside-5-glucoside, cyanidin-3-(*p*-coumaroyl)-sophoroside-5-glucoside, cyanidin-3-(feruloyl)-sophoroside-5-glucoside, cyanidin-3-(caffeoyl)(*p*-hydroxybenzoyl)-sophoroside-5-glucoside, cyanidin-3-(dicaffeoyl)-sophoroside-5-glucoside, peonidin-3-(*p*-hydroxybenzoyl)-sophoroside-5-glucoside, peonidin-3-(caffeoyl)-sophoroside-5-glucoside, peonidin-3-(*p*-coumaroyl)-sophoroside-5-glucoside, peonidin-3-(feruloyl)-sophoroside-5-glucoside, peonidin-3-(dicaffeoyl)-sophoroside-5-glucoside, peonidin-3-(caffeoyl)(*p*-hydroxybenzoyl)-sophoroside-5-glucoside, peonidin-3-(caffeoyl)(*p*-coumaroyl)-sophoroside-5-glucoside, peonidin-3-(caffeoyl)(feruloyl)-sophoroside-5-glucoside	*P*-Hydroxybenzoyl, *p*-coumaroyl, feruloyl, caffeoyl	Li et al. (2019)

Red cabbage	*Brassica oleracea* L. var. *Capitata* f. *rubra*	Cyanidin-3-(sinapoyl)-diglucoside-5-glucoside, cyanidin-3-(sinapoyl)-triglucoside-5-glucoside, cyanidin-3-(*p*-coumaroyl)-triglucoside-5-glucoside, cyanidin-3-(feruloyl)-triglucoside-5-glucoside, cyanidin-3-(sinapoyl)-triglucoside-5-glucoside, cyanidin-3-(feruloyl)(feruloyl)-triglucoside-5-glucoside, cyanidin-3-(feruloyl)(sinapoyl)-triglucoside-5-glucoside, cyanidin-3-(sinapoyl)(sinapoyl)-triglucoside-5-glucoside, cyanidin-3-(*p*-coumaroyl)-diglucoside-5-glucoside, cyanidin-3-(sinapoyl)-diglucoside-5-glucoside, cyanidin-3-(feruloyl)-diglucoside-5-glucoside, cyanidin-3-(sinapoyl)-glucoside-5-glucoside, cyanidin-3-(feruloyl)(feruloyl)-diglucoside-5-glucoside, cyanidin-3-(feruloyl)(sinapoyl)-triglucoside-5-glucoside, cyanidin-3-(feruloyl)(sinapoyl)-diglucoside-5-glucoside, cyanidin-3-(sinapoyl)(sinapoyl)-diglucoside-5-glucoside, cyanidin-3-(caffeoyl)-diglucoside-5-glucoside, cyanidin-3-(*p*-coumaroyl)-glucoside-5-glucoside, cyanidin-3-(sinapoyl)-glucoside-5-glucoside, cyanidin-3-(*p*-coumaroyl)(sinapoyl)-diglucoside-5-glucoside, cyanidin-3-(caffeoyl)(sinapoyl)-diglucoside-5-glucoside, cyanidin-3-(sinapoyl)(*p*-coumaroyl)-diglucoside-5-glucoside, cyanidin-3-(sinapoyl)(caffeoyl)-diglucoside-5-glucoside, peonidin-3-(*p*-coumaroyl)-diglucoside-5-glucoside, peonidin-3-(*p*-coumaroyl)(sinapoyl)-diglucose-5-glucoside	Sinapoyl, *p*-coumaroyl, feruloyl, caffeoyl	Fierri et al. (2023), Strauch et al. (2019)
Red and purple mustards	*Brassica juncea*	Cyanidin-3-(sinapoyl)-diglucoside-5-glucoside, cyanidin-3-(*p*-coumaroyl)(sinapoyl)-diglucoside-5-(malonyl)-glucoside, cyanidin-3-diferuloyl-sophoroside-5-glucoside, cyanidin-3-(feruloyl)-glucoside-5-(malonyl)-glucoside, cyanidin-3-(sinapoyl)(*p*-coumaroyl)-triglucoside-5-(malonyl)-glucoside, cyanidin-3-(sinapoyl)(*p*-coumaroyl)-triglucoside-5-(malonyl)-glucoside, cyanidin-3-(caffeoyl)(sinapoyl)-xylglucoside-5-(malonyl) glucoside, cyanidin-3-(*p*-coumaroyl)(sinapoyl)-diglucoside-5-*O*-(malonyl)-glucoside, cyanidin-3-(feruloyl)(feruloyl)-diglucoside-5-(malonyl)-glucoside, cyanidin-3-(sinapoyl)(feruloyl)-diglucoside-5-(malonyl)-glucoside, cyanidin-3-(sinapoyl)(feruloyl)-diglucoside-5-(malonyl)-glucoside, cyanidin-3-(*p*-coumaroyl)(sinapoyl)-diglucoside-5-(malonyl)-glucoside	Sinapoyl, malonyl, *p*-coumaroyl, caffeoyl, feruloyl	Sun et al. (2013)
Broccoli sprouts	*Brassica oleracea* 'Viola'	Cyanidin-3-*O*-(sinapoyl)-diglucoside-5-*O*-glucoside, cyanidin-3-*O*-(feruloyl)-diglucoside-5-*O*-glucoside, cyanidin-3-*O*-(sinapoyl)-triglucoside-5-*O*-glucoside, cyanidin-3-*O*-(*p*-coumaroyl)(sinapoyl)-diglucoside-5-*O*-glucoside, cyanidin-3-*O*-(sinapoyl)(feruloyl)-diglucoside-5-*O*-glucoside, cyanidin-3-*O*-(sinapoyl)(sinapoyl)-diglucoside-5-*O*-glucoside, cyanidin-3-*O*-(*p*-coumaroyl)(sinapoyl)-diglucoside-5-*O*-(malonyl)-glucoside, cyanidin-3-*O*-(sinapoyl)(feruloyl)-diglucoside-5-*O*-(malonyl)-glucoside, cyanidin-3-*O*-(sinapoyl)(sinapoyl)-diglucoside-5-*O*-(malonyl)-glucoside	*p*-Coumaroyl, sinapoyl, feruloyl, malonyl	Moreno et al. (2010)

(Continued)

TABLE 3.1 (Continued) Acylated ACN Composition of Common and Lesser-Known Edible Sources

SOURCE	*SCIENTIFIC NAME*	*ACYLATED ACN IDENTIFIED*	*ACYL GROUPS*	*REFERENCES*
Purple kohlrabi	*Brassica oleracea* var. *gongylodes*	Cyanidin-3-(*p*-coumaroyl)-sophoroside-5-(malonyl)-glucoside, cyanidin-3-(*p*-coumaroyl)(sinapoyl)-diglucoside-5-(malonyl)-glucoside, cyanidin-3-(feruloyl)-glucoside-5-(malonyl)-glucoside, cyanidn-3-(sinapoyl)-glucoside-5-(malonyl)-glucoside, cyanidin-3-diferuloyl-sophoroside-5-glucoside, cyanidin-3-(sinapoyl)(feruloyl)-sophoroside-5-glucoside, cyanidin-3-(feruloyl)(feruloyl)-diglucoside-5-(malonyl)-glucoside, cyanidin-3-(sinapoyl)(feruloyl)-diglucoside-5-(malonyl)-glucoside, cyanidin-3-(sinapoyl)(sinapoyl)-diglucoside-5-(malonyl)-glucoside	*p*-Coumaroyl, malonyl, sinapoyl, feruloyl	Sun et al. (2013)
Red mizuna	*Brassica rapa* var. *nipposinica* or *Brassica juncea* var. *japonica*	Cyanidin-3-(sinapoyl)-diglucoside-5-glucoside, cyanidin-3-(caffeoyl)(*p*-coumaroyl)diglucoside-5-glucoside, cyanidin-3-(glycopyranosyl-sinapoyl)-diglucoside-5-glucoside, cyanidin-3-(*p*-coumaroyl)(sinapoyl)-triglucoside-5-glucoside, cyanidin-3-(sinapoyl)-glucoside-5-glucoside, cyanidin-3-(*p*-coumaroyl)-diglucoside-5-glucoside, cyanidin-3-(sinapoyl)-diglucoside-5-glucoside, cyanidin-3-(*p*-coumaroyl)(sinapoyl)-diglucoside-5-glucoside, cyanidin-3-(feruloyl)(sinapoyl)-diglucoside-5-glucoside, cyanidin-3-(sinapoyl)(sinapoyl)-diglucoside-5-glucoside, cyanidin-3-(feruloyl)(sinapoyl)-diglucoside-5-glucoside	*p*-Coumaroyl, sinapoyl, caffeoyl, feruloyl	Park et al. (2020)
Red kale	*Brassica oleracea* var. *acephala*	Cyanidin-3-sinapoyl-diglucoside-5-glucoside, cyanidin-3-feruloyl-diglucoside-5-glucoside, cyanidin-3-sinapoyl-feruloyl-diglucoside-5-diglucoside, cyanidin-3-caffeoyl-feruloyl-diglucoside, cyanidin-3-sinapoyl-caffeoyl-diglucoside-5-glucoside, cyanidin-3-hydroxyferuloyl-feruloyl-diglucoside-5-glucoside, cyanidin-3-sinapoyl-*p*-coumaroyl-diglucoside-5-glucoside, cyanidin-3-sinapoyl-feruloyl-diglucoside-5-glucoside, cyanidin-3-disinapoyl-diglucoside-5-glucoside, cyanidin-3-di feruloyl-diglucoside-5-glucoside, cyanidin-3-feruloyl-sinapoyl-diglucoside-5-glucoside	Hydroxyferuloyl, feruloyl, *p*-coumaroyl, caffeoyl, sinapoyl	Olsen et al. (2010)
Purple cauliflower	Brassica oleracea L. var. botrytis	Cyanidin-3–(6-*p*-coumaroyl)-sophoroside-5-glucoside, cyanidin-3–(6-feruloyl)-sophoroside-5-glucoside, cyanidin-3–(6-*p*-coumaroyl)-sophoroside-5–(6-sinapoyl)-glucoside, cyanidin-3–(6-feruloyl)-sophoroside-5–(6-sinapoyl)-glucoside	*P*-coumaroyl, feruloyl, sinapoyl	Scalzo et al. (2008)
Thai and Filipino purple yam	*Dioscorea alata* L.	Cyanidin-3-*O*-(6-*O*-(6-*O*-(E)-sinapoyl-β-D-glucopyranosyl)-β-D-glucopyranosyl), cyanidin-3-*O*-(6-*O*-(6-*O*-(E)-sinapoyl-β-D-glucopyranosyl)-β-D-glucopyranosyl)-7-*O*-(6-*O*-(E)-sinapoyl-β-D-glucopyranosyl)-2′-*O*-(β-D-glucopyranosyl)	Sinapoyl	Srivichai & Hongsprabhas (2020), Yoshida et al. (1991)
Black carrot	*Daucus carota* L. ssp. *sativus* var. *atrorubens* Alef.	Cyanidin-3-*O*-xylosyl-feruloyl-glucosyl-galactoside, cyanidin-3-*O*-xylosyl-*p*-coumaroyl-glucosyl-galactoside, cyanidin-3-*O*-xylosyl-sinapoyl-glucosyl-galactoside, peonidin-3-*O*-xylosyl-sinapoyl-glucosyl-galactoside	Feruloyl, sinapoyl, *p*-coumaroyl	Koley et al. (2019)

Purple radish (root or leaf) "VRRAD-151"	*Raphanus sativus* L.	Cyanidin-3-*O*-sophoroside-5-*O*-malonylglucoside, cyanidin-3-*O*-xylosyl-*p*-coumaroyl-glucosyl-galactoside, cyanidin-3-*O*-xylosyl-feruloyl-glucosyl-galactoside, cyanidin-3-*O*-*p*-coumaroyl-sophoroside-5-*O*-malonyl-glucoside, cyanidin-3-*O*-caffeoyl-sophoroside-5-*O*-malonyl-glucoside, cyanidin-3-*O*-feruloyl-sophoroside-5-*O*-malonyl-glucoside, cyanidin-3-*O*-[2-*O*-(xylosyl)-6-*O*-(*p*-*O*-(glucosyl)-*p*-coumaroyl-glucoside]-5-*O*-glucoside, cyanidin-3-*O*-caffeoyl-*p*-coumaroyl-sophoroside-5-*O*-glucoside, cyanidin-3-*O*-*p*-coumaroyl-feruloyl-sophoroside-5-*O*-glucoside, cyanidin-3-*O*-caffeoyl-feruloyl-sophoroside-5-*O*-glucoside, cyanidin-3-*O*-di-*p*-coumaroyl-sophoroside-5-*O*-malonyl-glucoside, cyanidin-3-*O*-*p*-coumaroyl-triglucoside-5-*O*-malonyl-glucoside, cyanidin-3-*O*-*p*-coumaroyl-feruloyl-sophoroside-5-*O*-malonyl-glucoside, cyanidin-3-*O*-*p*-coumaroyl-caffeoyl-sophoroside-5-*O*-succinyl-glucoside, cyanidin-3-*O*-caffeoyl-triglucoside-5-*O*-malonyl-glucoside, cyanidin-3-*O*-caffeoyl-feruloyl-sophoroside-5-*O*-malonyl-glucoside, cyanidin-3-*O*-*p*-coumaroyl-feruloyl-sophoroside-5-*O*-malonyl-sophoroside, cyanidin-3-*O*-caffeoyl-feruloyl-sophoroside-5-*O*-malonyl-sophoroside, cyanidin-3-*O*-*p*-coumaroyl-sinapoyl-sophoroside-5-*O*-malonyl-sophoroside, pelargonidin-3-*O*-*p*-coumaroyl-diglucoside-5-*O*-glucoside, pelargonidin-3-*O*-feruloyl-diglucoside-5-*O*-glucoside, pelargonidin-3-*O*-caffeoyl-diglucoside-5-*O*-malonyl-glucoside, pelargonidin-3-*O*-caffeoyl-caffeoyl-diglucoside-5-*O*-malonyl-glucoside	Feruloyl, caffeoyl, *p*-coumaroyl, malonyl, succinyl	Koley et al. (2020)
Chinese red radish "Yan Zhi #2" "Tou Xin Hong" "Xin Ling Mei"	*Raphanus sativus* L.	Pelargonidin-3-(caffeoyl)-diglucoside-5-glucoside, pelargonidin-3-(caffeoyl)(caffeoyl)-diglucoside-5-glucoside, pelargonidin-3-(caffeoyl)-diglucoside-5-(malonyl)-glucoside, pelargonidin-3-(*p*-coumaroyl)-diglucoside-5-glucoside, pelargonidin-3-(feruloyl)-diglucoside-5-glucoside, pelargonidin-3-(caffeoyl)(feruloyl)-diglucoside-5-glucoside, pelargonidin-3-(caffeoyl)(*p*-coumaroyl)-diglucoside-5-glucoside, pelargonidin-3-(caffeoyl)(caffeoyl)-diglucoside-5-(malonyl)-glucoside, pelargonidin-3-(*p*-coumaroyl)-diglucoside-5-(malonyl)-glucoside, pelargonidin-3-(feruloyl)-diglucoside-5-(malonyl)-glucoside, pelargonidin-3-(feruloyl)(*p*-coumaroyl)-diglucoside-5-glucoside, pelargonidin-3-(feruloyl)(feruloyl)-diglucoside-5-glucoside, pelargonidin-3-(caffeoyl)(feruloyl)-diglucoside-5-(malonyl)-glucoside, pelargonidin-3-(caffeoyl)(*p*-coumaroyl)-diglucoside-5-(malonyl)-glucoside, pelargonidin-3-diglucoside-5-(malonyl)-glucoside, pelargonidin-3-(feruloyl)(feruloyl)-diglucoside-5-(malonyl)-glucoside, pelargonidin-3-(feruloyl)(*p*-coumaroyl)-diglucoside-5-(malonyl)-glucoside, pelargonidin-3-(*p*-coumaroyl)(*p*-coumaroyl)-diglucoside-5-(malonyl)-glucoside	Caffeoyl, *p*-coumaroyl, feruloyl, malonyl	Jing et al. (2012), W. Li et al. (2023)

(Continued)

TABLE 3.1 (Continued) Acylated ACN Composition of Common and Lesser-Known Edible Sources

SOURCE	*SCIENTIFIC NAME*	*ACYLATED ACN IDENTIFIED*	*ACYL GROUPS*	*REFERENCES*
Radish sprouts	*Raphanus sativus* cv Sango	Cyanidin-3-*O*-[2-*O*-(2-*O*-(E)-sinapoyl-6-*O*-(E)-feruloyl-β-D-glucopyranosyl)-β-D-glucopyranoside]–5-*O*-(β-D-glucopyranoside), cyanidin-3-*O*-[6-*O*-(E)-feruloyl-2-*O*-(2-*O*-(E)-sinapoyl-β-D-glucopyranosyl)-β-D-glucopyranoside]-5-*O*-(4-*O*-β-D-glucopyranosyl-6-*O*-malonoyl-β-D-glucopyranoside), cyanidin-3-*O*-[6-*O*-(E)-feruloyl-2-*O*-(2-*O*-(E)-feruloyl-β-D-glucopyranosyl)-β-D-glucopyranoside]–5-*O*-(4-*O*-β-D-glucopyranosyl-6-*O*-malonoyl-β-D-glucopyranoside), cyanidin-3-*O*-[6-*O*-(E)-feruloyl-2-*O*-(6-*O*-(E)-feruloyl-β-D-glucopyranosyl)-β-D-glucopyranoside]-5-*O*-(4-*O*-β-D-glucopyranosyl-6-*O*-malonoyl-β-D-glucopyranoside), cyanidin-3-*O*-[6-*O*-(E)-feruloyl-2-*O*-(2-*O*-(E)-sinapoyl-β-D-glucopyranosyl)-β-D-glucopyranoside]–5-*O*-(6-*O*-malonoyl-β-D-glucopyranoside), cyanidin-3-*O*-[2-*O*-(2-*O*-(E)-sinapoyl-β-D-glucopyranosyl)-β-D-glucopyranoside]–5-*O*-(β-D-glucopyranoside), cyanidin-3-*O*-[6-*O*-(E)-feruloyl-2-*O*-(2-*O*-(E)-sinapoyl-β-D-glucopyranosyl)-β-D-glucopyranoside]-5-*O*-(β-D-glucopyranoside), cyanidin-3-*O*-[2-*O*-(6-*O*-(E)-feruloyl-β-D-glucopyranosyl)-β-D-glucopyranoside]-5-*O*-(6-*O*-malonoyl-β-D-glucopyranoside), cyanidin-3-*O*-[6-*O*-(E)-feruloyl-2-*O*-(6-*O*-(E)-feruloyl-β-D-glucopyranosyl)-β-D-glucopyranoside]-5-*O*-(6-*O*-malonoyl-β-D-glucopyranoside)	Feruloyl, sinapoyl, malonyl	Matera et al. (2015)
Eggplant "Japanese type nasunin" (peel) "Transgenic eggplant" (fruit flesh)	*Solanum melongena* L.	Petunidin-3-(*p*-coumaroyl)-rutinoside-5-glucoside, petunidin 3-(feruloyl)-rutinoside-5-glucoside, delphinidin-3-(caffeoyl)-rutinoside-5-glucoside, delphinidin-3-(*p*-coumaroyl)-rutinoside-5-glucoside, malvidin-3-(*p*-coumaroyl)-rutinoside-5-glucoside, malvidin-3-(feruloyl)-rutinoside-5-glucoside, cyanidin-3-(*p*-coumaroyl)-rutinoside-5-glucoside	Caffeoyl, *p*-coumaroyl, feruloyl	Azuma et al. (2008), Zhang et al. (2016)
Red Chicory "Chioggia," "Treviso," "Treviso tardivo," "Verona"	*Cichorium intybus* L.	Delphinidin-3-*O*-(6″-*O*-malonyl)-glucoside-5-*O*-glucoside, cyanidin-3,5-di-*O*-(6″-*O*-malonyl)-glucoside	Malonyl	Carazzone et al. (2013)

Red onion "Krishnapuram"	*Allium cepa* L.	Cyanidin-3–(6″-malonyl)-glucoside, cyanidin-3–(3″-malonyl)-glucoside, cyanidin-3-malonyl-laminaribioside, peonidin-3-*O*-(6″-malonyl)-glucoside	Malonyl	Tedesco et al. (2015), Velasco González de Peredo et al. (2021), Vijayalakshmi et al. (2021)
Red-fleshed potato "Rosemary," "Red Emmalie," "Red Cardinal," "Herbie 26," "Rosalinde," "Highland," "Burgundy Red"	*Solanum tuberosum* L. cv	Pelargonidin-3-*O*-(*p*-coumaroyl)-rutinoside-5-*O*-glucoside, pelargonidin-3-*O*-(*p*-coumaroyl)-rutinoside, pelargonidin-3-*O*-(caffeoyl)-rutinoside-5-*O*-glucoside, pelargonidin-3-*O*-(feruloyl)-rutinoside-5-*O*-glucoside, peonidin-3-(*p*-coumaroyl)-rutinoside-5-glucoside, peonidin-3-feruoyl-rutinoside-5-glucoside	*p*-Coumaroyl, caffeoyl, feruloyl	Kita et al. (2013), Rytel et al. (2021), Sampaio et al. (2021)
Purple-fleshed potato "Purple," "Violetta," "Kefermarkter Blaue," "Salad Blue," "Vitelotte," "Valfi," "Blue Congo," "British Columbia Blue"	*Solanum tuberosum* L. cv	Petunidin-3-*O*-(caffeoyl)-rutinoside-5-*O*-glucoside, petunidin-3-*O*-(feruloyl)-rutinoside-5-*O*-glucoside, peonidin-3-*O*-(feruloyl)-rutinoside-5-*O*-glucoside, petunidin-3-*O*-(*p*-coumaroyl)-rutinoside, delphinidin-3-(*p*-coumaroyl)-rutinoside-5-glucoside, malvidin-3-(*p*-coumaroyl)-rutinoside-5-glucoside, malvidin-3-(feruloyl)-rutinoside-5-glucoside, malvidin-3-(*p*-coumaroyl)-rutinoside, cyanidin derivatives	*p*-Coumaroyl, caffeoyl, feruloyl	Kita et al. (2013), Lachman et al. (2009), Sampaio et al. (2021)
Marble-fleshed potato Shetland Black	*Solanum tuberosum* L. cv	Petunidin-3-*O*-(caffeoyl)-rutinoside-5-*O*-glucoside, petunidin-3-*O*-(*p*-coumaroyl)-rutinoside-5-*O*-glucoside, petunidin-3-*O*-(feruloyl)-rutinoside-5-*O*-glucoside, peonidin-3-*O*-(feruloyl)-rutinoside-5-*O*-glucoside	*p*-Coumaroyl, caffeoyl, feruloyl	Sampaio et al. (2021)
Purple pepper	*Capsicum annuum* L.	Delphinidin-3-*O*-(caffeoyl)-rutinoside-5-*O*-glucoside, delphinidin-3-*O*-(*cis*-*p*-coumaroyl)-rutinoside-5-*O*-glucoside, delphinidin-3-*O*-(*trans*-*p*-coumaroyl)-rutinoside-5-*O*-glucoside, delphinidin-3-*O*-(feruloyl)-rutinoside-5-*O*-hexoside, delphinidin-3-*O*-(*p*-coumaroyl)-hexoside, petunidin-3-*O*-(*p*-coumaroyl)-hexoside	Caffeoyl, *p*-coumaroyl, feruloyl	Sadilova et al. (2006)
Purple tomato transgenic plant "*Aft/Aft atv/atv*," "*Del/Ros1*," "Micro-Tom"	*Solanum lycopersicum*	Delphinidin-3-(*p*-coumaroyl)-rutinoside-5-glucoside, delphinidin-3-(caffeoyl)-rutinoside-5-glucoside, petunidin-3-(*p*-coumaroyl)-rutinoside-5-glucoside, petunidin-3-(caffeoyl)-rutinoside-5-glucoside, malvidin-3-(*p*-coumaroyl)-rutinoside-5-glucoside, malvidin-3-(caffeoyl)-rutinoside-5-glucoside, malvidin-3-(feruloyl)-rutinoside-5-glucoside	*p*-Coumaroyl, caffeoyl, feruloyl	Mathews et al. (2003), Mes et al. (2008), Su et al. (2016)

(Continued)

TABLE 3.1 (Continued) Acylated ACN Composition of Common and Lesser-Known Edible Sources

SOURCE	*SCIENTIFIC NAME*	*ACYLATED ACN IDENTIFIED*	*ACYL GROUPS*	*REFERENCES*
Red lettuce	*Lactuca sativa* L.	Cyanidin-3-*O*-(6-malonyl)-glucoside	Malonyl	Cheng et al. (2014)
Purple shamrock (leaves) "ssp. papilionacea"	*Oxalis triangularis*	Malvidin-3-*O*-(6-*O*-(4-*O*-malonyl-α-rhamnopyranosyl)-β-glucopyranoside)-5-*O*-β-glucopyranoside, malvidin-3-*O*-(6-*O*-α-rhamnopyranosyl-β-glucopyranoside)-5-*O*-(6-*O*-malonyl-β-glucopyranoside), malvidin-3-*O*-(6-*O*-(4-*O*-malonyl-α-rhamnopyranosyl)-β-glucopyranoside)-5-*O*-(6-*O*-malonyl-β-glucopyranoside), malvidin-3-*O*-(6-*O*-(4-*O*-malonyl-α-rhamnopyranosyl)-β-glucopyranoside, malvidin-3-*O*-(6-*O*-(Z)-*p*-coumaroyl-β-glucopyranoside)-5-*O*-β-glucopyranoside, malvidin-3-*O*-(6-*O*-(E)-*p*-coumaroyl-β-glucopyranoside)-5-*O*-β-glucopyranoside, petunidin-3-*O*-(malonyl-glucoside), pelargonidin-3-*O*-(caffeoyl-glucoside)-5-glucoside, pelargonidin-3-*O*-(*p*-coumaroyl-diglucoside)	Malonyl, *p*-coumaroyl, caffeoyl	Fossen et al. (2005), L. Li et al. (2023)
Perilla (Shiso) var. *acuta*	*Perilla frutescens*	Cyanidin-3-*O*-(6-*O*-*p*-coumaroyl)-glucoside-5-*O*-glucoside, cyanidin-3-*O*-(caffeoyl)-glucoside-5-*O*-glucoside, cyanidin-3-*O*-(caffeoyl)-glucoside-5-*O*-(malonyl)-glucoside, cyanidin-3-*O*-(feruloyl)-glucoside-5-*O*-glucoside, cyanidin-3-*O*-(*p*-coumaroyl)-glucoside-5-*O*-(malonyl)-glucoside	*p*-Coumaroyl, caffeoyl, feruloyl, malonyl	He et al. (2015)
Purple basil	*Ocimum basilicum* L.	Cyanidin-3-*O*-(6′-*p*-coumaroyl-6″-caffeoyl)-diglucoside, cyanidin-3-*O*-(6′-*p*-coumaroyl-6″-caffeoyl)-diglucoside-5-*O*-glucoside, cyanidin-3-*O*-(6′-*p*-coumaroyl)-*O*-glucoside-5-*O*-glucoside, cyanidin-3-*O*-(6′,6″-di-*p*-coumaroyl)-diglucoside-5-*O*-glucoside	*p*-Coumaroyl, caffeoyl	Tao et al. (2025)
Okinawan Spinach	*Gynura bicolor* DC.	Cyanidin-3-*O*-[β-D-glucopyranoside]-7-*O*-[6-*O*-(E)-(4-*O*-(6-*O*-(E)-caffeoyl-β-D-glucopyranosyl)-caffeoyl)-β-D-glucopyranoside]-3′-*O*-[6-*O*-(E)--caffeoyl-β-D-glucopyranoside], cyanidin-3-*O*-[6-*O*-malonyl-β-D-glucopyranoside]-7-*O*-[6-*O*-(E)-(4-*O*-(6-*O*-(E)-caffeoyl-β-D-glucopyranosyl)-caffeoyl)-β-D-glucopyranoside]-3′--*O*-[6-*O*-(E)-caffeoyl-β-D-glucopyranoside], cyanidin-3-*O*-[4, 6-di-*O*-malonyl-β-D-glucopyranoside]-7-*O*-[6-*O*-(E)-(4-*O*-(6-*O*-(E)-caffeoyl-β-D-glucopyranosyl)-caffeoyl)-β-D-glucopyranoside]-3′-*O*-[6-*O*-(E)-caffeoyl-β-D-glucopyranoside]	Caffeoyl, malonyl	Shimizu et al. (2010)
Grains and legumes				
Purple corn "cv Heizhenzhu," "self-ing and sibbing"	*Zea mays* L.	Cyanidin-3-*O*-malonyl-glucoside, pelargonidin-3-*O*-malonyl-glucoside, peonidin-3-*O*-malonyl-glucoside, cyanidin-3-succinyl-glucoside, cyanidin-3-dimalonyl-glucoside, pelargonidin-3-dimalonyl-glucoside, peonidin-3-dimalonyl-glucoside, cyanidin-3-malonyl-glucosyl-5-glucoside, cyanidin-3-succinyl-glucoside, cyanidin-3-(6″-(ethylmalonyl)-glucoside	Malonyl	C.-Y. Li et al. (2008), Massironi et al. (2025), Yang & Zhai (2010)

Black adzuki bean (seed coat) "Geomguseul cultivar"	*Vigna angularis*	Delphinidin-3-*O*-(*p*-coumaroyl)-glucoside, petunidin-3-*O*-(*p*-coumaroyl)-glucoside	*p*-Coumaroyl	Ha et al. (2021)
Red adzuki bean (seed coat)	*Vigna angularis*	Peonidin-3-(*p*-coumaroyl)-glucoside	*p*-Coumaroyl	Han et al. (2015)
Edible Flowers				
Morning glory	*Pharbitis nil*	Pelargonidin-3-*O*-(2-*O*-(6-*O*-(*trans*-3-*O*-(β-D-glucopyranosyl) caffeyl)-β-D-glucopyranosyl)-6-*O*-(trans-caffeyl)-β-D-glucopyranosyl)-5-*O*-β-D-glucopyranosyl	Caffeoyl	Dangles et al. (1993)
Cairo morning glory	*Ipomoea cairica*	Cyanidin-3-*O*-[2-*O*-(6-*O*-(E)-*p*-coumaroyl-β-d-glucopyranosyl)-6-*O*-(E)-caffeoyl-β-d-glucopyranoside]-5-*O*-β-D-glucopyranoside, cyanidin-3-*O*-[2-*O*-(2-*O*-(E)-*p*-coumaroyl-β-d-glucopyranosyl)-6-*O*-(E)-caffeoyl-β-d-glucopyranoside]-5-*O*-β-D-glucopyranoside, cyanidin-3-*O*-[2-*O*-(6-*O*-(E)-caffeoyl-β-d-glucopyranosyl)-6-*O*-(E)-caffeoyl-β-d-glucopyranoside]-5-*O*-β-D-glucopyranoside, cyanidin-3-*O*-[2-*O*-(2-*O*-(E)-caffeoyl-β-d-glucopyranosyl)-6-*O*-(E)-caffeoyl-β-d-glucopyranoside]-5-*O*-β-D-glucopyranoside, cyanidin-3-*O*-[(2-*O*-β-d-glucopyranosyl)-6-*O*-(E)-caffeoyl-β-d-glucopyranoside]-5-*O*-β-d-glucopyranoside	*p*-Coumaroyl, caffeoyl	Pomilio & Mercader (2018)
Ginger-leaf morning glory	*Ipomoea asarifolia* (Desv.) Roem. and Schult.	Cyanidin-3-*O*-[2-*O*-(6-*O*-E-caffeoyl-β-D-glucopyranosyl)]-{6-*O*-[4-*O*-(6-*O*-E-3,5-dihydroxycinnamoyl-β-D-glucopyranosyl)-E-caffeoyl]-β-D-glucopyranosyl}-5-*O*-β-D-glucopyranoside, cyanidin-3-*O*-[2-*O*-(6-*O*-E-*p*-coumaroyl-β-D-glucopyranosyl)]-{6-*O*-[4-*O*-(6-*O*-E-*p*-coumaroyl-β-D-glucopyranosyl)-E-caffeoyl]-β-D-glucopyranosyl}-5-*O*-β-D-glucopyranoside, cyanidin-3-*O*-[2-*O*-(6-*O*-E-caffeoyl-β-D-glucopyranosyl)]-{6-*O*-E-caffeoyl]-β-D-glucopyranosyl}-5-*O*-β-D-glucopyranoside, cyanidin-3-*O*-[2-*O*-(6-*O*-E-*p*-coumaroyl-β-D-glucopyranosyl)]-{6-*O*-E-caffeoyl]-β-D-glucopyranosyl}-5-*O*-β-D-glucopyranoside	*p*-Coumaroyl, caffeoyl and other hydroxycinnamoyl	Pale et al. (2003, 1998)
Petunia	*Petunia hybrida* Surfinia blue	Malvidin-3-*O*-[6-*O*-(4-*O*-E-caffeoyl-a-rhamnopyranosyl)-β-glucopyranoside]-5-*O*-β-glucopyranoside, malvidin-3-*O*-[6-*O*-(4-*O*-Z-*p*-coumaroyl-a-rhamnopyranosyl)-β-glucopyranoside]-5-*O*-β-glucopyranoside	Caffeoyl, *p*-coumaroyl	Slimestad et al. (1999)
Daisy	*Bellis perennis*	Cyanidin-3-*O*-[glucuronyl-(1→2)-(4″-malonylglucoside)], cyanidin-3-*O*-[glucuronyl-(1→2)-(6″-malonylglucoside)], cyanidin-3-*O*-[glucuronyl-(1→4)-(6″-malonylglucoside)], cyanidin-3-*O*-malonyl-glucoside, pelargonidin-3-*O*-malonyl-glucoside	Malonyl	Iwashina, (2015)

(Continued)

TABLE 3.1 (Continued) Acylated ACN Composition of Common and Lesser-Known Edible Sources

SOURCE	*SCIENTIFIC NAME*	*ACYLATED ACN IDENTIFIED*	*ACYL GROUPS*	*REFERENCES*
Camellia	*Camellia japonica*	Cyanidin-3-O-(6-O-(E)-p-coumaroyl)-β-glucopyranoside, cyanidin-3-O-(6-O-(E)-p-coumaroyl)-β-galactopyranoside, cyanidin-3-O-(6-O-(E)-caffeoyl)-β-glucopyranoside, cyanidin-3-O-(6-O-(E)-caffeoyl)-β-galactopyranoside	Caffeoyl, p-coumaroyl	Pereira et al. (2022)
Blue sailors	*Cichorium intybus*	Delphinidin-3,5-di-O-(6-O-malonyl-β-D-glucoside), delphinidin-3-O-(6-O-malonyl-β-D-glucoside)-5-O-β-D-glucoside, delphinidin-3-O-β-D-glucoside-5-O-(6-O-malonyl-β-D-glucoside)	Malonyl	Nørbæk et al. (2002)
Creeping Charlie	*Glechoma hederacea*	Delphinidin-3–(6″-p-coumaroylglucoside)-5-dimalonylglucoside, delphinidin-3-(6″-p-coumaroylglucoside)-5-(6‴-malonylglucoside), cyanidin-3-(6″-p-coumarylglucoside)-5-(4‴,6‴-dimalonylglucoside), cyanidin-3-(6″-malonylglucoside)-5-glucoside	p-Coumaroyl, malonyl	Saito & Harborne (1992)
Garden balsam "orange petals," "pink petals"	*Impatiens balsamina* L.	Pelargonidin-O-hexoside-O-acetyl-hexoside, malvidin-O-coumaroyl-hexoside-O-hexoside, pelargonidin-O-p-coumaroyl-hexoside-O-acetyl-hesoxide, peonidin-O-acetyl-hexoside-O-p-coumaroyl-hexoside, pelargonidin-O-p-coumaroyl-hexoside-O-acetyl-hexoside, malvidin-O-acetyl-hexoside-O-coumaroyl-hexoside, peonidin-O-acetyl-hexoside-O-coumaroyl-hexoside, malvidin-O-coumaroyl-hexoside, malvidin-O-acetyl-hexoside-O-coumaroyl-hexoside	p-Coumaroyl, acetyl	Pires, Pereira, Pereira et al. (2021)
Busy lizzie	*Impatiens walleriana*	Peonidin-O-acetyl-hexoside-O-p-coumaroyl-hexoside, pelargonidin-O-p-coumaroyl-hexoside-O-acetyl-hexoside, malvidin-O-acetyl-hexoside-O-p-coumaroyl-hexoside, malvidin-3-O-p-coumaroyl-hexoside-O-hexoside	p-Coumaroyl, acetyl	Pires, Pereira, Carocho et al. (2021)
Japanese honeysuckle "Chinensis (Wats.) Bak. (rFLJ)"	*Lonicera japonica* Thunb. var.	Cyanidin-3-(p-coumaroyl)-rutinoside-5-glucoside, cyanidin-3-(acetyl)-glucoside	p-Coumaroyl, acetyl	Yuan et al. (2014)
Carnations	*Dianthus caryophyllus*	Pelargonidin-3-malyl-glucoside, cyanidin-3-malyl-glucoside, macrocyclic malyl pelargonidin-3,5-diglucoside, macrocyclic malyl cyanidin-3,5-diglucoside	Malyl	Sasaki et al. (2013)
Carnations "cv. Moondust," "cv. Moonshadow"	*Dianthus caryophyllus*	Delphinidin-3-O-(6″-O-4-malyl-glucosyl)-5-O-glucoside, delphinidin-3,5-O-diglucoside-6″-O-4, 6‴-O-L-cyclic-malyl diester, delphinidin-3-O-(6″-O-4-malyl-glucosyl)-5-O-(6‴-O-L-malyl-glucoside)	Malyl	Fukui et al. (2003)
Orchids	*Dracula chimaera*, *Dracula cordobae* Luer	Cyanidin-3-O-(6″-O-malonyl-β-glucopyranoside), peonidin-3-O-(6″-O-malonyl-β-glucopyranoside)	Malonyl	Fossen & Øvstedal (2003)

Thyme	*Thymus* spp.	Cyanidin-3-(6″-malonylglucoside)-5-(6‴-malonylglucoside), cyanidin 3-(6″-malonylglucoside), cyanidin-3-(6″-malonyl-acetylglucoside), peonidin-3-(6″-malonylglucoside)-5-(6‴-malonylglucoside), pelargonidin-3-(6″-malonyl-acetylglucoside)	Malonyl, acetyl	Díaz-García et al. (2015)
Tulip	*Tulipa* spp.	Cyanidin-3-(2‴-acetylrutinoside), delphinidin-3-(2‴-acetylrutinoside), delphinidin-3-(3‴-acetylrutinoside), pelargonidin-3-(2‴-acetylrutinoside)	Acetyl	Nakayama et al. (2004)
Purple heart	*Tradescantia pallida* (Rose) D.R. Hunt cv *purpurea*	Cyanidin-tri(feruloyl)-pentoside-trihexoside, cyanidin-tri(feruloyl)(caffeoyl)-pentoside-trihexoside, cyanidin-tri(feruloyl)(coumaroyl)-pentoside-trihexoside, cyanidin-tetra(feruloyl)-pentoside-trihexoside	Caffeoyl, feruloyl, coumaroyl	Steingass et al. (2023)
Butterfly pea flower	*Clitoria ternatea*	Ternatin A1, ternatin A2, preternatin A3, ternatin B1, ternatin B2, ternatin B3, ternatin B4, ternatin C1, ternatin C2, Ternatin C4, ternatin D2, ternatin D1, ternatin D3, delphinidin-3-*O*-glucoside-3′-(6-*p*-coumaroyl)-*O*-glucoside-5′-(6-*p*-coumaroyl)-*O*-diglucoside, cyanidin-3-sambubioside, delphinidin-3-(6-succinyl)-*O*-glucoside-3′-(6-*p*-coumaroyl)-*O*-glucoside-5′-(6-*p*-coumaroyl)-*O*-diglucoside, delphinidin-3-*O*-glucoside-3′-(6-succinyl)-*O*-glucoside-5′-(6-*p*-coumaroyl)-*O*-glucoside	*p*-Coumaroyl, succinyl, malonyl	Multisona et al. (2023), Shen et al. (2016), Tao et al. (2025), Terahara et al. (1998), Zakaria et al. (2018)
Wild pansy	*Viola tricolor*	Delphinidin-3–(4′-*cis*-*p*-coumaroyl)-*O*-rutinoside-5-*O*-glucoside, delphinidin-3–(4′-*trans*-*p*-coumaroyl)-*O*-rutinoside-5-*O*-glucoside, cyanidin-3-(4′-*cis*-*p*-coumaroyl)-*O*-rutinoside-5-*O*-glucoside, delphinidin-3-(4′-*cis*-*p*-coumaroyl)-*O*-rutinoside, delphinidin-3-(4′-*trans*-*p*-coumaroyl)-*O*-rutinoside, delphinidin-3-(4′-*trans*-*p*-coumaroyl)-rutinoside-5-*O*-(6′-caffeoyl)-glucoside	*p*-Coumaroyl, caffeoyl	Tao et al. (2025), Teixeira et al. (2024)
Cornflower	*Centaurea cyanus*	Cyanidin-3-(6′-malonyl)-*O*-glucoside-5-*O*-glucoside, cyanidin-3-(6′-succinyl)-*O*-glucoside-5-*O*-glucoside, cyanidin-3-(6′-succinyl)-*O*-glucoside, cyanidin-3-(6″-succinylglucoside)-*O*-sophoroside	Malony, succinyl	Tao et al. (2025), Teixeira et al. (2024)
Cosmos	*Cosmos bipinnatus*	Cyanidin-3-*O*-(6″-*p*-coumaroyl)-glucoside, peonidin-3-*O*-(6″-*p*-coumaroyl)-glucoside, delphinidin-3-*O*-(6″-*p*-coumaroyl)-glucoside	*p*-Coumaroyl	Teixeira et al. (2024)
Wild bergamot, beebalm	*Monarda fistulosa*	Pelargonidin-3,5-diglucoside acylated with coumaric and malonic acids	Malony, coumaroyl	Davies & Mazza (1992)
Wood cranesbill	*Geranium sylvaticum*	Malvidin-3-(6-acetylglucoside)-5-glucoside	Acetyl	Andersen et al. (1995)
Blue garden cineraria	*Pericallis x hybrida*	Delphinidin-3-*O*-(6-*O*-malonyl-β-D-glucopyranosyl)-7-*O*-(6-*O*-(4-*O*-(6-*O*-caffeoyl-β-D-glucopyranosyl) caffeoyl)-β-D-glucopyranosyl)-3′-*O*-(6-*O*-caffeoyl-β-D-glucopyranosyl)	Malonyl, caffeoyl	Goto et al. (1984)

(Continued)

TABLE 3.1 (Continued) Acylated ACN Composition of Common and Lesser-Known Edible Sources

SOURCE	*SCIENTIFIC NAME*	*ACYLATED ACN IDENTIFIED*	*ACYL GROUPS*	*REFERENCES*
Common vervain	*Verbena hybrida*	Cyanidin-3,5-di-*O*-(6-*O*-acetyl-β-D-glucopyranopsides), pelargonidin-3,5-di-*O*-(6-*O*-acetyl-β-D-glucopyranopsides), pelargonidin-3-*O*-(6-*O*-malonyl-β-D-glucopyranopside)-5-*O*-(6-*O*-acetyl-β-D-glucopyranopside)	Acetyl, malonyl	Toki et al. (1995)
Cattleya	*Laeliocattleya* cv Mini Purple	Cyanidin-3-*O*-[6-*O*-(malonyl)-β-D-glucopyranoside-7-*O*-[6-*O*-(*trans-p*-coumaryl)-β-D-glucopyranoside]-3′-*O*-[6-*O*-[trans-4-*O*-[6-*O*-(*trans-p*-coumaryl)-β-D-glucopyranosyl]-*p*-coumaryl]-β-D-glucopyranoside]	*p*-Coumaroyl, malonyl	Tatsuzawa et al. (1994)
Garden lobelia	*Lobelia erinus*	Delphinidin-3-coumarylrutinosyl-5-malonyl-glucosyl-3′, 5′-di(caffeylglucosyl), delphinidin-3-coumarylrutinosyl-5-malonylglucosyl-3′-caffeoylglucosyl-5′-ferulylglucosyl	*p*-Coumaroyl, caffeoyl, feruloyl, malonyl	Kondo et al. (1989)
Bugleweed	*Ajuga reptans* L.	Delphinidin-3-(*p*-coumaroyl-feruloyl)-sophoroside-5-(malonyl)-glucoside, delphinidin-3-(diferuloyl)-sophoroside-5-(malonyl)-glucoside, cyanidin-3-(di-*p*-coumaroyl)-sophoroside-5-glucoside	*p*-Coumaroyl, feruloyl, malonyl	Terahara et al. (2001)
Night-scented stock	*Matthiola longipetala* spp. *bicornis*	Cyanidin-3-*O*-[2-*O*-(2-*O*-(*trans*-feruloyl)-β-xylosyl)-6-*O*-(*trans*-feruloyl)-β-glucoside]-5-*O*-[6-*O*-(malonyl)-β-glucoside], cyanidin-3-*O*-[2-*O*-(2-*O*-(*trans*-sinapoyl)-β-xylopyranosyl)-6-*O*-(*trans*-feruloyl)-β-glucopyranoside]-5-*O*-[6-*O*-(malonyl)-β-glucopyranoside], cyanidin-3-*O*-[2-*O*-(β-xylopyranosyl)-6-*O*-(*trans*-feruloyl)-β-glucopyranoside]-5-*O*-[6-*O*-(malonyl)-β-glucopyranoside]	Feruloyl, malonyl, sinapoyl	Tatsuzawa (2014)
Garden candytuft "Candycane Rose," "Candycane Red"	*Iberis umbellata* L.	Pelargonidin-3-*O*-[2-*O*-(glucosyl)-6-*O*-(*p*-coumaroyl)-glucoside]-5-*O*-[6-*O*-(malonyl)-glucoside], pelargonidin-3-*O*-[2-*O*-(2-*O*-(sinapoyl)-glucosyl)-6-*O*-(4-*O*-(glucosyl)-*p*-coumaroyl)-glucoside]-5-*O*-[6-*O*-(malonyl)-glucoside], pelargonidin-3-*O*-[2-*O*-(2-*O*-(feruloyl)-glucosyl)-6-*O*-(4-*O*-(glucosyl)-*p*-coumaroyl)-glucoside]-5-*O*-[6-*O*-(malonyl)-glucoside], pelargonidin-3-*O*-[2-*O*-(2-*O*-(sinapoyl)-glucosyl)-6-*O*-(4-*O*-(6-*O*-(feruloyl)-glucosyl)-*p*-coumaroyl)-glucoside]-5-*O*-[6-*O*-(malonyl)-glucoside], pelargonidin-3-*O*-[2-*O*-(2-*O*-(feruloyl)-glucosyl)-6-*O*-(4-*O*-(6-*O*-(feruloyl)-glucosyl)-*p*-coumaroyl)-glucoside]-5-*O*-[6-*O*-(malonyl)-glucoside], pelargonidin-3-*O*-[2-*O*-(2-*O*-(feruloyl)-glucosyl)-6-*O*-(4-*O*-(6-*O*-(4-*O*-(glucosyl)-feruloyl)-glucosyl)-*p*-coumaroyl)-glucoside]-5-*O*-[6-*O*--(malonyl)-glucoside]	*p*-Coumaroyl, sinapoyl, malonyl, feruloyl	Tatsuzawa (2019)
Herbal plant				
Boat lily (leaves)	*Rhoeo spathacea*	Cyanidin-3-(feruloyl)-arabinosylglucoside-7,3′-(diferuloyl)-diglucoside	Feruloyl	Tan et al. (2014)

Acylated ACN listed above were detected in the varieties or transgenic plants from the references provided and are not universally present in all varieties.

3.4 NON-EDIBLE SOURCES OF ACYLATED ANTHOCYANINS

Although edible sources have been widely explored and utilized for the extraction of acylated ACNs due to their safety and regulatory acceptance, numerous non-edible plants also contain structurally diverse and abundant acylated ACNs. However, these sources are generally unsuitable for direct application in food, pharmaceutical, or nutraceutical products due to safety concerns, lack of toxicological evaluation, or regulatory restrictions. A notable example is the tree leaves of certain maple species, which contain acylated ACNs in autumn, contributing to the seasonal foliage colors. Fujiwara et al. (2021) identified two different galloylated cyanidin-based ACNs, *i.e.,* cyanidin-3-*O*-[2″-*O*-(galloyl)-6″-*O*-(rhamnosyl)-β-glucoside] and cyanidin-3-*O*-[2″-*O*-(galloyl)-β-glucopyranoside] from fresh maple leaves (*Acer palmatum* cv 'Chishio'). Meanwhile, some ornamental plants are rich in ACNs, especially acylated ACNs, but the lack of systematic toxicology and food safety certification has led to them being considered non-edible plants. For instance, dimalonated ACNs, *i.e.,* pelargonidin-3-caffeoylglucoside-5-dimalonylglucoside and pelargonidin-3-*p*-coumaroylglucoside-5-dimalonylglucoside, have been identified in scarlet sage (*Salvia splendens*) flowers (Tomás-Barberán et al., 1987). Small amounts of acylated ACNs with one malonyl or acetyl group based on cyanidin and pelargonidin (cyanidin-3-(6″-acetylglucoside)-5-glucoside, cyanidin-3-(6″-malonylglucoside)-5-glucoside, and pelargonidin-3-(6″-malonylglucoside)) have been detected in poinsettia (*Euphorbia pulcherrima*) leaves (Slatnar et al., 2013). In addition, various wild or ornamental flowers such as certain cultivars of morning glory (*Ipomoea purpurea*) or petunia hybrida may accumulate high levels of acylated ACNs in their flowers or leaves. These plants are primarily cultivated for decorative purposes and may contain additional secondary metabolites that pose toxic risks to humans. Furthermore, their potential to accumulate heavy metals or pesticide residues from the environment makes them unreliable sources for safe pigment extraction. Besides, some transgenic sources designed to overexpress ACN biosynthesis genes have also been explored to boost acylated ACN production. However, ACNs from this source are still not commercially available in most countries due to the lack of formal approval by food safety authorities.

3.5 VARIATIONS IN ACYLATED ANTHOCYANIN PROFILES AMONG PLANT SPECIES

To date, ACNs production and distribution are complex biosynthesis process that has been extensively studied at the metabolic and molecular levels (Zhang et al., 2016). ACNs exhibit considerable structural diversity among different plant species, mainly due to variations in their aglycone types, glycosylation patterns, and acylation modifications. These structural differences contribute not only to the wide spectrum of colors observed in plant tissues ranging from red and purple to blue but also to their stability, antioxidant capacity, and potential health benefits. ACN glycosyl acylation is typically the final step in ACNs biosynthesis. Differences in acyltransferase activity are key to the biosynthesis of diverse acylated ACNs (Luo et al., 2022). ACN profiles and content vary significantly between species and even within varieties of the same species. ACNs accumulation and profile can show significant differences, which are influenced by several factors such as varietal diversity, genetic regulation, phytohormones, developmental stage, grafting, light, season, soil type, environmental conditions, maturity, ripening stage, stress levels, and cultural practices.

3.6 CONCLUSION AND FUTURE PERSPECTIVES

Currently, ACNs are unstable and easily affected by degradation in harsh environments. Acylation of the sugar moieties with aliphatic acids and aromatic acids is an efficient strategy to improve the overall environmental resistance of ACNs. Oliveira et al. (2019) isolated two acylated ACNs, *i.e.*, peonidin-3-(6′-hydroxybenzoyl)-sophoroside-5-glucoside and peonidin-3-(6′-hydroxybenzoyl-6″-caffeoyl)-sophoroside-5-glucoside, from purple-fleshed sweet potatoes, and these two acylated ACNs showed high resistance to pH variations compared to the parent ACNs, with improved ability to retain red and blue colorants at acidic and basic pH. Acylated ACNs have been shown to be promising alternatives to the use of synthetic dyes in food systems. Depending on the ACNs chemical structure and pH of the food matrix, a wide variety of hues can be achieved. Acylated ACNs in edible sources impart desirable color and stability. For example, acylated pelargonidin derivatives from radish and potato can closely match the color characteristics of orange-red color under acidic conditions. Acylated cyanidin derivatives from black carrots or red cabbage can provide hues ranging from deep red to purple, depending on the pH (Giusti & Wrolstad, 2003). Their attractive color and higher stability make acylated ACNs possess wide application in beverages, fruit preparations, dairy products, ice creams, and confectionaries. Apart from being used as food colorants, epidemiological evidence reveals that ACNs intake is associated with a lower incidence of chronic and degenerative diseases, and the excellent antioxidant properties of acylated ACNs also indicate that they may be used as functionalizing agents to prevent cardiovascular disease, neurological diseases, cancer, inflammation, and diabetes (Luo et al., 2022). Z.-C Zhang et al. (2019) found that the combination of acylated ACNs obtained from purple sweet potato and allopurinol was more effective in reducing serum uric acid levels and kidney damage in hyperuricemic mice than allopurinol alone. In addition to the above applications in food and pharmaceutical industries, acylated ACNs have also been reported to apply in food packing and cosmetic sectors. Acylated ACNs are widely distributed in both edible and non-edible plants, with significant variations in their profiles across species. Their enhanced stability and diverse structures make them valuable for various applications. This chapter provides a comprehensive overview of the widespread occurrence of acylated ACNs in both edible and non-edible plants. From common fruits like grapes and blueberries to unique plants like shiso and butterfly pea flower, acylated ACNs offer a palette of stable, vibrant pigments. Enhanced research efforts in this domain could lead to innovative uses of acylated ACNs in various industries. Although acylated ACNs have shown improved stability and enhanced fat solubility, there are still challenges and limitations in obtaining acylated ACNs more efficiently and applying acylated ACNs more widely in the food industry. Besides, further exploration into the metabolic pathways responsible for acylation may unlock new methods for enhancing ACNs stability, paving the way for novel uses of these natural pigments in various sectors. Additionally, further research is needed to explore their potential health benefits and industrial uses.

REFERENCES

Alappat, B., & Alappat, J. (2020). Anthocyanin pigments: Beyond aesthetics. *Molecules*, *25*(23), 5500. https://www.mdpi.com/1420-3049/25/23/5500

Andersen, Ø. M., & Jordheim, M. (2010). Anthocyanins. *Encyclopedia of Life Sciences (ELS)*. https://doi.org/10.1002/9780470015902.a0001909.pub2

Andersen, Ø. M., Viksund, R. I., & Pedersen, A. T. (1995). Malvidin 3-(6-acetylglucoside)-5-glucoside and other anthocyanins from flowers of *Geranium sylvaticum*. *Phytochemistry*, *38*(6), 1513–1517. https://doi.org/10.1016/0031-9422(94)00898-4

Andreeva, V. (2020). Identification of red grape varieties by anthocyanin profile. *BIO Web of Conferences*, 25, 02020. https://doi.org/10.1051/bioconf/20202502020Aubert, C., & Chalot, G. (2020). Physicochemical characteristics, vitamin C, and polyphenolic composition of four European commercial blood-flesh peach cultivars (*Prunus persica* L. Batsch). *Journal of Food Composition and Analysis*, *86*, 103337. https://doi.org/10.1016/j.jfca.2019.103337

Azuma, K., Ohyama, A., Ippoushi, K., Ichiyanagi, T., Takeuchi, A., Saito, T., & Fukuoka, H. (2008). Structures and antioxidant activity of anthocyanins in many accessions of eggplant and its related species. *Journal of Agricultural and Food Chemistry*, *56*(21), 10154–10159. https://doi.org/10.1021/jf801322m

Bakker, J., Bridle, P., & Bellworthy, S. J. (1994). Strawberry juice colour: A study of the quantitative and qualitative pigment composition of juices from 39 genotypes. *Journal of the Science of Food and Agriculture*, *64*(1), 31–37.

Carazzone, C., Mascherpa, D., Gazzani, G., & Papetti, A. (2013). Identification of phenolic constituents in red chicory salads (*Cichorium intybus*) by high-performance liquid chromatography with diode array detection and electrospray ionisation tandem mass spectrometry. *Food Chemistry*, *138*(2), 1062–1071. https://doi.org/10.1016/j.foodchem.2012.11.060

Cebadera-Miranda, L., Domínguez, L., Dias, M. I., Barros, L., Ferreira, I. C. F. R., Igual, M., . . . Cámara, M. (2019). Sanguinello and tarocco (*Citrus sinensis* [L.] Osbeck): Bioactive compounds and colour appearance of blood oranges. *Food Chemistry*, *270*, 395–402. https://doi.org/10.1016/j.foodchem.2018.07.094

Cheng, D. M., Pogrebnyak, N., Kuhn, P., Poulev, A., Waterman, C., Rojas-Silva, P., . . . Raskin, I. (2014). Polyphenol-rich Rutgers Scarlet Lettuce improves glucose metabolism and liver lipid accumulation in diet-induced obese C57BL/6 mice. *Nutrition*, *30*(7, Supplement), S52–S58. https://doi.org/10.1016/j.nut.2014.02.022

Costa, E., Cosme, F., Jordão, A. M., & Mendes-Faia, A. (2014). Anthocyanin profile and antioxidant activity from 24 grape varieties cultivated in two Portuguese wine regions. *Oeno One*, *48*(1), 51–62.

Dangles, O., Saito, N., & Brouillard, R. (1993). Kinetic and thermodynamic control of flavylium hydration in the pelargonidin-cinnamic acid complexation: Origin of the extraordinary flower color diversity of *Pharbitis nil*. *Journal of the American Chemical Society*, *115*(8), 3125–3132.

Davies, A. J., & Mazza, G. (1992). Separation and characterization of anthocyanins of *Monarda fistulosa* by high-performance liquid chromatography. *Journal of Agricultural and Food Chemistry*, *40*(8), 1341–1345. https://doi.org/10.1021/jf00020a009

de Pascual-Teresa, S., & Sanchez-Ballesta, M. T. (2008). Anthocyanins: From plant to health. *Phytochemistry Reviews*, *7*, 281–299.

Díaz-García, M. C., Castellar, M. R., Obón, J. M., Obón, C., Alcaraz, F., & Rivera, D. (2015). Production of an anthocyanin-rich food colourant from *Thymus moroderi* and its application in foods. *Journal of the Science of Food and Agriculture*, *95*(6), 1283–1293.

Fierri, I., De Marchi, L., Chignola, R., Rossin, G., Bellumori, M., Perbellini, A., . . . Zoccatelli, G. (2023). Nanoencapsulation of anthocyanins from red cabbage (*Brassica oleracea* L. var. *capitata* f. *rubra*) through coacervation of whey protein isolate and apple high methoxyl pectin. *Antioxidants*, *12*(9), 1757. https://www.mdpi.com/2076-3921/12/9/1757

Fossen, T., & Øvstedal, D. O. (2003). Anthocyanins from flowers of the orchids *Dracula chimaera* and *D. cordobae*. *Phytochemistry*, *63*(7), 783–787. https://doi.org/10.1016/S0031-9422(03)00339-X

Fossen, T., Rayyan, S., Holmberg, M. H., & Andersen, Ø. M. (2005). Acylated anthocyanins from leaves of *Oxalis triangularis*. *Phytochemistry*, *66*(10), 1133–1140.

Fujiwara, Y., Miwa, M., Nagatsu, A., & Honma, A. (2021). Identification of maple anthocyanin and its antiproliferative activity against LLC, T47D and C3H10T1/2 cells. *Anti-Cancer Agents in Medicinal Chemistry (Formerly Current Medicinal Chemistry-Anti-Cancer Agents)*, *21*(7), 894–901.

Fukui, Y., Tanaka, Y., Kusumi, T., Iwashita, T., & Nomoto, K. (2003). A rationale for the shift in colour towards blue in transgenic carnation flowers expressing the flavonoid 3′,5′-hydroxylase gene. *Phytochemistry*, *63*(1), 15–23. https://doi.org/10.1016/S0031-9422(02)00684-2

Gayan Chandrajith, V. G., Lim, Y. Y., & Choo, W. S. (2021). Sources and relative stabilities of acylated and nonacylated anthocyanins in beverage systems. *Journal of Food Science and Technology*, *59*(3), 831–845. https://doi.org/10.1007/s13197-021-05054-z

Giusti, M. M., & Wrolstad, R. E. (2003). Acylated anthocyanins from edible sources and their applications in food systems. *Biochemical Engineering Journal*, *14*(3), 217–225. https://doi.org/10.1016/S1369-703X(02)00221-8

Goto, T., Kondo, T., Kawai, T., & Tamura, H. (1984). Structure of cinerarin, a tetra-acylated anthocyanin isolated from the blue garden cineraria, *Senecio cruentus*. *Tetrahedron Letters*, *25*(52), 6021–6024. https://doi.org/10.1016/S0040-4039(01)81749-4

Ha, T. J., Park, J. E., Lee, K.-S., Seo, W. D., Song, S.-B., Lee, M.-H., . . . Lee, J. H. (2021). Identification of anthocyanin compositions in black seed coated Korean adzuki bean (*Vigna angularis*) by NMR and UPLC-Q-Orbitrap-MS/MS and screening for their antioxidant properties using different solvent systems. *Food Chemistry*, *346*, 128882. https://doi.org/10.1016/j.foodchem.2020.128882

Han, K.-H., Kitano-Okada, T., Seo, J.-M., Kim, S.-J., Sasaki, K., Shimada, K.-I., & Fukushima, M. (2015). Characterisation of anthocyanins and proanthocyanidins of adzuki bean extracts and their antioxidant activity. *Journal of Functional Foods*, *14*, 692–701. https://doi.org/10.1016/j.jff.2015.02.018

Harborne, J. B., & Grayer, R. J. (1988). The anthocyanins. *The Flavonoids: Advances in Research Since 1980*. Springer, pp. 1–20.

He, Y.-K., Yao, Y.-Y., & Chang, Y.-N. (2015). Characterization of anthocyanins in *Perilla frutescens* var. *acuta* extract by advanced UPLC-ESI-IT-TOF-MS n method and their anticancer bioactivity. *Molecules*, *20*(5), 9155–9169.

Hong, V., & Wrolstad, R. E. (1990). Use of HPLC separation/photodiode array detection for characterization of anthocyanins. *Journal of Agricultural and Food Chemistry*, *38*(3), 708–715.

Iwashina, T. (2015). Contribution to flower colors of flavonoids including anthocyanins: A review. *Natural Product Communications*, *10*(3), https://doi.org/10.1177/1934578X1501000335

Jang, Y. P., Zhou, J., Nakanishi, K., & Sparrow, J. R. (2005). Anthocyanins protect against A2E photooxidation and membrane permeabilization in retinal pigment epithelial cells¶. *Photochemistry and Photobiology*, *81*(3), 529–536.

Jing, P., Zhao, S.-J., Ruan, S.-Y., Xie, Z.-H., Dong, Y., & Yu, L. (2012). Anthocyanin and glucosinolate occurrences in the roots of Chinese red radish (*Raphanus sativus* L.), and their stability to heat and pH. *Food Chemistry*, *133*(4), 1569–1576. https://doi.org/10.1016/j.foodchem.2012.02.051

Jokioja, J., Yang, B., & Linderborg, K. M. (2021). Acylated anthocyanins: A review on their bioavailability and effects on postprandial carbohydrate metabolism and inflammation. *Comprehensive Reviews in Food Science and Food Safety*, *20*(6), 5570–5615.

Jordheim, M., Måge, F., & Andersen, Ø. M. (2007). Anthocyanins in berries of ribes including gooseberry cultivars with a high content of acylated pigments. *Journal of Agricultural and Food Chemistry*, *55*(14), 5529–5535. https://doi.org/10.1021/jf0709000

Kay, C. D., Mazza, G., Holub, B. J., & Wang, J. (2004). Anthocyanin metabolites in human urine and serum. *British Journal of Nutrition*, *91*(6), 933–942.

Kidøy, L., Nygård, A. M., Andersen, Ø. M., Pedersen, A. T., Aksnes, D. W., & Kiremire, B. T. (1997). Anthocyanins in fruits of *Passiflora edulis* and *P. suberosa*. *Journal of Food Composition and Analysis*, *10*(1), 49–54. https://doi.org/10.1006/jfca.1996.0514

Kita, A., Bąkowska-Barczak, A., Hamouz, K., Kułakowska, K., & Lisińska, G. (2013). The effect of frying on anthocyanin stability and antioxidant activity of crisps from red- and purple-fleshed potatoes (*Solanum tuberosum* L.). *Journal of Food Composition and Analysis*, *32*(2), 169–175. https://doi.org/10.1016/j.jfca.2013.09.006

Koley, T. K., Khan, Z., Oulkar, D., Singh, B. K., Maurya, A., Singh, B., & Banerjee, K. (2020). High resolution LC-MS characterization of phenolic compounds and the evaluation of antioxidant properties of a tropical purple radish genotype. *Arabian Journal of Chemistry*, *13*(1), 1355–1366. https://doi.org/10.1016/j.arabjc.2017.11.007

Koley, T. K., Srivastava, S., Tripathi, Y., Banerjee, K., Oulkar, D., Goon, A., . . . Singh, B. (2019). High-resolution LCMS profiling of phenolic compounds of Indian black carrot and evaluation of its effect on antioxidant defense and glucose metabolism in animal model. *Agricultural Research*, *8*, 481–489.

Kondo, T., Yamashiki, J., Kawahori, K., & Goto, T. (1989). Structure of lobelinin A and B, novel anthocyanins acylated with three and four different organic acids, respectively. *Tetrahedron Letters*, *30*(44), 6055–6058.

Kraithong, S., Liu, Y., Suwanangul, S., Sangsawad, P., Theppawong, A., & Bunyameen, N. (2025). A comprehensive review of the impact of anthocyanins from purple/black rice on starch and protein digestibility, gut microbiota modulation, and their applications in food products. *Food Chemistry*, *473*, 143007. https://doi.org/10.1016/j.foodchem.2025.143007

Kumari, P., & Bhargava, B. (2021). Phytochemicals from edible flowers: Opening a new arena for healthy lifestyle. *Journal of Functional Foods*, *78*, 104375.

Kőrösi, L., Molnár, S., Teszlák, P., Dörnyei, Á., Maul, E., Töpfer, R., . . . Röckel, F. (2022). Comparative study on grape berry anthocyanins of various teinturier varieties. *Foods*, *11*(22), 3668. https://www.mdpi.com/2304-8158/11/22/3668

Lachman, J., Hamouz, K., Šulc, M., Orsák, M., Pivec, V., Hejtmánková, A., . . .Čepl, J. (2009). Cultivar differences of total anthocyanins and anthocyanidins in red and purple-fleshed potatoes and their relation to antioxidant activity. *Food Chemistry*, *114*(3), 836–843. https://doi.org/10.1016/j.foodchem.2008.10.029

Lee, M. J., Park, J. S., Choi, D. S., & Jung, M. Y. (2013). Characterization and quantitation of anthocyanins in purple-fleshed sweet potatoes cultivated in Korea by HPLC-DAD and HPLC-ESI-QTOF-MS/MS. *Journal of Agricultural and Food Chemistry*, *61*(12), 3148–3158. https://doi.org/10.1021/jf3055455

Li, C.-Y., Kim, H.-W., Won, S. R., Min, H.-K., Park, K.-J., Park, J.-Y., . . . Rhee, H.-I. (2008). Corn husk as a potential source of anthocyanins. *Journal of Agricultural and Food Chemistry*, *56*(23), 11413–11416.

Li, G., Lin, Z., Zhang, H., Liu, Z., Xu, Y., Xu, G., . . . Qiu, Y. (2019). Anthocyanin accumulation in the leaves of the purple sweet potato (*Ipomoea batatas* L.) cultivars. *Molecules*, *24*(20), 3743.

Li, J., Li, X.-D., Zhang, Y., Zheng, Z.-D., Qu, Z.-Y., Liu, M., . . . Qu, L. (2013). Identification and thermal stability of purple-fleshed sweet potato anthocyanins in aqueous solutions with various pH values and fruit juices. *Food Chemistry*, *136*(3), 1429–1434. https://doi.org/10.1016/j.foodchem.2012.09.054

Li, L., Xia, L., Xiao, F., Xiao, Y., Liu, L., Jiang, S., & Wang, H. (2023). Colorimetric active carboxymethyl chitosan/oxidized sodium alginate-*Oxalis triangularis* ssp. *papilionacea* anthocyanins film@gelatin/zein-linalool membrane for milk freshness monitoring and preservation. *Food Chemistry*, *405*, 134994. https://doi.org/10.1016/j.foodchem.2022.134994

Li, N., Simon, J. E., & Wu, Q. (2024). Determination of anthocyanins, organic acids, and phenolic acids in hibiscus market products using LC/UV/MS. *Journal of Food Science*, *89*(2), 1098–1113.

Li, W., Zhang, Y., Deng, H., Yuan, H., Fan, X., Yang, H., & Tan, S. (2023). In vitro and in vivo bioaccessibility, antioxidant activity, and color of red radish anthocyanins as influenced by different drying methods. *Food Chemistry: X*, *18*, 100633. https://doi.org/10.1016/j.fochx.2023.100633

Liu, P., Li, W., Hu, Z., Qin, X., & Liu, G. (2020). Isolation, purification, identification, and stability of anthocyanins from *Lycium ruthenicum* Murr. *LWT*, *126*, 109334. https://doi.org/10.1016/j.lwt.2020.109334

Luo, X. e., Wang, R., Wang, J., Li, Y., Luo, H., Chen, S., . . . Han, Z. (2022). Acylation of anthocyanins and their applications in the food industry: Mechanisms and recent research advances. *Foods*, *11*(14), 2166. https://www.mdpi.com/2304-8158/11/14/2166

Massironi, A., Toccaceli, M., Marinelli, A., Maggioni, D., Scapuzzi, C., Emide, D., . . . Marzorati, S. (2025). Extraction and encapsulation of purple corn cob anthocyanins into starch- and zein-based delivery systems. *Journal of Drug Delivery Science and Technology*, *105*, 106624. https://doi.org/10.1016/j.jddst.2025.106624

Matera, R., Gabbanini, S., Berretti, S., Amorati, R., De Nicola, G. R., Iori, R., & Valgimigli, L. (2015). Acylated anthocyanins from sprouts of *Raphanus sativus* cv. Sango: Isolation, structure elucidation and antioxidant activity. *Food Chemistry*, *166*, 397–406. https://doi.org/10.1016/j.foodchem.2014.06.056

Mathews, H., Clendennen, S. K., Caldwell, C. G., Liu, X. L., Connors, K., Matheis, N., . . . Lightner, J. (2003). Activation tagging in tomato identifies a transcriptional regulator of anthocyanin biosynthesis, modification, and transport. *The Plant Cell*, *15*(8), 1689–1703.

Mes, P. J., Boches, P., Myers, J. R., & Durst, R. (2008). Characterization of tomatoes expressing anthocyanin in the fruit. *Journal of the American Society for Horticultural Science*, *133*(2), 262–269.

Mikulic-Petkovsek, M., Schmitzer, V., Slatnar, A., Todorovic, B., Veberic, R., Stampar, F., & Ivancic, A. (2014). Investigation of anthocyanin profile of four elderberry species and interspecific hybrids. *Journal of Agricultural and Food Chemistry*, *62*(24), 5573–5580. https://doi.org/10.1021/jf5011947

Moreno, D. A., Pérez-Balibrea, S., Ferreres, F., Gil-Izquierdo, Á., & García-Viguera, C. (2010). Acylated anthocyanins in broccoli sprouts. *Food Chemistry*, *123*(2), 358–363. https://doi.org/10.1016/j.foodchem.2010.04.044

Multisona, R. R., Shirodkar, S., Arnold, M., & Gramza-Michalowska, A. (2023). *Clitoria ternatea* flower and its bioactive compounds: Potential use as microencapsulated ingredient for functional foods. *Applied Sciences*, *13*(4), 2134. https://www.mdpi.com/2076-3417/13/4/2134

Nakayama, M., Okada, M., Taya-Kizu, M., Urashima, O., Kan, Y., Fukui, Y., & Koshioka, M. (2004). Coloration and anthocyanin profile in tulip flowers. *Japan Agricultural Research Quarterly: JARQ*, *38*(3), 185–190.

Nørbæk, R., Nielsen, K., & Kondo, T. (2002). Anthocyanins from flowers of *Cichorium intybus*. *Phytochemistry*, *60*(4), 357–359.

Oliveira, H., Basilio, N., Pina, F., Fernandes, I., de Freitas, V., & Mateus, N. (2019). Purple-fleshed sweet potato acylated anthocyanins: Equilibrium network and photophysical properties. *Food Chemistry*, *288*, 386–394.

Olsen, H., Aaby, K., & Borge, G. I. A. (2010). Characterization, quantification, and yearly variation of the naturally occurring polyphenols in a common red variety of curly kale (*Brassica oleracea* L. convar. *acephala* var. *sabellica* cv. 'Redbor'). *Journal of Agricultural and Food Chemistry*, *58*(21), 11346–11354. https://doi.org/10.1021/jf102131g

Pale, E., Kouda-Bonafos, M., Nacro, M., Vanhaelen, M., & Vanhaelen-Fastré, R. (2003). Two triacylated and tetraglucosylated anthocyanins from *Ipomoea asarifolia* flowers. *Phytochemistry*, *64*(8), 1395–1399. https://doi.org/10.1016/j.phytochem.2003.08.011

Pale, E., Nacro, M., Vanhaelen, M., Vanhaelen-Fastré, R., & Ottinger, R. (1998). Acylated anthocyanins from the flowers of *Ipomoea asarifolia*. *Phytochemistry*, *48*(8), 1433–1437. https://doi.org/10.1016/S0031-9422(97)00805-4

Park, C. H., Bong, S. J., Lim, C. J., Kim, J. K., & Park, S. U. (2020). Transcriptome analysis and metabolic profiling of green and red mizuna (*Brassica rapa* L. var. *japonica*). *Foods*, *9*(8), 1079.

Peng, J., Jia, Y., Du, X., Wang, Y., Yang, Z., & Li, K. (2019). Study of physicochemical stability of anthocyanin extracts from black peanut skin and their digestion enzyme and adipogenesis inhibitory activities. *LWT*, *107*, 107–116. https://doi.org/10.1016/j.lwt.2019.03.016

Peng, Y., Yan, Y., Wan, P., Dong, W., Huang, K., Ran, L., . . . Cao, Y. (2020). Effects of long-term intake of anthocyanins from *Lycium ruthenicum* Murray on the organism health and gut microbiota in vivo. *Food Research International*, *130*, 108952. https://doi.org/10.1016/j.foodres.2019.108952

Pereira, A. G., Garcia-Perez, P., Cassani, L., Chamorro, F., Cao, H., Barba, F. J., . . . Prieto, M. A. (2022). Camellia japonica: A phytochemical perspective and current applications facing its industrial exploitation. *Food Chemistry: X*, *13*, 100258. https://doi.org/10.1016/j.fochx.2022.100258

Pina, F., Oliveira, J., & de Freitas, V. (2015). Anthocyanins and derivatives are more than flavylium cations. *Tetrahedron*, *71*(20), 3107–3114. https://doi.org/10.1016/j.tet.2014.09.051

Pires, E. D. O. Jr., Pereira, E., Carocho, M., Pereira, C., Dias, M. I., Calhelha, R. C., . . . Barros, L. (2021). Study on the potential application of *Impatiens balsamina* L. flowers extract as a natural colouring ingredient in a pastry product. *International Journal of Environmental Research and Public Health, 18*(17), 9062. https://www.mdpi.com/1660-4601/18/17/9062

Pires, E. D. O. Jr., Pereira, E., Pereira, C., Dias, M. I., Calhelha, R. C., Ćirić, A., . . . Caleja, C. (2021). Chemical composition and bioactive characterisation of *Impatiens walleriana*. *Molecules*, *26*(5), 1347.

Pomilio, A. B., & Mercader, A. G. (2018). Natural acylated anthocyanins and other related flavonoids: Structure elucidation of *Ipomoea cairica* compounds and QSAR studies including multidrug resistance. *Studies in Natural Products Chemistry*, *55*, 293–322.

Riaz, M., Zia-Ul-Haq, M., & Saad, B. (2016). *Anthocyanins and human health: Biomolecular and therapeutic aspects* (vol. 10). Springer, pp. 978–983.

Rytel, E., Tajner-Czopek, A., Kita, A., Tkaczyńska, A., Kucharska, A. Z., & Sokół-Łętowska, A. (2021). The influence of the production process on the anthocyanin content and composition in dried potato cubes, chips, and french fries made from red-fleshed potatoes. *Applied Sciences*, *11*(3), 1104. https://www.mdpi.com/2076-3417/11/3/1104

Sadilova, E., Stintzing, F. C., & Carle, R. (2006). Anthocyanins, colour and antioxidant properties of eggplant (*Solanum melongena* L.) and violet pepper (*Capsicum annuum* L.) peel extracts. *Zeitschrift für Naturforschung C*, *61*(7–8), 527–535.

Saito, N., & Harborne, J. B. (1992). Correlations between anthocyanin type, pollinator and flower colour in the Labiatae. *Phytochemistry*, *31*(9), 3009–3015.

Sampaio, S. L., Lonchamp, J., Dias, M. I., Liddle, C., Petropoulos, S. A., Glamočlija, J., . . . Barros, L. (2021). Anthocyanin-rich extracts from purple and red potatoes as natural colourants: Bioactive properties, application in a soft drink formulation and sensory analysis. *Food Chemistry*, *342*, 128526. https://doi.org/10.1016/j.foodchem.2020.128526

Sang, J., Zhang, Y., Sang, J., & Li, C.-q. (2019). Anthocyanins from *Nitraria tangutorun*: Qualitative and quantitative analyses, antioxidant and anti-inflammatory activities and their stabilities as affected by some phenolic acids. *Journal of Food Measurement and Characterization*, *13*, 421–430.

Santos-Buelga, C., Mateus, N., & De Freitas, V. (2014). Anthocyanins. Plant pigments and beyond. *Journal of Agricultural and Food Chemistry*, *62*(29), 6879–6884. https://doi.org/10.1021/jf501950s

Sasaki, N., Matsuba, Y., Abe, Y., Okamura, M., Momose, M., Umemoto, N., . . . Ozeki, Y. (2013). Recent advances in understanding the anthocyanin modification steps in carnation flowers. *Scientia Horticulturae*, *163*, 37–45. https://doi.org/10.1016/j.scienta.2013.07.029

Sasaki, N., Nishizaki, Y., Ozeki, Y., & Miyahara, T. (2014). The role of acyl-glucose in anthocyanin modifications. *Molecules*, *19*(11), 18747–18766.

Scalzo, R. L., Genna, A., Branca, F., Chedin, M., & Chassaigne, H. (2008). Anthocyanin composition of cauliflower (*Brassica oleracea* L. var. *botrytis*) and cabbage (*B. oleracea* L. var. *capitata*) and its stability in relation to thermal treatments. *Food Chemistry*, *107*(1), 136–144. https://doi.org/10.1016/j.foodchem.2007.07.072

Shen, Y., Du, L., Zeng, H., Zhang, X., Prinyawiwatkul, W., Alonso-Marenco, J. R., & Xu, Z. (2016). Butterfly pea (*Clitoria ternatea*) seed and petal extracts decreased HEp-2 carcinoma cell viability. *International Journal of Food Science and Technology*, *51*(8), 1860–1868.

Shimizu, Y., Imada, T., Zhang, H., Tanaka, R., Ohno, T., & Shimomura, K. (2010). Identification of novel polyacylated anthocyanins from Gynura bicolor leaves and their antioxidative activity. *Food Science and Technology Research*, *16*(5), 479–486.

Slatnar, A., Mikulic-Petkovsek, M., Veberic, R., Stampar, F., & Schmitzer, V. (2013). Anthocyanin and chlorophyll content during poinsettia bract development. *Scientia Horticulturae*, *150*, 142–145. https://doi.org/10.1016/j.scienta.2012.10.014

Slimestad, R., Aaberg, A., & Andersen, Ø. M. (1999). Acylated anthocyanins from petunia flowers. *Phytochemistry*, *50*(6), 1081–1086. https://doi.org/10.1016/S0031-9422(98)00647-5

Slimestad, R., & Solheim, H. (2002). Anthocyanins from black currants (*Ribes nigrum* L.). *Journal of Agricultural and Food Chemistry*, *50*(11), 3228–3231. https://doi.org/10.1021/jf011581u

Srivichai, S., & Hongsprabhas, P. (2020). Profiling anthocyanins in Thai purple yams (*Dioscorea alata* L.). *International Journal of Food Science, 2020*(1), 1594291.

Steingass, C. B., Burkhardt, J., Bäumer, V., Kumar, K., Mibus-Schoppe, H., Zinkernagel, J., . . . Schweiggert, R. (2023). Characterisation of acylated anthocyanins from red cabbage, purple sweet potato, and *Tradescantia pallida* leaves as natural food colourants by HPLC-DAD-ESI(+)-QTOF-MS/MS and ESI(+)-MSn analysis. *Food Chemistry*, *416*, 135601. https://doi.org/10.1016/j.foodchem.2023.135601

Strauch, R. C., Mengist, M. F., Pan, K., Yousef, G. G., Iorizzo, M., Brown, A. F., & Lila, M. A. (2019). Variation in anthocyanin profiles of 27 genotypes of red cabbage over two growing seasons. *Food Chemistry*, *301*, 125289. https://doi.org/10.1016/j.foodchem.2019.125289

Su, X., Xu, J., Rhodes, D., Shen, Y., Song, W., Katz, B., . . . Wang, W. (2016). Identification and quantification of anthocyanins in transgenic purple tomato. *Food Chemistry*, *202*, 184–188. https://doi.org/10.1016/j.foodchem.2016.01.128

Sun, J., Xiao, Z., Lin, L.-Z., Lester, G. E., Wang, Q., Harnly, J. M., & Chen, P. (2013). Profiling polyphenols in five *Brassica* species microgreens by UHPLC-PDA-ESI/HRMS n. *Journal of Agricultural and Food Chemistry*, *61*(46), 10960–10970.

Tan, J. B. L., Lim, Y. Y., & Lee, S. M. (2014). *Rhoeo spathacea* (Swartz) Stearn leaves, a potential natural food colorant. *Journal of Functional Foods*, *7*, 443–451. https://doi.org/10.1016/j.jff.2014.01.012

Tao, W., De Luca, L., Mateus, N., de Freitas, V., Bordiga, M., & Oliveira, H. (2025). A reliable, green and fast two-step solid-phase extraction methodology for the obtention of anthocyanin-rich extracts from different sources. *Journal of Food Composition and Analysis*, *140*, 107232. https://doi.org/10.1016/j.jfca.2025.107232

Tatsuzawa, F. (2014). Acylated cyanidin 3-sambubioside-5-glucosides from the purple-violet flowers of *Matthiola longipetala* subsp. *bicornis* (Sm) P. W. Ball. (Brassicaceae). *Phytochemistry Letters*, *9*, 17–21. https://doi.org/10.1016/j.phytol.2014.03.011

Tatsuzawa, F. (2019). Acylated pelargonidin glycosides from the red-purple flowers of *Iberis umbellata* L. and the red flowers of *Erysimum × cheiri* (L.) Crantz (Brassicaceae). *Phytochemistry*, *159*, 108–118. https://doi.org/10.1016/j.phytochem.2018.12.010

Tatsuzawa, F., Saito, N., Yokoi, M., Shigihara, A., & Honda, T. (1994). An acylated cyanidin glycoside in the red-purple flowers of x *Laeliocattleya* cv mini purple. *Phytochemistry*, *37*(4), 1179–1183. https://doi.org/10.1016/S0031-9422(00)89553-9

Tedesco, I., Carbone, V., Spagnuolo, C., Minasi, P., & Russo, G. L. (2015). Identification and quantification of flavonoids from two Southern Italian cultivars of *Allium cepa* L., Tropea (Red Onion) and Montoro (Copper Onion), and their capacity to protect human erythrocytes from oxidative stress. *Journal of Agricultural and Food Chemistry*, *63*(21), 5229–5238. https://doi.org/10.1021/acs.jafc.5b01206

Teixeira, M., De Luca, L., Faria, A., Bordiga, M., de Freitas, V., Mateus, N., & Oliveira, H. (2024). First insights on the bioaccessibility and absorption of anthocyanins from edible flowers: Wild pansy, cosmos, and cornflower. *Pharmaceuticals*, *17*(2), 191. https://www.mdpi.com/1424-8247/17/2/191

Teixeira, M., Tao, W., Fernandes, A., Faria, A., Ferreira, I. M. P. L. V. O., He, J., . . . Oliveira, H. (2023). Anthocyanin-rich edible flowers, current understanding of a potential new trend in dietary patterns. *Trends in Food Science & Technology*, *138*, 708–725. https://doi.org/10.1016/j.tifs.2023.07.010

Terahara, N., Callebaut, A., Ohba, R., Nagata, T., Ohnishi-Kameyama, M., & Suzuki, M. (2001). Acylated anthocyanidin 3-sophoroside-5-glucosides from *Ajuga reptans* flowers and the corresponding cell cultures. *Phytochemistry*, *58*(3), 493–500. https://doi.org/10.1016/S0031-9422(01)00172-8

Terahara, N., Toki, K., Saito, N., Honda, T., Matsui, T., & Osajima, Y. (1998). Eight new anthocyanins, ternatins C1–C5 and D3 and preternatins A3 and C4 from young *Clitoria ternatea* flowers. *Journal of Natural Products*, *61*(11), 1361–1367.

Tian, B., Zhao, J., Zhang, M., Chen, Z., Ma, Q., Liu, H., . . . Li, J. (2021). *Lycium ruthenicum* anthocyanins attenuate high-fat diet-induced colonic barrier dysfunction and inflammation in mice by modulating the gut microbiota. *Molecular Nutrition & Food Research*, *65*(8), 2000745.

Toki, K., Saito, N., Kuwano, H., Terahara, N., & Honda, T. (1995). Acylated anthocyanins in verbena flowers. *Phytochemistry*, *38*(2), 515–518. https://doi.org/10.1016/0031-9422(94)00698-S

Tomás-Barberán, F. A., Harborne, J. B., & Self, R. (1987). Dimalonated anthocyanins from the flowers of *Salvia splendens* and *S. coccinea*. *Phytochemistry*, *26*(10), 2759–2760. https://doi.org/10.1016/S0031-9422(00)83586-4

Truong, V.-D., Deighton, N., Thompson, R. T., McFeeters, R. F., Dean, L. O., Pecota, K. V., & Yencho, G. C. (2010). Characterization of anthocyanins and anthocyanidins in purple-fleshed sweetpotatoes by HPLC-DAD/ESI-MS/MS. *Journal of Agricultural and Food Chemistry*, *58*(1), 404–410. https://doi.org/10.1021/jf902799a

Tsuda, T. (2012). Dietary anthocyanin-rich plants: Biochemical basis and recent progress in health benefits studies. *Molecular Nutrition & Food Research*, *56*(1), 159–170.

Velasco González de Peredo, A., Vázquez Espinosa, M., Espada Bellido, E., Ferreiro González, M., Carrera Fernández, C. A., Fernández Barbero, G., & Palma Lovillo, M. (2021). Development of optimized ultrasound-assisted extraction methods for the recovery of total phenolic compounds and anthocyanins from onion bulbs. *Antioxidants*, *10*(11), 1755. https://doi.org/10.3390/antiox10111755

Vijayalakshmi, G., Raja, M. M., Naik, M. L., Carbone, V., Russo, G., & Khan, P. S. V. (2021). Determination of antioxidant capacity and flavonoid composition of onion (*Allium cepa* L.) landrace 'Krishnapuram' bulb using HPLC-ESI-ITMS. *Journal of Biosciences*, *46*, 1–7.

Wang, L., Lan, W., & Chen, D. (2024). Blueberry (Vaccinium spp.) anthocyanins and their functions, stability, bioavailability, and applications. *Foods*, *13*(17), 2851. https://www.mdpi.com/2304-8158/13/17/2851

Wang, L.-S., & Stoner, G. D. (2008). Anthocyanins and their role in cancer prevention. *Cancer Letters*, *269*(2), 281–290.

Wu, X., Beecher, G. R., Holden, J. M., Haytowitz, D. B., Gebhardt, S. E., & Prior, R. L. (2006). Concentrations of anthocyanins in common foods in the United States and estimation of normal consumption. *Journal of Agricultural and Food Chemistry*, *54*(11), 4069–4075. https://doi.org/10.1021/jf060300l

Xu, J., Su, X., Lim, S., Griffin, J., Carey, E., Katz, B., . . . Wang, W. (2015). Characterisation and stability of anthocyanins in purple-fleshed sweet potato P40. *Food Chemistry*, *186*, 90–96. https://doi.org/10.1016/j.foodchem.2014.08.123

Yang, C., Sun, N., Qin, X., Liu, Y., Sui, M., Zhang, Y., . . . Shen, X. (2024). Analysis of flavonoid metabolism of compounds in succulent fruits and leaves of three different colors of Rosaceae. *Scientific Reports*, *14*(1), 4933.

Yang, Z., & Zhai, W. (2010). Optimization of microwave-assisted extraction of anthocyanins from purple corn (*Zea mays* L.) cob and identification with HPLC-MS. *Innovative Food Science & Emerging Technologies*, *11*(3), 470–476. https://doi.org/10.1016/j.ifset.2010.03.003

Yoshida, K., Kondo, T., Kameda, K., Kawakishi, S., Lubag, A. J. M., Mendoza, E. M., & Goto, T. (1991). Structures of alatanin A, B and C isolated from edible purple yam *Dioscorea alata*. *Tetrahedron Letters*, *32*(40), 5575–5578.

Yoshida, Y., Koyama, N., & Tamura, H. (2002). Color and anthocyanin composition of strawberry fruit: Changes during fruit development and differences among cultivars, with special reference to the occurrence of pelargonidin 3-malonylglucoside. *Journal of the Japanese Society for Horticultural Science*, *71*(3), 355–361.

Yuan, Y., Yang, J., Yu, X., Huang, L., & Lin, S. (2014). Anthocyanins from buds of *Lonicera japonica* Thunb. var. *chinensis* (Wats.) Bak. *Food Research International*, *62*, 812–818.

Zakaria, N., Okello, E., Howes, M. J., Birch-Machin, M., & Bowman, A. (2018). In vitro protective effects of an aqueous extract of *Clitoria ternatea* L. flower against hydrogen peroxide-induced cytotoxicity and UV-induced mtDNA damage in human keratinocytes. *Phytotherapy Research*, *32*(6), 1064–1072.

Zhang, L., Zhou, J., Liu, H., Khan, M. A., Huang, K., & Gu, Z. (2012). Compositions of anthocyanins in blackberry juice and their thermal degradation in relation to antioxidant activity. *European Food Research and Technology*, *235*(4), 637–645. https://doi.org/10.1007/s00217-012-1796-6

Zhang, X., Liu, Y., Yong, H., Qin, Y., Liu, J., & Liu, J. (2019). Development of multifunctional food packaging films based on chitosan, TiO2 nanoparticles and anthocyanin-rich black plum peel extract. *Food Hydrocolloids*, *94*, 80–92. https://doi.org/10.1016/j.foodhyd.2019.03.009

Zhang, Y., Chu, G., Hu, Z., Gao, Q., Cui, B., Tian, S., . . . Chen, G. (2016). Genetically engineered anthocyanin pathway for high health-promoting pigment production in eggplant. *Molecular Breeding*, *36*, 1–14.

Zhang, Z.-C., Zhou, Q., Yang, Y., Wang, Y., & Zhang, J.-L. (2019). Highly acylated anthocyanins from purple sweet potato (*Ipomoea batatas* L.) alleviate hyperuricemia and kidney inflammation in hyperuricemic mice: possible attenuation effects on allopurinol. *Journal of Agricultural and Food Chemistry*, *67*(22), 6202–6211.

Zhao, C.-L., Yu, Y.-Q., Chen, Z.-J., Wen, G.-S., Wei, F.-G., Zheng, Q., . . . Xiao, X.-L. (2017). Stability-increasing effects of anthocyanin glycosyl acylation. *Food Chemistry*, *214*, 119–128. https://doi.org/10.1016/j.foodchem.2016.07.073

Extraction and Purification of Acylated Anthocyanins

4

Adele Papetti and Raffaella Colombo

4.1 INTRODUCTION

Anthocyanins (ACNs) are secondary metabolites widely distributed in nature and possess many different healthy properties that make these compounds useful as food supplements in addition to being used as ingredients for the dye industry. Therefore, the isolation and extraction of ACNs, including their acylated derivatives (AACNs), from the natural matrices is a key step and requires subsequent purification to obtain concentrated and stable extracts. AACNs are generally more stable than the corresponding ACNs due to the presence of one or more acyl residues on their skeleton. Over the years, different approaches have been set up to extract AACNs, and today the conventional solid to liquid method is overcome, even if still used, especially on a large scale; in this case, a purification step is required before the extract is concentrated. As regards the extraction solvents, generally aqueous or hydroalcoholic mixtures are used, and acidification (using hydrochloric, malic, tartaric, ascorbic, citric, or acetic acid) adjusts the pH, increasing the extraction yield and stability. The need to have sustainable and green methods has led to the development of different processes based on the use of water, ethanol, or natural deep eutectic solvents (NaDES). Ultrasound-assisted extraction (UAE), microwave-assisted extraction (MAE), and pressurized liquid extraction (PLE) are the currently most used approach. As the development of such methods needs the set-up of different experimental parameters, the design of experiment (DoE) approach is a useful tool to reduce time, cost, solvent, and consumption (Ongkowijoyo et al., 2018; Ghareaghajlou et al., 2021; Nunes et al., 2022; Tena & Asuero, 2022; Rodríguez-Mena et al., 2023; Airouyuwa et al., 2024; Yücetepe et al., 2024; Yun et al., 2024; Xue et al., 2024).

In this chapter, both the conventional and the most recent green methods will be discussed, with a section dedicated to each of them. In each section, the key parameters and the applications present in the literature of the last decade will be considered. The advantages and disadvantages of each approach will also be discussed.

DOI: 10.1201/9781003519591-5

4.2 CONVENTIONAL EXTRACTION METHODS

In recent years, the conventional solid–liquid extraction has already been used to separate AACNs from many food/plant matrices. Different extraction solvent mixtures, temperatures, and solid-to-liquid (S/L) ratios have been used. Generally, when high temperatures are applied, thermal degradation of ACNs can occur following a first-order kinetic; higher stability can be reached using an acidic medium, and the degradation is highly correlated with the degree of acylation. However, hot water extraction can be considered a feasible method because under suitable extraction conditions (time and temperature), the extracts are rich in AACNs and safe for food use, as in the case of AACNs in grape skin (Yu et al., 2024), red cabbage (De Marchi et al., 2024), and purple sweet potato (Steingass et al., 2023; Pereira et al., 2024). Few papers reported the use of hydroalcoholic mixtures without any acidification; it requires different temperature and time conditions set up according to the matrix, as in the case of AACNs extracted from red grape skin by-products; however, in this case, a higher yield than the corresponding extract obtained using only water at high temperature was obtained (Baron et al., 2021). Another example is the use of an ethanol–water mixture to obtain AACNs from corn flour: an ethanol:water mixture of 60:40 (v/v) was used for 4 and 16 h under dark conditions at room temperature, obtaining high yields (Piazza et al., 2024).

Ethanol was also used as a single solvent with Britton–Robinson buffer to extract AACNs from potato, and then the extract was purified using macroporous adsorbent resin D101. Different fractions rich in AACNs have been obtained using a gradient elution of EtOH:water and tested for their anti-tyrosinase activity (Ouyang et al., 2024). Another similar example regarded black seed-coated Korean adzuki bean (*Vigna angularis*) (Ha et al., 2021).

More common is the use of an acidified aqueous or hydroalcoholic solution. Many examples are present in the literature; for example, this type of solvent mixture has been used for extracting acylated derivatives of peonidin from purple-flashed sweet potato, which possesses a higher resistance to pH variations compared to their parent anthocyanin, peonidin-3-*O*-glucoside (He et al., 2016; Oliveira et al., 2019), or for the extraction of AACNs from *Raphanus sativus* cv. Sango seeds or sprouts (Matera et al., 2012, 2015). A similar solvent extraction composition was also used with grape pomace after an investigation about different ethanol percentages, pH values, and temperatures by DoE (Moutinho et al., 2023). Other examples of acidified alcoholic solutions used as extracting solvents are related to purple jam (Miyata et al., 2022) and purple- and orange-fleshed sweet potatoes (Bennett et al., 2021). de Aguiar Cipriano et al. (2015) successfully adopted a preheating step before solid–liquid extraction and the addition of polyphenol oxidase inhibitors to avoid texture changes when ACNs are used as pigments. Different temperatures were investigated to pretreat samples, ranging from 30 to 90°C, for 10 min; after this step, the samples were extracted for 1 h with water acidified (pH 2.0) with citric acid at 70°C under stirring. The addition of oxalic acid and sodium borate to a standard aqueous citric acid solution as a chemical modifier led to the chelation of metals and increased the ionic strength of the extraction solution. So, a relatively stable AACN extract was obtained that is useful for application in food and beverages (de Aguiar Cipriano et al., 2015).

To avoid the use of high temperature, it is also possible to operate at lower temperatures with an enzyme-assisted process to reduce extraction time and improve extraction efficiency (see Table 4.1), as reported for AACNs from *Lycium ruthenicum* Murr. In this case, pectinase was used and optimized by applying a single-factor experimental design. Therefore, an aqueous hydrochloric acid solution (HCl 0.1%, pH 2) was added with an enzyme at 38°C for 37 min, and the extraction yield was higher than that registered using NaDES (19.51 ± 0.21 mg/g). The extract was then purified using macroporous resin XDA-6 (Shen et al., 2020). Differently, 20% (w/w) of α-amylase, cellulase, and pectinase was added to colored potatoes for the extraction of AACNs using 50%–70% ethanol (*v*/v) as solvent, and the effect of a wide range of temperature (30–70°C) for 10–90 min was investigated. The optimal selected condition was 75.43% ethanol for 39.96 min at 51.80°C (Han et al., 2024).

TABLE 4.1 Main Applications of Conventional Extraction Methods

SAMPLE	*EXTRACTION SOLVENT*	*EXTRACTION CONDITIONS*	*AACN EXTRACTED*	*REFERENCES*
Grape skin	H_2O	30–120 min, 30–90°C, S/L 4:200 (w/v)	n.i.	Yu et al. (2024)
Red cabbage, purple sweet potato	Citric acid solution 0.5%, pH 2.3	Homogenization for 45 sec in a hand-held mixer	Cy-3-(hex-*p*-coum)-dihex-5-hex, Cy-3-(hex-fer)-dihex-5-hex, Cy-3-(hex-sin)-dihex-5-hex, Cy-3-(caf)-dihex-5-hex, Cy-3-(coum)-dihex-5-hex, Cy-3-(fer)-dihex-5-hex, Cy-3-(sin)-dihex-5-hex, Cy-(sin)-trihex, Cy-(sin)-tetrahex, Cy-3-(coum)(sin)-dihex-5-hex, Cy-(fer)(sin)-dihex-5-hex, Cy-(sin) (sin)-dihex-5-hex, Cy-3-(*p*-hb)-dihex-5-hex, Cy-3-(caf)-dihex-5-hex, Peo-3-(*p*hb)-dihex-5-hex, Peo-3-(caf)-dihex-5-hex, Cy-3-(caf)(caf)-dihex-5-hex, Cy-3-(caf) (*p*hb)-dihex-5-hex, Cy-3-(caf)-dihex-5-hex, Cy-3-(caf)(fer)-dihex-5-hex, Cy-3-(caf)(coum)-dihex-5-hex, Peo-3-(caf)(caf)-dihex-5-hex, Peo-3-(caf)(*p*hb)-dihex-5-hex, Peo-3-(caf)-dihex-5-hex, Peo-3-(caf)(fer)-dihex-5-hex, Peo-3-(caf)(coum)-dihex-5-hex	Steingass et al. (2023)
Lycium ruthenicum Murr.	Water acidified HCl 0.1% (pH 2) + pectinase	37 min, 38°C, pH 2.0, S/L 1:20 (w/v)	Pt-3-*O*-(glu-*trans*-*p*-coum)-rut-5-*O*-glu, Dp-3-*O*-(*p*-coum)-glu, Pt-3-*O*-(caf)-rut-5-*O*-glu, Dp-3-*O*-(*p*-coum)-rut-5-*O*-glu, Pt-3-*O*-(*trans*-*p*-coum)-rut-5-*O*-glu, Mv-3-*O*-(*p*-coum)-rut-5-*O*-glu, Pt-3-*O*-(*p*-coum)-rut isomer	Shen et al. (2020)

(Continued)

TABLE 4.1 (Continued) Main Applications of Conventional Extraction Methods

SAMPLE	*EXTRACTION SOLVENT*	*EXTRACTION CONDITIONS*	*AACN EXTRACTED*	*REFERENCES*
Pigmented corn	EtOH : H_2O 60:40 (v/v)	4 or 16 h, room temperature, S/L 1:10 (w/v)	Cy-3,6-(mal)-glu, Pg-3,6-(mal)-glu, Pn-3,6-(mal)-glu	Piazza et al. (2024)
Red cabbage	H_2O	80°C, pH 3, S/L 1:20 (w/v)	Cy-3-(sin)-diglu-5-glu, Cy-3-(caf)(*p*-coum)-diglu-5-glu, Cy-3-(fer)-triglu-5-glu, Cy-3-(sin)-triglu-5-glu, Cy-3-(fer)(sin)-triglu-5-glu, Cy-3-(sin)(*p*-coum)-triglu-5-glu, Cy-3-(sin)(sin)-triglu-5-glu, Cy-3-(*p*-coum)-diglu-5-glu, Cy-3-digluc-5-(sin)-glu, Cy-3-(fer)-diglu-5-glu, Cy-3-(fer)(fer)-diglu-5-glu, Cy-3-(fer)(sin)-diglu-5-glu, Cy-3-(sin)-(sin)-diglu-5-glu	Pereira et al. (2024)
Potato	70% Ethanol–Britton–Robinson buffer		Pn-3-*O*-*trans*-(caf)-rut-5-*O*-glu, Pn-3-*O*-(fer)-rut-5-*O*-glu, Pt-3-*O*-*trans*-(*p*-coum)-rut, Pt-3-*O*-*trans*-(*p*-coum)-rut, Pn-3-*O*-*trans*-(*p*-coum)-rut-5-*O*-glu, Mv-3-*O*-*trans*-(*p*-coum)-rut-5-*O*-glu, Pt-3-*O*-*trans*-(caf)-rut-5-*O*-glu, Dp-3-*O*-*trans*-(*p*-coum)-rut-5-*O*-glu, Mv-3-*O*-*trans*-(caf)-rut-5-*O*-glu, Pn-3-*O*-*trans*-(caf)-rut-5-*O*-glu, Pt-3-*O*-(fer)-rut-5-*O*-glu, Mv-3-*O*-(fer)-rut-5-*O*-glu, Pn-3-*O*-(fer)-rut-5-*O*-glu	Ouyang et al. (2024)
Purple-fleshed sweet potato	H_2O acidified with 5% formic acid	12 h, 40°C, S/L 1:20 (w/v)	Pn-3-(6′-*p*hb)-soph-5-glu, Pn-3-(6′-*p*hb-6″-caf)-soph-5-glu	Oliveira et al. (2019)
Grape pomace	EtOH : H_2O 70:3 v/v + 3.5% HCl	30 min, 60°C, S/L 1:40 (w/v)	Pt-3-(6″-coum)-glu	Moutinho et al. (2023)

Purple yam	MeOH : H_2O 50:50 (v/v) acidified with citric acid buffer (pH 2.5 or 7.0)	30–90°C, S/L 2:100 (w/v)	Cy-3-(caf)(fer)-soph-5-glu, Pn-3-(caf)(fer)-soph-5-glu, Pt-3-(coum)-rut-5-glu	Miyata et al. (2022)
Raphanus sativus cv. Sango seeds	Aqueous TFA solution (0.5%) (v/v)	5 min, S/L 1:24 (w/v)	Cy-3-*O*-[6-*O*-fer-2-*O*-(2-*O*-fer-glu)-glu]-5-*O*-(4-*O*-glu-6-*O*-mal-glu), Cy-3-*O*-[6-*O*-fer-2-*O*-(6-*O*-fer-glu)-glu]-5-*O*-(4-*O*-glu-6-*O*-mal-glu), Cy-3-*O*-[6-*O*-fer-2-*O*-(2-*O*-sin-glu)-glu]-5-*O*-(6-*O*-mal-glu), Cy-3-*O*-[2-*O*-(2-*O*-sin-glu)-glu]-5-*O*-glu), Cy-3-*O*-[6-*O*-fer-2-*O*-(2-*O*-sin-glu)-glu]-5-*O*-(glu), Cy-3-*O*-[2-*O*-(6-*O*-fer-glu)-glu]-5-*O*-(6-*O*-mal-glu), Cy-3-*O*-[6-*O*-(E)-fer-2-*O*-(6-*O*-fer-glu)-glu]-5-*O*-(6-*O*-mal-glu)	Matera et al. (2015)
Purple-fleshed sweet potatoes	EtOH acidified with 1% formic acid (v/v)	2 h, 25°C, S/L 0.5:10 (w/v)	Cy-3-(caf)(*p*hb)-soph, Cy-3-(caf)(*p*hb)-soph-5-glu, Cy-3-(caf)(van)-soph-5-glu, Pn-3-(caf)(van)-soph, Pn-3-(caf)(van)-soph-5-glu, Cy-3-(dicaf)-soph-5-glu, Pn-3-(dicaf)-soph-5-glu	He et al. (2016)
Colored potatoes	EtOH : H_2O 75.43:24.57 (v/v) acidified with HCl pH 2.5 + 20% enzymes	39.96 min, 51.80°C, S/L 1:20	Dp-3-*O*-5-*O*-(6-*O*-coum)-diglu, Mv-3-*O*-(6-*O*-*p*-coum)-glu, Pg-3-*O*-[2-*O*-glu-6-*O*-*p*-coum-glu]-5-*O*-glu, Pg-3-*O*-5-*O*-(6-*O*-coum)-diglu, Pg-3-*O*-[6-*O*-fer-2-*O*-glu-glu]-5-*O*-glu, Pg-3-*O*-(6-*O*-*p*-coum)-glu	Han et al. (2024)
Purple carrot concentrate	Ethyl acetate	L/L 1:1	Cy-3-*O*-(6-*O*-(4-*O*-*p*-coum)-glu)-xyl-galac, Cy-3-*O*-(6-*O*-fer-glu)-xyl-galac, Cy-3-*O*-(6-*O*-sin-glu)-xyl-galac	Hahm et al. (2021)
Black seed-coated Korean adzuki bean	EiHO:H_2O:40(v/v) acidified with 1% HCl	3 days, room temperature, S/L 2:5 (w/v)	Dp-3-*O*-(*p*-coum)-glu, Pt-3-*O*-(*p*-coum)-glu	Ha et al. (2021)

(Continued)

TABLE 4.1 (Continued) Main Applications of Conventional Extraction Methods

SAMPLE	*EXTRACTION SOLVENT*	*EXTRACTION CONDITIONS*	*AACN EXTRACTED*	*REFERENCES*
Red cabbage	Formic acid solution, pH 3.3	1 h, room temperature, S/L 1:1.5 (w/v)	Cy-3-(sin)-diglu-5-glu, Cy-3-(glu-sin)-diglu-5-glu, Cy-3-(glu-*p*-coum)-diglu-5-glu, Cy-3-(glu-fer)-diglu-5-glu, Cy-3-(glu-sin)-diglu-5-glu, Cy-3-(glu)(*p*-coum)(sin)-diglu-5-glu, Cy-3-(glu)(fer)(sin)-diglu-5-glu, Cy-3-(*p*-coum)-diglu-5-glu, Cy-3-(syn)-diglu-5-glu, Cy-3-(*p*-coum)(sin)-diglu-5-glu, Cy3-(*p*-fer)(sin)-diglu-5-glu, Cy-3-(sin)(sin)-diglu-5-glu, Cy-3-(caf)-diglu-5-glu, Cy-3-(glu)(sin)(sin)-diglu-5-glu, Cy-3-(sin)-glu-5-glu	De Marchi et al. (2024)
Purple sweet potatoes	H_2O acidified with citric acid pH 2.0	1 h, 70°C, S/L 1:6 (w/v)	Cy-3-(6″-*p*hb-soph)-5-glu, Cy-3-(6″-fer-soph)-5-glu, Cy-3-(6″-caf-soph)-5-glu, Cy-3-(6″-caf-6‴-*p*hb-soph)-5-glu, Pn-3-(6″-caf-soph)-5-glu, Cy-3-(6″-caf-6‴-fer-soph)-5-glu Pn-3-(6″-caf-6‴-*p*hb-soph)-5-glu	de Aguiar Cipriano et al. (2015)
Purple sweet potatoes	Citric acid aqueous solution pH 2.0 + SPE for purification	1 h, 70°C, S/L 1:6 (w/v)	Cy-3-(6″-*p*hb-soph)-5-glu, Cy-3-(6″-fer-soph)- 5-glu, Cy-3-(6″-caf-6‴-*p*hb-soph-5-glu, Pn-3-(6″-caf-soph)-5-glu, Cy-3-(6″-caf-6‴-fer-soph)-5-glu, Pn-(6″-caf-6‴-*p*hb-soph)-5-glu	de Aguiar Cipriano et al. (2022)

Purple and orange-fleshed sweet potatoes	EtOH : H_2O 75:25 (v/v) acidified with 10% acetic acid	60 min, S/L 10:5 (w/v)	Cy-3-*p*hb-soph-5-glu, Cy-3-caf-soph-5-glu, Pn-3-*p*hb-soph-5-glu, Pn-3-caf-soph-5-glu, Cy-3-fer-soph-5-glu, Pn-3-fer-soph-5-glu, Cy-3-caf-soph-5-glu, Cy-3-caf-*p*hb-soph-5-glu, Cy-3-dicaf-soph-5-glu, Pn-3-caf-soph-5-glu, Pn-3-caf-soph-5-glu, Cy-3-(caf)(fer)-soph-5-glu, Pn-3-dicaf-soph-5-glu, Pn-3-(caf)(*p*hb)-soph-5-glu, Pn-3-(caf)(fer)-soph-5-glu, Pn-3-(caf)(fer)-soph-5-glu	Bennett et al. (2021)
Red grape skin by-products	EtOH : H_2O 70:30 (v/v)	24 h, room temperature or 1 h 50°C, S/L 1:20 (w/v)	Pn-3-(6″-ace)-glu Mv-3-(6″-ace)-glu Dp-3-(6″-coum)-glu Mv-3-(6″-caf)-glu Pt-3-(6″-coum)-glu Pn-3-(6″-coum)-glu Mv-3-(6″-coum)-glu	Baron et al. (2021)

n.i., not identified; Cy, cyanidin; Pn, peonidin; Pg, pelargonidin; Dp, delphinidin; Pt, petunidin; Mv, malvidin; Hex, hexoside; Glu, glucoside; Soph, sophoroside; Rut, rutinoside; Xil, xiloside; Galac, galactoside; Fer, feruoloyl; Sin, sinapoyl; Coum, coumaroyl; Caf, caffeoyl; phb, p-hydroxybenzoyl; Mal, malonyl; Ace, acetyl; Van, vanilloyl.

Another approach consists in the direct extraction of AACNs from a liquid and/or homogenate matrix using solid phase extraction (SPE). AACNs from purple sweet potatoes were isolated from the homogenate after dilution SPE using methanol. After removing the solvent, the dried material was dissolved in citric acid buffer, pH 3.0, before the chemical analysis for AACN identification. This approach is useful to purify samples and separate ACNs from other polyphenols, and it is widely used (de Aguiar Cipriano et al., 2022).

Less common is the use of liquid–liquid extraction. An example is the use of such methodology for extracting a purple carrot concentrate, using ethyl acetate in the ratio 1:1 (v/v). The obtained extract needed further purification of the aqueous phase, which was performed using a Sephadex LH20 column. AACNs were eluted with 30% methanol and chemically characterized (Hahma et al., 2021).

4.3 ULTRASOUND-ASSISTED EXTRACTION METHOD

UAE is a green extraction approach widely used in the recent years to extract natural bioactive compounds, including AACNs. Ultrasound waves generate thermal, mechanical, and cavitational effects that damage cell walls, enhancing the extraction efficiency. In fact, the mechanical effect improves solvent penetration into the matrix, thus increasing the surface area of contact between the solvent and matrix particles; this effect is combined with cavitational effects that consist of the collapse of bubbles generated by ultrasonic waves on the surface of the matrix particles. High temperature and pressure are produced and generate shock waves, leading to micro-fractures on cell walls. UAE is a technique suitable for reducing energy consumption while obtaining high extraction yields; it is also ideal for the extraction of high-molecular-mass compounds. For ACNs, the parameters that must be generally set up are ultrasound power, time, and temperature; pH; and S/L ratio.

In the literature, many different optimized methods have been reported indicating the increasing use of this approach, which represents the most commonly used approach today to extract AACNs by green methods (Lepaus et al., 2023; Xue et al., 2024).

DoE is generally used to reduce the number of experiments, thus obtaining the optimized working conditions in a shorter time, as reported for the extraction of AACNs from purple sweet potatoes. UAE was used as a pretreatment step in AACN extraction, and time, temperature, pH, and ethanol percentage were optimized by an experimental design using response surface methodology (RSM). A comparison with the conventional extraction performed under the same experimental conditions, but without ultrasound support, was made. In addition to the total ACN content, total polyphenol content, total protein content, and the energy consumption associated with the extraction process were considered as response factors in RSM. The highest AACN extraction yield was obtained by applying UAE for 40 min before extracting with 80% ethanol for 120 min at 80°C and pH 2.5. A reduction in UAE time and extraction temperature induced a decrease in the yields, confirming what was previously reported by Barba et al. (2015) who highlighted how UAE coupled with mild temperature could produce higher results using an aqueous solvent rather than a hydroalcoholic mixture. However, the final conditions used were slightly different because the optimization considered all the response factors and not only the ACN content: UAE pretreatment for 40 min, 58% EtOH, extraction at 80°C, 120 min, pH 2.5. The obtained ACN yield was 0.293 mg/g, and 13 AACNs have been extracted, of which the main representatives were peonidin-3-caffeoyl-*p*-hydroxybenzoylsophoroside-5-glucoside, peonidin-3-(6″-caffeoyl-6‴-feruloylsophoroside)-5-glucoside, cyanidin-3-caffeoyl-*p*-hydroxybenzoylsophoroside-5-glucoside, and peonidin-3-phydroxybenzoylsophoroside-5-glucoside. The results obtained suggested the possibility to scale up the process as more convenient due to a reduction in energy use (Zhu et al., 2017). Using the same approach, other optimized parameters, consisting in a S/L ratio of 1:15, 78.9% of ethanol, 63.8°C, about 48 min, but pH 7.0, were identified for the extraction of AACNs from grape pomace, obtaining a yield of 193.547 mg/100 g.

Ultrasonication was also successfully applied to extract 14 AACNs from six edible *Dendrobium officinalis* samples derived from different artificial cultivations. Fresh stems or dry material from *D. officinalis* is widely used in China in food or food supplements as they are rich in bioactive compounds (Yang et al., 2023). A hydroalcoholic mixture solvent acidified with 1% hydrochloric acid was used to extract the stem material at room temperature in a short time (20 min) leading to an extract highly concentrated in cyanidin-3-*O*-[6-*O*-sinapoyl-2-*O*-(6-*O*-sinapoyl-glucosyl)-glucoside]-5-*O*-glucoside after a purification step consisting in the use of macroporous resin AB-8 (Zhu et al., 2024).

The use of UAE increased the extraction yield of ACN from grape pomace (193.547 mg/100 g), increasing also the number of compounds extracted; in fact, delphinidin-acetylglucoside and cyanidin-coumaroylglucoside were identified in the extract for the first time. Time affected the yield obtained: the yield increased up to 50 min, after which it decreased due to ACN degradation by oxidation, hydrolysis, and polymerization. The increase in pH value from 3 to 7 and the increase in ethanol concentration up to 70% were also investigated, registering a better extraction at pH 7 and 70% of ethanol; conversely, a further increase from 70% to 90% did not affect the yields, probably due to the enhancement of solvent permeability, which could dissolve compounds with similar polarity as ACNs (Zhou & Yang, 2022). On the same matrix, UAE was used as a pretreatment step to isolate monomeric forms simultaneously in a single step using semi-preparative high-performance liquid chromatography (HPLC) with a high purity grade (≥90%). Acidified MeOH was used, and the extract was purified on Amberlite XAD-7HP resin. The main advantage of this approach is that both nonacylated ACNs and AACNs and their *cis* and *trans* isomers were extracted with a simple, but extremely efficient method, which did not require any other pre-fractionation. As stated by Zhou and Yang (2022), cyanidin, delphynidin, petunidin, peonidin, and malvidin acetylglucoside were extracted as well as the same coumaroyl derivatives. In addition, peonidin and malvidin caffeoylglucoside were detected for the first time (Zhao et al., 2020).

Improved extraction of cyanidin-, petunidin-, malvidin-3-(6″-acetylglucoside); petunidin 3-(6″-acetylglucoside); malvidin-3-(6″-*p*-caffeyglucoside); and cyanidin-, petunidin, and peonidin-3-(6″-*O*-*p*-coumaroyl-glucoside) was also obtained for wine lees, a by-product of the wine industry, after the optimization of six variables, i.e., ethanol percentage, pH, S/L, temperature, amplitude, and cycle, through a Box–Behnken design (BBD). In this case, the percentage of ethanol in the solvent mixture, extraction temperature, and amplitude were the significant factors affecting both ACN extraction and extract antioxidant activity. Furthermore, the developed method had high repeatability and intermediate precision (variation coefficient (CV) lower than 5%). This protocol is easily applicable to recycle by-products derived from the winemaking industry to obtain bioactive compounds useful in medicine, pharmaceutical, and nutraceutical fields (Umsza-Guez et al., 2023). Conversely, for wine lees, when UAE was applied as a pretreatment before conventional extraction, it did not enhance the AACN extraction, probably because ultrasound waves affect only the external mass transfer and not the internal mass transfer, thus limiting the AACN extraction; however, even if the yield is similar to that obtained without this pretreatment, this extraction approach has the advantage to consistently reduce the time (Romero-Díez et al., 2023).

Other applications of UAE, all based on the optimization by DoE, regard *Nitraria tangutorum* Bobr. seed meal. Generally, temperature, ethanol percentage, and extraction time are the significant factors affecting the yield; in particular, the maximum yield (especially when the aim is to extract polyphenols and not only AACNs) have been obtained by applying high temperature for a short time. The effect of ethanol percentage is significant as the yield increases by increasing the solvent, but there is an optimal concentration after which the yield decreases by further increasing the solvent (Sang et al., 2017). A methanolic solution was also used to extract AACNs from pigmented corn using a BBD to optimize the extraction conditions. The method was validated with high precision (CV < 5% for both repeatability and intermediate precision) and a high level of accuracy (>90% of recovery) (Nurkhasanah et al., 2023).

In the last decade, ACNs from black carrots (*Daucus carota* ssp. *sativus* var. *atrorubens* Alef.) have been widely used in different foods to impart bright red color, and therefore, it is important to set up a suitable extraction method to obtain a stable extract. After the inactivation of polyphenol oxidase and peroxidase due to the low pH and the added methanol, samples were extracted with high yield using water/methanol/formic acid (75/20/5, v/v/v, pH 2.3) and probe-sonicated for 20 s at 75% amplitude using an ultrasound homogenizer (three extraction cycles). It is important to cool samples during UAE to avoid heating, which could degrade AACNs. This approach is simple and allows the brilliant color to be maintained (Gras et al., 2015), as also confirmed by Agcam et al. (2017) for black carrot pomace. In this case, a thermosonication process has been developed, and an RSM with a central composite design was successfully used to optimize ultrasound energy density and temperature, which were 183.1 J/g and 50°C, respectively.

An interesting research aimed to use only acidified water as solvent to extract AACNs. After effective optimization of extraction time (1–5 min), power density (8.3–16.7 W/mL), and pulse interval (0–2 s), it was possible to obtain extracts containing more than 2.5 mg/g of dried material. Another important factor affecting the extraction yield was the pretreatment of samples; in fact, when the samples were oven-dried (5 h, 60°C), a higher amount of AACNs was obtained compared to using freeze-drying.

Other recent applications of optimized UAE regard the extraction of AANCs present in onion bulbs (González-de-Peredo et al., 2021) and purple potatoes (Carrera et al., 2021). All the methods had high repeatability and intermediate precision.

A more recent approach in the extraction of ACNs, including the acylated forms, is the use of NaDES, which represent a green and effective alternative to the conventional solvent mixture. They are components present in nature and consist of a hydrogen bond acceptor (i.e., choline chloride) and a hydrogen bond donor (i.e., 1,2-propanediol or glycerol). NaDES have the advantage of directly interacting with the target compounds or destroying the cell wall to promote the release of the target compound from the plant matrix, thus improving the extraction yield. Other advantages of their use are low cost, high thermal stability, and environmental friendliness (Hikmawanti et al., 2021; Santos et al., 2022). The main drawbacks could be the high viscosity of such solvents due to the formation of a hydrogen bond network between the donor and acceptor, which could limit the extraction due to mass transfer problems. In fact, viscosity increases with the increase in hydrogen bonds and the number of carboxyl, hydroxyl, carbonyl, and amino groups, and it is closely correlated with the number of hydrogen bonds and van der Waals and electrostatic forces. Another important factor affecting NaDES stability is the molar ratio between the donor and acceptor. In fact, when a large number of hydroxyl groups in choline chloride are present (occurring at 1:4 or 1:5 ratio), glycerol hydroxyl groups cannot bind totally to choline chloride groups, and this leads to a loss in stability. In addition, an increase in the glycerol ratio leads to an increase in solvent viscosity and a decrease in ACN diffusion (Pires et al., 2024). Considering all the above mentioned, a double or even higher extraction yield than those usually obtained with conventional methods have been obtained for grape skin AACNs. In fact, citric acid and D-(+)-maltose at 4:1 molar ratio have been selected, and after the optimization of UAE parameters (using RSM), delphinidin 3-(6″-*p*-coumarylglucoside), cyanidin 3-(6″-acetlyglucoside)-5-glucoside, delphinidin-3-(6-*O*-*p*-coumarylglucoside)-5-glucoside, cyanidin 3-(6″-caffeylglucoside)-5-glucoside, cyanidin 3-galactoside-5-(6-*p*-coumarylglucoside), cyanidin 3-(4‴-caffeylrutinoside), cyanidin 3-glucoside-5-(6-*p*-coumaroylglucoside), petunidin 3-(6″-β-coumarylglucoside)-5-glucoside, cyanidin 3-(6″-caffeylglucoside) or delphinidin 3-*O*-β-D-(6-*O*-(E)-*p*-coumaryl)galactopyranoside, malvidin-3-(6-*O*-*p*-coumarylglucoside)-5-glucoside or malvidin 3-*O*-(6-*O*-(Z)-*p*-coumalonyl-β-glucopyranoside)-5-*O*-β-glucopyranoside, peonidin 3-(6″-*p*-coumarylglucoside)-5-glucoside, and peonidin 3-(6″-caffeylgucoside) have been putatively identified (Jeong et al., 2015).

This approach has been also successfully applied to extract malvidin-, delphinidin-, and peonidin-3-*O*-5-*O*-(6-*O*-coumaroyl)-diglucoside, and malvidin-, delphinidin-, peonidin-, and petunidin-3-*O*-(6-*O*-*p*-coumaroyl)-glucoside from *Vitis davidii* Foex. After an initial screening performed using seven different NaDES, choline chloride-glycerol was selected for the experimental design considering the total ACN content as the response factor. A purification step using macroporous resin AB-8 was necessary to eliminate glycerol before liquid chromatography-mass spectrometry analysis; proteins, sugars and polar compounds were also eluted away. The extract obtained under the optimized conditions was not only rich in ACNs but also more stable than those obtained using the classical acidified organic solvent under stress conditions (high temperature and pH value, and light exposure) (Zhang et al., 2024).

UAE was also applied to extract AACNs from purple sweet potatoes after an enzymatic hydrolysis step performed using ethanolic solutions of cellulose, pectinase, and papain. The optimal conditions were defined by single-factor experiments, Plackett–Burman experimental design, and RSM. In single-factor experiments, the effects of ethanol concentration (60%–100%), S/L (1:5–1:25 g/mL), total enzyme amount (0.5%–2.5%), enzyme ratio (cellulase: pectinase: papain from 2:2:1 to 1:1:1), hydrolysis temperature (30–70°C), pH (4.0–6.0), hydrolysis time (0.5–2.5 h), UAE temperature (30%–70%), UAE time (10–50 min), and number of UAE cycles (1–3) were investigated by changing one factor while keeping the others fixed. Then, Plackett–Burman experimental design was used to investigate the effects of six parameters on AACN extraction, and finally, a BBD was used to investigate the associations of the four factors mainly affecting the extraction, i.e., ethanol percentage, UAE and hydrolysis temperatures, and total enzyme amount. The best extraction conditions were obtained using 78% ethanolic solution containing cellulase: pectinase: papain 2:2:1 (w:w:w), performing the hydrolysis at 41°C, pH 4.5 for 1.5 h, and then applying two UAE cycles at 48°C for 20 min. This approach based on the UAE enzymatic method can be successfully used at the industrial level as it allows the rapid and effective extraction of AACNs (Wang et al., 2022). A similar approach was also used to extract AACNs from lees. β-Glucanase and a pectolytic enzyme (polygalacturonase, pectin methyl esterase, and pectin-lyase mixture) were used, thus increasing the extraction of AACNs with respect to the application of UAE alone. This method is particularly useful to monitor the aging on lees because it favors the release of polysaccharides, thus increasing the recovery of AACNs lost during vinification (del Fresno et al., 2018; Osete-Alcaraz et al., 2022).

A pectinase-assisted extraction method has been developed to increase the release of AACNs from mulberry (*Morus alba* L.) fruit cell walls. The optimal conditions predicted by RSM analysis indicated the requirement of 3.5% pectinase (v/w) and 1.5% citric acid (w/w) for 113 min at 50°C. The results suggested that the enzymatic treatment increased AACN permeability and solubility because pectinase degraded the complex pectin matrices, and it could be easily used in the food industry as an alternative and efficient process (Kim et al., 2021).

A recent new approach using UAE was used to extract AACNs from vegetables and fruits (purple cabbages, purple onions, purple sweet potatoes, black beans, eggplants, strawberries, blueberries, cherries, raspberries, and mulberries). It consisted in the use of a high-speed shear homogenizer for 5 min at 10,000 rpm and 25°C followed by a first extraction performed at constant temperature for 1 h; a second extraction with UAE was coupled for 30 min. SPE was used to purify the obtained solutions to remove interfering compounds (sugars, organic acids, etc.) before the identification of compounds. This approach remarkably improved the efficiency of ACNs including petunidin-3-*O*-(*p*-coumaroyl) glucoside, thus contributing to the development of new methodologies (Ren, Zhang et al., 2024).

In 2025, the possibility to use bio-based solvents such as ethanol, isopropanol, propylene-glycol, and ethylene-glycol for extracting AACNs from grape pomace by UAE was investigated. Using a solvent concentration of 50% v/v at 50°C and a S/L ratio of 1:10 and 1:50 g/mL, at higher ultrasonic intensities, resulted in fast analyte extraction in 5 min. Among the tested solvents, ethanol showed the highest ACN extraction, even if propylene-glycol had the highest eco-scale scores defined as "Excellent" with very low energy consumption. Therefore, the results highlighted how the integration of UAE with bio-based solvents could be considered a sustainable and efficient method for valorizing grape pomace, thus promoting the circular economy (Córdova et al., 2025).

Ultrasound has been also used under unspecified experimental conditions to extract AACNs from *Lycium ruthenicum* Murr. fruits. Extract purification through passage on macroporous resin AB-8 was necessary to isolate eight AACNs, putatively identified as delphinidin-3-*O*-*p*-coumaroyl-glucoside-5-*O*-glucoside, petunidin-3-*O*-*p*-coumaroyl-glucoside-5-*O*-glucoside, petunidin-3-*O*-(*trans*-*p*-coumaroylrutinoside)-5-*O*-glucoside,petunidin-3-*O*-(*cis*-*p*coumaroylrutinoside)-5-*O*-glucoside,petunidin-3-*O*-(*p*-coumaroylrutinoside)-7-*O*-glucoside, delphinidin-3-*O*-(*p*-coumaroylrutinoside)-5-*O*-glucoside, malvidin-3-*O*-(*p*-coumaroylrutinoside)-5-*O*-glucoside, and petunidin-3-*O*-*p*-coumaroylrutinoside (Ren, Tan et al., 2024).

Some research compared different extraction methods, as in the case of AACN extraction from *Licyum ruthenicum* (black goji berries). In fact, after the optimization of hot water extraction conditions, the effects of UAE, microwave-assisted extraction (MAE), and pectinase were investigated. The application of a blending step deeply improved the extraction yield, which was higher than that obtained using sub-critical water extraction at 170°C for 55 min at 0.3 mL/min flow (Wang, Yan et al., 2018) probably because the process reduces the sample material particle size, helping the mass transfer. UAE applied to the optimized hot water extracts (S/L 1:15 (w/v), 50°C, 30 min) produced the lowest extraction yield among the three approaches; in contrast, pectinase-assisted extraction resulted in the highest extraction yield, especially of petunidin-3-*O*-(*trans*-*p*-coumaroyl)-rutinoside-5-*O*-glucoside, when pectinase 3% (w/v) was used because pectinase degrades cell wall material and facilitates the diffusion of intracellular compounds during extraction (Gamage et al., 2023).

Extraction of AACNs from red grape skins using UAE and PLE was compared. The stability of peonidin-3-acetylglucoside, malvidin-3-acetylglucoside, malvidin-3-caffeoylglucoside, petunidin-3-*p*-coumaroylglucoside, and malvidin-3-*p*-coumaroylglucoside (*trans*) were investigated under the different experimental conditions. AACNs were stable up to 75 and 100°C when UAE or PLE was applied, respectively, and their stability was always higher than that registered for the nonacylated compounds, with differences more marked at the highest temperature and when higher levels of oxygen were present in the extraction media (Liazid et al., 2014).

The efficiency of AACN extraction from blackberries, blueberries, and grumixama by-products using UAE combined with PLE was investigated. Effectively the combination of the two approaches increased the AACN extraction, even if UAE, also using hydroalcoholic solvents, was highly efficient for extracting ACNs (Da Fonseca Machado et al., 2017).

Another recent interesting example is the set-up of an eco-friendly extraction method for grape pomace AACNs on a larger scale. First, eight different NaDES were investigated for their extraction efficiency and stabilizing capacity; viscosity was the main factor affecting the process for organic acid-based NaDES as the yield increased as solvent viscosity decreased. Considering these factors, choline chloride: citric acid was selected also for its lower cost, which is an important factor in scale-up development. Different extraction approaches have been studied such as low-frequency-ultrasound irradiation, alone or simultaneously with multimode-microwave, and the highest ACN (including AACN) extraction yield was obtained by applying simultaneous ultrasound/microwave-assisted extraction (UMAE). The process parameters were then optimized using RSM-based experimental design. In particular, the water content in NaDES, MAE power, and extraction time were the most influencing parameters due to their effect on viscosity and extraction efficiency. This set-up has the great advantage of being scaled up using a sequential ultrasound–microwave irradiation system in batch reactors. A purification step for ACN recovery is necessary, and simple resin adsorption on a glass column packed with Sepabeads SP207 macroporous resin was used to recover the analytes from NaDES (Panića et al., 2019).

Another research aiming to scale-up the extraction of ACNs from *Pyrus communis* 'Starkrimson' fruit peel compared UAE with conventional extraction methods (see Table 4.2). Under UAE optimized conditions, cyanidin-3-galactoside was the ACN extracted in higher amounts, but cyanidin 3-*O*-(6-caffeoyl-glucoside) and petunidin 3-*O*-(6″-*O*-coumaryl)-glucoside were also quantitatively extracted. The only factor with a negative impact on the extraction yield was viscosity, which could have a high impact during the scaling-up of the UAE process; however, the extraction yield was higher under batch scale-up compared to continuous operation (Belwal et al., 2019).

An innovative approach was developed by Da Fonseca Machado (2015). It consisted of an optimized blackberry by-product ethanolic extract obtained using UAE coupled with a co-precipitation process with polyvinylpyrrolidone, a polymer useful to encapsulate and thus stabilize the extract (particularly useful as it is soluble in water and in many solvents) through the supercritical antisolvent (SAS) process. SAS is a very promising technique to produce nano- and microparticles without residual solvent, and it generally uses carbon dioxide as the SAS. The use of CO_2 has many advantages, from the possibility to easily set up temperature, pressure, injection nozzle size, antisolvent, and solution flow rates to optimize the extraction selectivity and yield, due to its features such as inertness, green properties, non-toxic, non-flammability, and low cost. In addition, its moderate critical temperature makes it suitable for thermally unstable compounds such as AACNs. This research highlighted that using low temperature, polymer concentration, and solution flow rate, particles characterized by small diameter, homogeneous particle size distribution, and low ethanol residue and moisture were obtained.

TABLE 4.2 Main Applications of UAE Methods

SAMPLE	*EXTRACTION SOLVENT*	*EXTRACTION CONDITIONS*	*AACN EXTRACTED*	*REFERENCES*
Dendrobium officinalis stem	EtOH : H_2O 85:15 acidified with 1% HCl	20 min, room temperature, 400 W, 40 kHz	Cy-3-*O*-(6-*O*-fer-2-*O*-glu-glu)-5-*O*-glu, Cy-3-*O*-(6-*O*-sin-2-*O*-glu-glu)-5-*O*-glu, Cy-3-*O*-[6-*O*-3,4-dihbphenylace-2-*O*-(6-*O*-sin-glu)-glu]-5-*O*-glu, Cy-3-*O*-[6-*O*-caf-2-*O*-(6-*O*-sin-glu)-glu]-5-*O*-glu, Dp-3-*O*-[6-*O*-sin-2-*O*-(6-*O*-sin-glu)-glu]-5-*O*-glu, Cy-3-*O*-[6-*O*-fer-2-*O*-(6-*O*-sin-glu)-glu]-5-*O*-glu, Cy-3-*O*-[6-*O*-fer-2-*O*-(6-*O*-fer-glu)-glu]-5-*O*-glu, Cy-3-*O*-[6-*O*-sin-2-*O*-(6-*O*-sin-glu)-glu]-5-*O*-glu, Cy-3-*O*-[6-*O*-sin-2-*O*-(6-*O*-fer-glu)-glu]-5-*O*-(6-*O*-mal-glu), Cy-3-*O*-[6-*O*-sin-2-*O*-(6-*O*-sin-glu)-glu]-5-*O*-(6-*O*-mal-glu), Cy-3-*O*-[6-*O*-sin-2-*O*-(6-*O*-fer-glu)-glu]-5-*O*-(6-*O*-succinyl-glu), Cy-3-*O*-(6-*O*-sin-2-*O*-glu)-5-*O*-(6-*O*-succinyl-glu) Cy-3-*O*-[6-*O*-sin-2-*O*-(6-*O*-sin-glu)-glu]-5-*O*-(6-*O*-succinyl-glu)	S. Zhu et al. (2024)
Fresh purple sweet potato	EtOH : H_2O 58:42 acidified with 4% HCl drop by drop	40 min, room temperature, 178 W, 45 kHz, S/L 1:20 (w/W), pH 2.5	Pn-3-(caf)-(*p*hb)-soph-5-glu, Pn-3–(6″-caf-6‴-fer-soph)-5-glu, Cy-3-(caf)(*p*hb)-soph-5-glu, Pn-3-(*p*hb)-soph-5-glu, Cy-3-(*p*hb)-soph-5-glu, Cy-3-(6″-fer-soph)-5-glu, Pn-3-(6″-fer-soph)-5-glu, Cy-3-(6″-caf-soph)-5-glu, Cy-3-(6″-caf-6‴-fer-soph)-5-glu, Pn-3-(6′-*p*hb-6″-fer-soph)-5-glu	Z. Zhu et al. (2017)
Grape pomace	EtOH : H_2O 79.8:20.2 (v/v)	48 min, 63.8°C, S/L 1:15, pH 7.0	Cy-ace-glu, Pn-ace-glu, Mv-ace-glu, Dp-ace-glu, Pt-ace-glu, Cy-cum-glu, Pn-coum-glu, Dp-coum-glu, Mv-coum-glu	Zhou & Yang (2022)

TABLE 4.2 (Continued) Main Applications of UAE Methods

SAMPLE	*EXTRACTION SOLVENT*	*EXTRACTION CONDITIONS*	*AACN EXTRACTED*	*REFERENCES*
Grape pomace	MeOH acidified with 2% formic acid	10 min, 25–35°C, 59 kHz, S/L1:10 (w/v)	Cy-ace-glu, Pn-ace-glu, Mv-ace-glu, Dp-ace-glu, Pt-ace-glu, Cy-cum-glu, Pn-coum-glu, Dp-coum-glu, Mv-coum-glu, Mv-caf-glu	Zhao et al. (2020)
Vitis davidii Foex	Choline chloride-glycerol 1:2 molar ratio	49.64°C, 431.67 W, S/L 1:34.46 (w/v)	Mv-*O*-5-*O*-(6-*O*-coum)-diglu Dp-*O*-5-*O*-(6-*O*-coum)-diglu, Pn-3-*O*-5-*O*-(6-*O*-coum)-diglu, Mv-3-*O*-(6-*O*-*p*-coum)-glu, Dp-3-*O*-(6-*O*-*p*-coum)-glu, Pn-3-*O*-(6-*O*-*p*-coum)-glu, Pt-3-*O*-(6-*O*-*p*-coum)-glu	Zhang et al. (2024)
Purple sweet potato	EtOH : H_2O 78:22 containing 1.3% enzymes (cellulase: pectinase: papain 2:2:1)	48°C, S/L 1:15 (w/v), pH 4.5	Pn-3-caf-soph-5-glu	Wang et al. (2022)
Licyum ruthenicum (black goji berries)	Water added to 3% pectinase	30 min, 50°C, 60 Hz, 230 W, S/L 1:15 (w/v)	Pt-3-*O*-rut-(*trans*-*p*-coum)-5-*O*-glu, Pt-3-*O*-rut-(coum-(4-*O*-glu))-5-*O*-glu, Dp-3-*O*-rut-(coum)-5-*O*-glu	Gamage et al. (2023)
Wine lees	EtOH : H_2O 69:40	15 min, 37°C, pH 2, 35 W, 0.6 s^{-1} cycle values, S/L 0.57:25 (w/v)	Cy-3-(6″-ace-glu), Pt-3-(6″-ace-glu), Cy-3-(6″-ace-glu), Mv-3-(6″-ace-glu), Mv-3-(6″-*p*-caf-glu), Cy-3-(6″-*O*-*p*-coum-glu), Pt-3-(6″-*O*-*p*-coum-glu), Pn-3-(6″-*O*-*p*-coum-glu)	Umsza-Guez et al. (2023)

Nitraria tangutorun Bobr. seed meal	EtOH : H_2O 51.15:48.85 acidified with 0.1% HCl (v/v)	32.73 min; 70°C; 30 kHz, 300 W, S/L 1:15(w/v)	Dp-3-*O*-(caf)-diglu, Cy-3-*O*-(caf)-diglu, Cy-3-*O*-(*cis*-*p*-coum)-diglu, Cy-3-*O*-(*trans*-*p*-coum)-diglu, Pg-3-*O*-(*p*-coum)-diglu, Dp-3-*O*-(*p*-coum)-hex	Sang et al. (2017)
Wine lees	EtOH : H_2O 50:50 acidified with HCl (v/v)	5 min, 25°C, S/L 1:10 (w/v), pH 2.5	Dp-3-*O*-(6″-*p*-ace-glu), 10-carboxypyranoMv-3-6″-*p*-coum-glu, Cy-3-*O*-(6″-*p*-ace-glu), Mv-3-*O*-6″-ace-glu, Dp-3-*O*-(6″-*p*-coum-glu), Pt-3-*O*-(6″-*p*-coum-glu), Mv-3-*O*-(6″-*p*-coum-glu)	Romero-Díez et al. (2019)
Purple cabbages, purple onions, purple sweet potatoes, black beans, eggplants, strawberries, blueberries, cherries, raspberries, and mulberries	EtOH : H_2O 70:30 or 80:20 (v/v)	High-speed shear homogenization extraction (5 min, 10,000 rpm, 25°C, shaker for 60 min) 30 min, 40–50°C, 40 kHz, 160 W, pH 3–4; two cycles	Pt-3-*O*-(*p*-coum)-glu	Ren, Tan et al. (2024)
Lycium ruthenicum Murr. fruits	EtOH : H_2O 70:30 (v/v)	15 min, S/L 1:100	Dp-3-*O*-*p*-coum-glu-5-*O*-glu, Pt-3-*O*-*p*-coum-glu-5-*O*-glu, Pt-3-*O*-(*trans*-*p*-coum-rut)-5-*O*-glu, Pt-3-*O*-(*cis*-*p*-coum-rut)-5-*O*-glu, Pt-3-*O*-(*p*-coum-rut)-7-*O*-glu, Dp-3-*O*-(*p*-coum-rut)-5-*O*-glu, Mv-3-*O*-(*p*-coum-rut)-5-*O*-glu, Pt-3-*O*-*p*-coum-rut	Ren, Zhang et al. (2024)
Grape pomace	Choline chloride: citric acid 2:1	10 min, simultaneous UAE (30 W) and MAE (300 W)	Pn-3-*O*-ace-glu, Mv-3-*O*-ace-glu, Cy-3-(6-*O*-*p*-coum)-glu, Pn-3-(6-*O*-*p*-coum)-glu, Mv-3-(6-*O*-*p*-coum)-glu	Panić et al. (2019)

(Continued)

TABLE 4.2 (Continued) Main Applications of UAE Methods

SAMPLE	*EXTRACTION SOLVENT*	*EXTRACTION CONDITIONS*	*AACN EXTRACTED*	*REFERENCES*
Lees	EtOH : H_2O 12:88, pH 3.6, containing pectolytic enzyme (0.03 mL/L) and glucanase (0.05 g/L)	120 min, 400 W, 24 kHz	n.i.	del Fresno et al. (2018), Osete-Alcaraz et al. (2022)
Corn cob	MeOH: 36:64 (v/v), pH 7	10 min, 70°C, S/L 1:25 (w/v), 73% power	Cy-3-mal-glu, Pg-3-mal-glu, Pn-3-mal-glu	Nurkhasanah et al. (2023)
Red grape skin	H_2O	20 min, S/L 1.5:9 (w/v) 400 W, PLE 20 min, S/L 8 g (50 mL), 100°C	Pn-3-ace-glu, Mv-3-ace-glu, Mv-3-caf-glu, Pt-3-*p*-coum-glu, Mv-3-*p*-coum-glu	Liazid et al. (2014)
Grape skin	Citric acid and D-(+) maltose, 4:1 molar ratio	9 min, S/L 0.1/0.83 (w/v)	Dp-3-(6″-*p*-coum-glu), Cy-3-(6″-ace-glu)-5-glu, Cy-3-glu-5-(6-ace-glu), Dp-3-(6-*O*-*p*-coum-glu)-5-glu, Cy-3-(6″-caf-glu)-5-glu, Cy-3-galac-5-(6-*p*-coum-glu), Cy-3-(4‴-caf-rut), Cy-3-glu-5-(6-*p*-coum-glu), Pt-3-(6″-β-coum-glu)-5-glu, Cy-3-(6″-caf-glu), Mv-3-(6-*O*-*p*-coum-glu)-5-glu, Pn-3-(6″-*p*-coum-glu)-5-glu, Pn-3-(6″-caf-glu)	Jeong et al. (2015)
Black carrots	Water/methanol/formic acid (75/20/5, v/v/v), pH 2.3	20 s, 75% amplitude, S/L 0.05:20 (w/v)	Cy-3-*O*-sin-xyl-glu-galac, Cy-3-*O*-fer-xyl-glu-galac, Cy-3-*O*-*p*-coum-xyl-glu-galac	Gras et al. (2015)
Black carrot pomace	H_2O	102.4 W, 24 kHz, 183.1 J/g, 50°C S/L 1:3 (w/v)	Cy-3-xyl-glu-galac sinapic acid, Cy-3-xyl-glu-galac-ferulic acid, Cy-3-xyl-glu-galac-coumaric acid	Agcam et al. (2017)

Onion bulbs	MeOH : H_2O 57:43 (v/v), pH 2.0	0.64 s cycle, S/L 0.2:15 (w/v), 60°C, 90% amplitude, 20 Hz	Cy-3-*O*-(3″-mal-glu), Cy-3-*O*-(6″-mal-glu), Cy-3-*O*-(6″-mal-laminaribioside), Pn-3-*O*-(6″-mal-glu)	González-de-Peredo et al. (2021)
Purple potato	MeOH : H_2O 60:40 (v/v), pH 2.9	15 min, 0.5 s cycle, S/L 0.5:20 (w/v), 70°C, 70% amplitude	Pt-3-(*p*-coum)-rut-5-glu, Pn-3-(*p*-coum)-rut-5-glu, Mv-3-(*p*-coum)-rut-5-glu, Mv-3-(fer)-rut-5-glu, Pg-3-(*p*-coum)-rut-5-glu	Carrera et al. (2021)
Blueberry, blackberry, and grumixana by-products	EtOH : H_2O 50:50 and 70:30 (v/v), pH 2.0	90 min, 80°C, S/L 2:45 (w/v), 580 W, 37 kHz; 8 min, 80°C, S/L 5:4 (w/v), 580 W, 37 kHz*	Cy-3-*O*-ace-hex, Cy-3-*O*-(6″-mal)-glu, Cy-3-*O*-(6″-dioxalil)-glu	Da Fonseca Machado et al. (2017)
Grape pomace	Acidified water (HCl), pH 1.5	5 min, 25°C, 500 W, S/L 1.5:30 (w/v)	Dp-coum-diglu, Mv-ace-glu, Dp-coum-glu, Pt-coum-diglu, Pt-coum-glu, Mv-coum-glu, Pn-coum-glu	Angioletti Decker et al. (2024)
Pyrus communis 'Starkrimson' fruit peel	EtOH : H_2O 57:43 (v/v), acidified with 3% trifluoroacetic acid	11 min, 162 W, 71°C, S/L 1:30/w/v	Cyanidin 3-*O*-(6-caffeoyl-glucoside), petunidin 3-*O*-(6″-*O*-coumaryl)-glucoside	Belwal et al. (2019)
Blackberry by-products	Absolute ethanol	10 min, room temperature, 150 W, 37 kHz, S/L 1.5:20 (w/v)	Cy-3-*O*-mal-glu, Cy-3-*O*-dioxalyl-glu	Da Fonseca Machadoa et al. (2015)

* when UAE was combined with PLEn.i., not identified, but quantified as total AACNs; Cy, cyanidin; Pn, peonidin; Pg, pelargonidin; Dp, delphinidin; Pt, petunidin; Mv, malvidin;Hex, hexoside; Glu, glucoside; Soph, sophoroside; Rut, rutinoside; Xil, xiloside; Galac, galactoside; Fer, feruoloyl; Sin, sinapoyl; Coum, coumaroyl; Caf, caffeoyl; phb, p-hydroxybenzoyl; Mal, malonyl; Ace, acetyl; Van, vanilloyl.

4.4 MICROWAVE-ASSISTED EXTRACTION METHODS

MAE is another suitable method to extract bioactive compounds in a short time with increased recovery. It is based on rapid heating of the matrix by microwave radiation, which causes the breakdown of cell walls and the subsequent release of bioactive compounds. The heating is due to the microwave absorption, which excites molecules to upper rotational levels, thus dissipating the excess energy in the form of heat (see Table 4.3). The extraction process starts with the separation of the dissolved molecules from the matrix, followed by an increase in pressure and temperature to induce the diffusion of the solvent mixture into the sample. Therefore, the key parameters to be set up are S/L ratio, extraction solvent mixture composition, temperature, microwave power, frequency, irradiation time, and pressure. Obviously, the extraction solvent has to absorb microwave energy; hence, its selection is critical for obtaining a high extraction yield. Therefore, solvents characterized by high dielectric constant are ideal, such as water and ethanol, which are the most used. Another advantage of the use of MAE is the low amount of solvent used, so this method could be considered a sustainable approach. In addition, it saves energy due to its very short processing time. Conversely, a potential disadvantage of this approach is the possibility to degrade AACNs due to the negative effect of microwave energy on such compounds; to avoid this effect, the experimental conditions must be optimized (Tena & Asuero, 2022; Diaconeasa et al., 2023).

MAE was successfully applied to extract AACNs from onion bulbs in a short time with good recovery, high repeatability, and intermediate precision by using a DoE approach. The optimal experimental conditions were 62% methanol as a solvent, pH 2.0, 56°C, and S/L ratio 0.2:13 (w/v). The obtained yield was higher than that previously reported in the literature, also for UAE extraction (González-de-Peredo et al., 2022). Another interesting example of MAE application concerns a by-product of winemaking – grape marc. Optimized conventional extraction was compared with MAE. Different solvents were tested using the conventional extraction approach, and 50% ethanol acidified with HCl was the most convenient to maximize AACN extraction after eight extraction cycles. Sample pretreatment was also investigated considering different drying methods before a single extraction cycle, wherein freeze-drying was identified to be better than heat-drying at 70°C. Using the same extraction solvent, MAE increased the yield of AACNs, also preserving the extract composition. Citric acid was used in substitution of HCl, reaching about 90% of AACN extraction yield, and this has the advantage to make the protocol a sustainable green approach for industrial applications (Yang & Kilmartin, 2025).

Sometimes, after MAE extraction, a purification step was performed, as for purple heart radish AACNs. The new process was developed by combining MAE with macroporous resin purification performed by eluting AACNs with 80% ethanol acidified with 0.1% HCl, after an initial acidification of the extract. An extract of higher grade purity was obtained, and 12 AACNs were detected (Yuan et al., 2024).

Recently, a new sustainable and economical extraction methodology based on the use of MAE hydrodiffusion and gravity (MHG) was developed to extract strawberry and raspberry AACNs. This approach consists of an oven in which microwave heating and gravity at atmospheric pressure are combined. In this system, the irradiation of electromagnetic waves penetrates the matrix, and both the molecular interactions and heating of the medium favor the extraction because the cell wall of the matrix is broken, and the extract is collected by gravity at the bottom of the equipment. The so-obtained extract is very concentrated (Périno et al., 2016; Farias et al., 2021). A disadvantage could be the fact that sometimes after MHG, the matrix contains bioactive compounds; therefore, it could be necessary to carry out new extractions using a green solvent to avoid environmental problems. Because this residue has a low moisture content, it must be hydrated before a new MHG to exhaustively extract other compounds. Extraction yields higher than 60% have been obtained, and when the residues have been re-extracted, it was possible to exhaustively extract the ACNs from both fruits (90%). An advantage is the possibility to reuse the residue for other applications as this approach has low environmental impact (Farias et al., 2022).

TABLE 4.3 Main Applications of MAE methods

SAMPLE	*EXTRACTION SOLVENT*	*EXTRACTION CONDITIONS*	*AACN EXTRACTED*	*REFERENCES*
Strawberries and raspberries	MHG without solvent + MHG after pretreatment with H_2O or acidified water, pH 1.5 (citric acid)	15 min, 500 W, and 12 min at 600 W, 10°C, 200 g	Pn-3-(mal)-hex, Pn-3-(ace)-hex	Farias et al. (2022)
Onion bulb	MeOH:H_2O 62:38 (v/v), pH 2.0	2 min, 56°C, S/L 0.2:13 (w/v)	Cy-3-*O*-(3″-mal-glu), Cy-3-*O*-(6″-mal-glu), Cy-3-*O*-(6″-mal-laminaribioside), Pn-3-*O*-(6″-mal-glu)	Gonzalez-de-Peredo et al. (2022)
Merlot grape marc	EtOH:H_2O 50:50 (v/v) acidified with HCl or citric acid	60 min, 60°C, 500 W, S/L 20:40 (w/v) or 8.4:40 (w dried material/v)	Dp-3-(ace)-glu, Cy-3-(ace)-glu, Pt-3-(ace)-glu, Pn-3-(ace)-glu, Mv-3-(ace)-glu, Pt-3-(*p*-coum)-glu, Pn-3-(*p*-coum)-glu, Mv-3-(*p*-coum)-glu	Yang & Kilmartin (2025)
Purple-heart radish	EtOH:H_2O 60:40 (v/v)	10 min, 70°C, S/L 0.5:30 (w dried material/v)	Pg-3-soph-5-(mal)-galac, Pg-3-(mal)-soph-5-glu, Pg-3-(caf)-soph-5-galac, Pg-3-(caf)-soph-5-glu, Pg-3-(*p*-coum)-soph-5-galac, Pg-3-(fer)-soph-5-galac, Pg-3-(caf)-soph-5-(mal)-galac, Pg-3-(*p*-coum)-soph-5-glu, Pg-3-(*p*-coum)-soph-5-(mal)-glu, Pg-3-(ferl)-soph-5-glu, Pg-3-(caf)-soph-5-(mal)-glu, Pg-3-(fer)(caf)-soph-5-(mal)-glu	Yuan et al. (2024)

Cy, cyanidin; Pn, peonidin; Pg, pelargonidin; Dp, delphinidin; Pt, petunidin; Mv, malvidin;
Hex, hexoside; Glu, glucoside; Soph, sophoroside; Rut, rutinoside; Xil, xiloside; Galac, galactoside;
Fer, feruoloyl; Sin, sinapoyl; Coum, coumaroyl; Caf, caffeoyl; phb, p-hydroxybenzoyl; Mal, malonyl; Ace, acetyl; Van, vanilloyl.

4.5 PRESSURIZED LIQUID EXTRACTION METHODS

PLE is a useful technique to extract naturally bioactive compounds from plants, but its use is limited by the high cost of the equipment needed (see Table 4.4). In fact, this method is based on the use of a solvent at high temperature (50–150 °C) and pressure (3–21 MPa); therefore, a suitable extractor system is required. So, PLE is a technique based on the use of pressurized solvents that remain in a liquid state well above their boiling points, thus allowing better analyte solubilization and desorption kinetics from the matrix. In the extractor, sand and diatomaceous earth are used as dispersants and are added to reduce solvent channeling and particle clumping. The main advantage of this technique is the consistent reduction of time and solvent volume for each extraction cycle (Yun et al., 2024). High-pressure carbon dioxide is considered a green nonthermal technique because the extraction is performed by carbon dioxide under high pressure (5–50 MPa) and mild temperature (5–80 °C). Also in this case, the high pressure of carbon dioxide facilitates the recovery of bioactive molecules by disrupting the cell membrane, accelerating mass transfer, and causing an electrolyte balance disorder at the intracellular level (Khan et al., 2019). Specific equipment is also required for this extraction method. Supercritical carbon dioxide is a promising fluid generated at pressure higher than 7.38 MPa and temperature lower than 31.1°C (Arumugham et al., 2021).

PLE is ideal for extracting AANCs at high temperatures (40–200°C) in a short time (3–20 min) in a closed and inert environment under high pressure (3.3–20.3 MPa). In fact, PLE was used to extract ACNs from red wine grape skin. Different solvents have been tested, and the highest AACN extraction yield was obtained using a ternary mixture containing methanol/acetone/water acidified with HCl (40:40:20:0.1) at 60°C, 10.1 MPa, and performing three cycles of extraction for 5 min each (Ju & Howard, 2003).

If the extraction solvent is water at high temperature (150–300°C) and pressure, the method is referred to subcritical water extraction (SWE), which is a newly developed extraction approach. In this method, AACN extraction is carried out using water heated to a high temperature (100–374 °C) under a pressure below 22.1 MPa. Water is a polar solvent at room temperature that can shift its polarity as the temperature increases under pressure. Therefore, when the water temperature reaches 250 °C, water acts as a low-polarity solvent like ethanol. It is evident that in such conditions, water can act as ethanol in the extraction process, without the need for toxic, flammable, and environmentally unfriendly solvents (Rizkiyah et al., 2023). The extraction of bioactive substances including AACNs is highly variable depending on the operating conditions (pressure, temperature, flow rate, and particle size) (Yücetepe et al., 2024).

As previously reported, PLE was coupled with UAE to increase the extractability of AACNs from blueberries, blackberries, and grumixama by-products. When applied as the only process, special attention was paid to achieve a mass ratio of 18 kg solvent/kg fresh residue by working on the solvent's density. However, a lower yield was obtained when the newly proposed UAE + PLE technique used, which was higher than that obtained using only UAE (Da Fonseca Machado et al., 2017).

An innovative system based on pressurized water extraction that is able to directly produce powdered particles of black carrot extract was developed in 2017 by Uzel et al. by investigating the effects of temperature and pressure on the ACNs (and AACNs) and phenolic compound extraction yield. The process was also performed using a secondary co-solvent (ethanol). This innovative system consists of a supercritical water delivery unit, with a pre-heater, an extraction vessel, and an oven. A nozzle and a collector form the crystallizing particle unit. The pre-heating coil inside the oven assures the right experimental water temperature before the water enters the extractor, while a back-pressure regulator controls the extraction pressure. The main advantage of this system is the selectivity of SWE in extracting active compounds by simply changing the experimental temperature. Different ratios of ethanol to subcritical water have been investigated as well as the effect of temperature (40–100°C) on AACN yield by keeping the pressure (10 or 20 MPa) and water flow rate constant. The pre-expansion temperatures and pressures were set to be equal to the extraction temperatures and pressures. The temperature increase can improve the saturated

TABLE 4.4 Main Applications of PLE Methods and Enzyme-Assisted Methods

SAMPLE	*EXTRACTION SOLVENT*	*EXTRACTION CONDITIONS*	*AACN EXTRACTED*	*REFERENCES*
Mulberry fruits	3.5% pectinase (v/w) and 1.5% citric acid (w/w)	113 min, 50°C, S/L 1:5 (w/v)	Dp-3-(6-*p*-coum)-glu	Kim et al. (2021)
Dried red grape skin	PLE 0.1% HCl in methanol/ acetone/water; solvent mixture (40:40:20, v/v/v), pH 1.9	3 x 5 min cycles, 0.2 g mixed with 30 g sea sand, 60°C, 10.1 MPa	Mv-3-ace-glu, Dp-3-(*p*-coum)-glu, Mv-3-(*p*-caf)-glu, Cy-3-(*p*-coum)-glu, Pt-3-(*p*-coum)-glu, Mv-3-(*p*-coum)-glu	(Ju & Howard, 2003)
Blueberry, blackberry, grumixama by-products	PLE $EtOH:H_2O$ acidified with citric acid 50:50 or 70:30 (v/v), pH 2.0	30 min, 10.0 MPa, 80°C, S/L 1:18 (w/w)	Cy-3-*O*-(ace)-hex, Cy-3-*O*-(6″-mal)-glu, Cy-3-*O*-(6″-dioxalil)-glu	Da Fonseca et al. (2017)
Lycium ruthenicum Murr.	SWE	55 min, 3 mL/min, 170°C, 6.8 Mpa, 10 g	Pt-3-*O*-[6-*O*-(4-*O*-(4-*O*-glu)-*trans*-*p*-coum)-rham-glu]-5-*O*-[glu], Pt-3-*O*-[6-*O*-(4-*O*-(4-*O*-(glu)-*cis*-*p*-coum)-rham)-glu]-5-*O*-[glu], Pt-3-*O*-[6-*O*-(4-*O*-*p*-caf-rham)-glu]-5-*O*-[glu], Dp-3-*O*-[6-*O*-(4-*O*-*p*-coum-rham)-glu]-5-*O*-[glu], Pt-3-*O*-[6-*O*-(4-*O*-*trans*-*p*-coum-rham)-glu]-5-*O*-[glu], Pt-3-*O*-[6-*O*-(4-*O*-*cis*-*p*-coum-rham)-glu]-5-*O*-[glu], Mv-3-*O*-[6-*O*-(4-*O*-*p*-coum-rham)-glu]-5-*O*-glu	Wang Y et al. (2018)
Black carrot	PLE H_2O, acidified H_2O, EtOH	40–100°C, 10–20 MPa	n.i.	Uzel (2017)
Nitraria sibirica Pall. fruits	SWE	45 min, 140°C, flow rate 7 mL/ min	Dp-3-*O*-(*trans*-*p*-coum)-glu-5-*O*-glu, Pt-3-*O*-(*p*-coum-rut)-5-*O*-glu), Cy-3-*O*-(*trans*-*p*-coum)-diglu, Pn-3-*O*-(*trans*-*p*-coum)-diglu. Pt-3-*O*-(*cis*-*p*-coum)-glu, Pt-3-*O*-(*trans*-*p*-coum)-glu, Cy-3-*O*-(*trans*-*p*-coum)-glu-5-*O*-glu	Ren, Dong et al. (2024)

n.i., not identified, but quantified as total acylated anthocyanins; Cy, cyanidin; Pn, peonidin; Pg, pelargonidin; Dp, delphinidin; Pt, petunidin; Mv, malvidin;
Hex, hexoside; Glu, glucoside; Soph, sophoroside; Rut, rutinoside; Xil, xiloside; Galac, galactoside; Rham, rhamnoside; Fer, feruoloyl; Sin, sinapoyl; Coum, coumaroyl; Caf, caffeoyl; phb, p-hydroxybenzoyl; Mal, malonyl; Ace, acetyl; Van, vanilloyl.

vapor pressure of the solute, but a high temperature can reduce the density of the subcritical water and therefore reduce or increase the solubility of the solute, depending on the physicochemical characteristics of the sample. The results indicated 80°C at 10 MPa as the most efficient temperature for extraction with a pre-expansion temperature in the range 50–60°C. These conditions yielded an average particle size of 3 mm.

More recently, SWE was set up to obtain an extract from *Nitraria sibirica* Pall. fruits rich in ACNs and able to interact with α-glucosidase in a stable configuration, thus representing a promising α-glucosidase inhibitor for diabetes mellitus (Ren, Dong et al., 2024). RSM combined with the BBD was also performed in this research to evaluate the effect of temperature, extraction time, and flow rate on AACN yield. The extraction performed under optimized conditions (45 min, 140°C, and flow rate 7 mL/min) had a higher AACN yield than that obtained by applying conventional processes using methanol or hot water. The reason could be the fact that water viscosity, dielectric constant, and surface tension change under a subcritical state, thus improving the efficiency and selectivity of the extraction. In addition, water at high temperature possesses high superficial fluid velocity through the extraction cell and the mass transfer of AACNs is enhanced, increasing the solubility of ACN in water (Kuasnei et al., 2022). These results confirm the great advantage of SWE – a green approach. Following the same approach, AACNs have been extracted from *Lycium ruthenicum* Murr. in 55 min using a flow rate of 3 mL/min at 170°C. The high yield obtained (26.33%) confirmed that SWE is more efficient than the conventional extraction performed using UAE with a hydroalcoholic solvent or hot water (Wang Y et al., 2018).

4.6 ENZYME-ASSISTED EXTRACTION

Enzyme-assisted extraction is based on the use of suitable enzymes that can disrupt the cell wall, thus increasing the contact of molecules to be extracted with the solvent (Asanka Sanjeewa et al., 2023). Thanks to this phenomenon, the extraction efficiency of ACNs from fruits, vegetables, and various plant sources generally noticeably increases (Ngoc Nhon et al., 2022). Cellulases, hemicellulases, pectinases, proteases, and α-amylases are some of the enzymes generally used in this method (Gonzalez et al., 2022). The main advantages of this extraction technique are the low solvent volume, short extraction time, and high quality of the final product. Additionally, enzyme-assisted extraction limits the use of chemical substances in extraction processes, leading to the reduction of energy requirements. Therefore, this approach can be considered an environmentally friendly extraction technique (Costa et al., 2020). In enzymatic extraction, extraction efficiency is affected by parameters such as enzyme type/concentration, extraction time, extraction temperature, and solvent ratio. Conversely, the main disadvantages are the high cost, difficulty of reuse, and complex set-up (Yücetepe et al., 2024). A good compromise has been reached using cellulase and α-amylase to prepare a magnetic nanobiocatalyst by immobilizing the enzymes on amino-functionalized magnetic nanoparticles. This approach improved the thermal and pH stability of the enzymes and also their catalytic activity. In fact, the enzymes have been used to extract ACNs from black rice (including AACNs), and after six reuse cycles, they maintained about 65%–70% of their activity. The optimum temperature for extracting the black rice AACNs using the nanobiocatalyst was 30°C (lower than the value found for the free enzyme), and the maximum extraction capacity was at pH 5.0. The method has a considerable potential for reducing the costs, suggesting broad applications in the food industry (Yi et al., 2021).

Recently, enzyme-assisted extraction has also been used to extract ACNs from blackcurrant, and the results have been compared with those obtained by using UAE. No significant differences in extraction yield has been registered between the two approaches (Gonzalez et al., 2022).

REFERENCES

Agcam, E., Akyıldız, A., Balasubramaniam, V.M. (2017). Optimization of anthocyanins extraction from black carrot pomace with thermosonication. *Food Chemistry*, 237, 461–470.

Airouyuwa, J.O., Sivapragasam, N., Redha, A.A., Maqsood, S. (2024). Sustainable green extraction of anthocyanins and carotenoids using deep eutectic solvents (DES): A review of recent developments. *Food Chemistry*, 448, 139061.

Angioletti Decker, B.L., de Godoy Alves Filho, E., Alexandre e Silva, L.M., Riceli Vasconcelos Ribeiro, P., Sousa de Brito, E., Fernandes, F.A.N., Vidal Fonteles, T., Rodrigues, S. (2024). Ultrasound-assisted extraction of anthocyanins from grape pomace using acidified water: Assessing total monomeric anthocyanin and specific anthocyanin contents. *Food Research International*, 194, 114910.

Arumugham, T., Rambabu, K., Hasan, S.W., Show, P.L., Rinklebe, J., Banat, F. (2021). Supercritical carbon dioxide extraction of plant phytochemicals for biological and environmental applications–a review. *Chemosphere*, 271, 129525.

Asanka Sanjeewa, K.K., Herath, K.H.I.N.M., Young-Sang Kim, You-Jin Jeon, Se-Kwon Kim (2023). Enzyme-assisted extraction of bioactive compounds from seaweeds and microalgae. *Trends in Analytical Chemistry*, 167, 117266.

Barba, F.J., Brianceau, S., Turk, M., Boussetta, N., Vorobiev, E. (2015). Effect of alternative physical treatments (ultrasounds, pulsed electric fields, and high voltage electrical discharges) on selective recovery of bio-compounds from fermented grape pomace. *Food and Bioprocess Technology*, 8(5), 1139–1148.

Baron, G., Ferrario, G., Marinello, C., Carini, M., Morazzoni, P., Aldini, G. (2021). Effect of extraction solvent and temperature on polyphenol profiles, antioxidant and anti-inflammatory effects of red grape skin by-product. *Molecules*, 26, 5454.

Belwal, T., Huang, H., Li, L., Duan, Z., Zhang, X., Aalim, A., Luo, Z. (2019). Optimization model for ultrasonic-assisted and scale-up extraction of anthocyanins from *Pyrus communis* 'Starkrimson' fruit peel. *Food Chemistry*, 297, 124993.

Bennett, A.A., Mahood, E.H., Fan, K., Moghe, G.D. (2021). Untargeted metabolomics of purple and orange-fleshed sweet potatoes reveals a large structural diversity of anthocyanins and flavonoids. *Scientific Reports*, 11, 16408.

Carrera, C., Aliaño-González, M.J., Valaityte, M., Ferreiro-González, M., Barbero, G.F., Palma, M. (2021). A novel ultrasound-assisted extraction method for the analysis of anthocyanins in potatoes (*Solanum tuberosum* L.). *Antioxidants*, 10, 1375.

Córdova, A., Catalán, S., Carrasco, V., Farias, F.O., Trentin, J., López, J., Salazar, F., Mussagy, C.U. (2025). Sustainable assessment of ultrasound-assisted extraction of anthocyanins with bio-based solvents for upgrading grape pomace Cabernet Sauvignon derived from a winemaking process. *Ultrasonics Sonochemistry*, 112, 107201.

Costa, J.R., Tonon, R.V., Cabral, L., Gottschalk, L., Pastrana, L., Pintado, M.E. (2020). Valorization of agricultural lignocellulosic plant byproducts through enzymatic and enzyme-assisted extraction of high-value-added compounds: A review. *Sustainable Chemistry & Engineering*, 8, 13112–13125.

Da Fonseca Machado, A.P., Duarte Pereira, A.L., Fernández Barbero, G., Martínez, J. (2017). Recovery of anthocyanins from residues of *Rubus fruticosus*, *Vaccinium myrtillus* and *Eugenia brasiliensis* by ultrasound assisted extraction, pressurized liquid extraction and their combination. *Food Chemistry*, 231, 1–10.

Da Fonseca Machado, A.P., Pasquel-Reátegui, J.L., Fernández Barbero, G., Martínez, J. (2015). Pressurized liquid extraction of bioactive compounds from blackberry (*Rubus fruticosus* L.) residues: A comparison with conventional methods. *Food Research International*, 77, 675–683.

de Aguiar Cipriano, P., Ekici, L., Barnes, R.C., Gomes, C., Talcott, S.T. (2015). Pre-heating and polyphenol oxidase inhibition impact on extraction of purple sweet potato anthocyanins. *Food Chemistry*, 180, 227–234.

de Aguiar Cipriano, P., Kim, H., Fang, C., Venancio, V.P., Mertens-Talcott, S.U., Talcott, S.T. (2022). In vitro digestion, absorption and biological activities of acylated anthocyanins from purple sweet potatoes (*Ipomoea batatas*). *Food Chemistry*, 374, 131076.

del Fresno, J.M., Loira, I., Morata, A., González, C., Suárez-Lepe, J.A., Cuerda, R. (2018). Application of ultrasound to improve lees ageing processes in red wines. *Food Chemistry*, 261, 157–163.

De Marchi, D., Salemi, L., Bellumori, M., Chignola, R., Mainente, F., Santisteban Soto, D.V., Fierri, I., Ciulu, M., Zoccatelli, G. (2024). Thermal degradation of red cabbage (*Brassica oleracea* L. var. *Capitata* f. *rubra*) anthocyanins in a water model extract under accelerated shelf-life testing. *Food Chemistry*, 440, 138272.

Diaconeasa, Z., Iuhas, C.I., Ayvaz, H., Mortas, M., Farcas, A., Mihai, M., Danciu, C., Stanila, A. (2023). Anthocyanins from agro-industrial food waste: Geographical approach and methods of recovery – a review. *Plants*, 12, 74.

Farias, C.A.A., Moraes, D.P., Lazzaretti, M., Ferreira, D.F., Zabot, G.L., Barin, J.S., Ballus, C.A., Barcia, M.T. (2021). Microwave hydro diffusion and gravity as pretreatment for grape dehydration with simultaneous obtaining of high phenolic grape extract. *Food Chemistry*, 337, 127723.

Farias, C.A.A., Moraes, D.P., Neuenfeldt, N.H., Zabot, G.L., Emanuelli, T., Barin, J.S., Ballus, C.A., Barcia, M.T. (2022). Microwave hydrodiffusion and gravity model with a unique hydration strategy for exhaustive extraction of anthocyanins from strawberries and raspberries. *Food Chemistry*, 383, 132446.

Gamage, G.C.V., Choo, W.S. (2023). Effect of hot water, ultrasound, microwave, and pectinase-assisted extraction of anthocyanins from black goji berry for food application. *Heliyon*, 9, e14426.

Ghareaghajlou, N., Hallaj-Nezhadi, S., Ghasempour, Z. (2021). Red cabbage anthocyanins: Stability, extraction, biological activities and applications in food systems. *Food Chemistry*, 365, 130482.

Gonzalez, J.A., Carrera, C., Barbero, G.F., Palma, M. (2022). A comparison study between ultrasound-assisted and enzyme-assisted extraction of anthocyanins from blackcurrant (*Ribes nigrum* L.). *Food Chemistry*, X, 13.

González-de-Peredo, A.V., Vázquez-Espinosa, M., Espada-Bellido, E., Ferreiro-González, M., Carrera, C., Barbero, G.F., Palma, M. (2021). Development of optimized ultrasound-assisted extraction methods for the recovery of total phenolic compounds and anthocyanins from onion bulbs. *Antioxidants*, 10, 1755.

González-de-Peredo, A.V., Vázquez-Espinosa, M., Espada-Bellido, E., Ferreiro-González, M., Carrera, C., Barbero, G.F., Palma, M. (2022). Extraction of antioxidant compounds from onion bulb (*Allium cepa* L.) using individual and simultaneous microwave-assisted extraction methods. *Antioxidants*, 11, 846.

Gras, C.C., Carle, R., Schweiggert, R.M. (2015). Determination of anthocyanins from black carrots by UHPLC-PDA after ultrasound-assisted extraction. *Journal of Food Composition and Analysis*, 44, 170–177.

Ha, T.J., Park, J.E., Lee, K.-S, Seo, W.D., Song, S.-B., Lee, M.-H., Kim, S., Kim, J.-I, Oh, E., Pae, S.-B., Kwak, D.-Y., Lee, J.H. (2021). Identification of anthocyanin compositions in black seed coated Korean adzuki bean (*Vigna angularis*) by NMR and UPLC-Q-Orbitrap-MS/MS and screening for their antioxidant properties using different solvent systems. *Food Chemistry*, 346, 128882.

Hahma, T.H., Tanakab, M., Nguyenc, H.-N., Tsutsumid, A., Aizawad, K., Matsuia, T. (2021). Matrix-assisted laser desorption/ionization mass spectrometry-guided visualization analysis of intestinal absorption of acylated anthocyanins in Sprague-Dawley rats. *Food Chemistry*, 334, 127586.

Han, H., Li, R., Jin, X., Li, Y., Chen, Q., He, C., Wang, M. (2024). Metabolomic analysis, extraction, purification and stability of the anthocyanins from colored potatoes. *Food Chemistry: X*, 22, 101423.

He, W., Zeng, M., Chen, J., Jiao, Y., Niu, F., Tao, G., Zhang, S., Qin, F., He, Z. (2016). Identification and quantitation of anthocyanins in purple-fleshed sweet potatoes cultivated in China by UPLC-PDA and UPLC-QTOFMS/MS. *Journal of Agricultural and Food Chemistry*, 64, 171–177.

Hikmawanti, N.P.E., Ramadon, D., Jantan, I., Munìm, A. (2021) Natural deep eutectic solvents (NADES): Phytochemical extraction performance enhancer for pharmaceutical and nutraceutical product development. *Plants*, 10(10), 2091.

Jeong, K.M., Zhao, J., Jin, Y., Heo, S. R., Han, S.Y., Yoo, D.E., Lee, J. (2015). Highly efficient extraction of anthocyanins from grape skin using deep eutectic solvents as green and tunable media. *Archives of Pharmacal Research*, 38, 2143–2152.

Ju, Z.Y., Howard, L.R. (2003). Effects of solvent and temperature on pressurized liquid extraction of anthocyanins and total phenolics from dried red grape skin. *Journal of Agricultural and Food Chemistry*, 51, 5207–5213.

Khan, S.A., Aslam, R., Makroo, H.A. (2019). High pressure extraction and its application in the extraction of bioactive compounds: A review. *Journal of Food Process Engineering*, 42, e12896.

Kim, M., Nam, D.-G., Choe, J.-S., Hwang, K.-A., Choi, A.-J. (2021). Optimization of pectinase-assisted extraction condition of mulberry (*Morus alba* L.) fruit using response surface methodology and its effect on anthocyanin synthesis pathway-related metabolites. *Journal of Food Science*, 86, 3926–3938.

Kuasnei, M., Wojeicchowski, J.P., Santos, N.H., Pinto, V.Z., Ferreira, S.R.S., Zielinski, A.A.F. (2022). Modifiers based on deep eutectic mixtures: A case study for the extraction of anthocyanins from black bean hulls using high pressure fluid technology. *Journal of Supercritical Fluids*, 191, 105761.

Lepaus, B.M., Valiati, B.S., Machado, B.G., Domingos, M.M., Silva, M.N., Faria-Silva, L., Bernardes, P.C., Oliveira, D.D.S., de São José, J.F.B. (2023). Impact of ultrasound processing on the nutritional components of fruit and vegetable juices. *Trends in Food Science & Technology*, 138, 752–765.

Liazid, A., Barbero, G.F., Azaroual, L., Palma, M., Barroso, C.G. (2014). Stability of anthocyanins from red grape skins under pressurized liquid extraction and ultrasound-assisted extraction conditions. *Molecules*, 19, 21034–21043.

Matera, R., Gabbanini, S., Berretti, S., Amorati, R., De Nicola, G.R., Iori, R., Valgimigli, L. (2015). Acylated anthocyanins from sprouts of *Raphanus sativus* cv. Sango: Isolation, structure elucidation and antioxidant activity. *Food Chemistry*, 166, 397–406.

Matera, R., Gabbanini, S., De Nicola, G.R., Iori, R., Petrillo, G., Valgimigli, L. (2012). Identification and analysis of isothiocyanates and new acylated anthocyanins in the juice of *Raphanus sativus* cv. Sango sprouts. *Food Chemistry*, 133, 563–572.

Miyata, R., Sano, H., Hoshino, S., Kumazawa, S. (2022). Thermostability and catechol-O-methyltransferase inhibitory activity of acylated anthocyanins from purple yam. *Bioscience, Biotechnology, and Biochemistry*, 86(7), 916–921.

Moutinho, J., Gouvinhas, I., Domínguez-Perles, R., Barros, A. (2023). Optimization of the extraction methodology of grape pomace polyphenols for food applications. *Molecules*, 28, 3885.

Ngoc Nhon, H.T., Diem My, N.T., Tuong Vi, V.N., Kim Lien, P.T., Thao Minh, N.T., Doan Duy, L.N., Hong Anh, L.T., Anh Dao, D.T. (2022). Enhancement of extraction effectiveness and stability of anthocyanin from *Hibiscus sabdariffa* L. *Journal of Agriculture and Food Research*, 10, 100408.

Nunes, A.N., Borges, A., Matias, A.A., Rosário Bronze, M., Oliveira, J. (2022). Alternative extraction and downstream purification processes for anthocyanins. *Molecules*, 27, 368.

Nurkhasanah, A., Fardad, T., Carrera, C., Setyaningsih, W., Palma, M. (2023). Ultrasound-assisted anthocyanins extraction from pigmented corn: Optimization using response surface methodology. *Methods and Protocols*, 6, 69.

Oliveira, H., Basílio, N., Pina, F., Fernandes, I., de Freitas, V., Mateus, N. (2019). Purple-fleshed sweet potato acylated anthocyanins: Equilibrium network and photophysical properties. *Food Chemistry*, 288, 386–394.

Ongkowijoyo, P., Luna-Vital, D.A., Gonzalez de Mejia, E. (2018). Extraction techniques and analysis of anthocyanins from food sources by mass spectrometry: An update. *Food Chemistry*, 250, 113–126.

Osete-Alcaraz, A., Bautista-Ortín, A.B., Pérez-Porras, P., Gómez-Plaza, E. (2022). The application of ultrasound and enzymes could be promising tools for recovering polyphenols during the aging on lees process in red winemaking. *Foods*, 11, 19.

Ouyang, J., Hu, N., Wang, H. (2024). Isolation, purification and tyrosinase inhibitory activity of anthocyanins and their novel degradation compounds from Solanum tuberosum L. *Molecules*, 29, 1492.

Panića, M., Gunjević, V., Cravotto, G., Radojčić Redovnikovića, I. (2019). Enabling technologies for the extraction of grape-pomace anthocyanins using natural deep eutectic solvents in up-to-half-litre batches extraction of grape-pomace anthocyanins using NADES. *Food Chemistry*, 300, 125185.

Pereira, A.R., Cruz Fernandez, V., Delerue-Matos, C., de Freitas, V., Mateus, N., Oliveira, J. (2024). Exploring acylated anthocyanin-based extracts as a natural alternative to synthetic food dyes: Stability and application insights. *Food Chemistry*, 461, 140945.

Périno, S., Pierson, J.T., Ruiz, K., Cravotto, G., Chemat, F. (2016). Laboratory to pilot scale: Microwave extraction for polyphenols lettuce. *Food Chemistry*, 204, 108–114.

Piazza, S., Bani, C., Colombo, F., Mercogliano, F., Pozzoli, C., Martinelli, G., Petroni, K., Pilu, S.R., Sonzogni, S., Fumagalli, M., Sangiovanni, E., Restani, P., Dell'Agli, M., Di Lorenzo, C. (2024). Pigmented corn as a gluten-free source of polyphenols with anti-inflammatory and antioxidant properties in CaCo-2 cells. *Food Research International*, 191, 114640.

Pires, I.V., Meller da Silva, L.H., da Cruz Rodrigues, A.M., Saldaña, M.D.A. (2024). Natural deep eutectic solvents for anthocyanin extraction from agricultural sources: Process parameters, economic and environmental analysis, and industrial challenges. *Comprehensive Review in Food Science and Food Safety*, 23, e70057.

Ren, L., Dong, Q., Liu, Z., Wang, Y., Tan, N., Wang, H., Hu, N. (2024). Optimization of subcritical water extraction, UPLC-triple-TOF-MS/MS analysis, antioxidant and α-glucosidase inhibitory activities of anthocyanins from *Nitraria sibirica* Pall. *Fruits. Food Chemistry X*, 23, 101626.

Ren, L., Tan, N., Ouyang, J., Wang, R., Tie, F., Dong, Q., Wang, H., Hu, N. (2024). Hypoglycaemic activity of the anthocyanin enriched fraction of *Lycium ruthenicum* Murr. fruits and its ingredient identification via UPLC-triple-TOF-MS/MS. *Food Chemistry*, 461, 140837.

Ren, Y., Zhang, S., Zhao, B., Qian, Y., Cheng, X., Chen, C., Liu, H., Zhang, C. (2024). Enhancing anthocyanin extraction efficiency in vegetables and fruits: A high-speed shear homogenization technology. *Journal of Science and Food Agriculture*, 104, 8975–8990.

Rizkiyah, D.N., Putra, N.R., Idham, Z., Aziz, A.H.A., Che Yunus, M.A., Veza, I., Irianto, Chinedu Mamah, S., Qomariyah, L. (2023). Recovery of anthocyanins from *Hibiscus sabdariffa* L. using a combination of supercritical carbon dioxide extraction and subcritical water extraction. *Processes*, 11, 751.

Rodríguez-Mena, A., Ochoa-Martínez, L.A., González-Herrera, S.M., Rutiaga-Quiñones, O.M., González-Laredo, R.F., Olmedilla-Alonso, B. (2023). Natural pigments of plant origin: Classification, extraction and application in foods. *Food Chemistry*, 398, 133908.

Romero-Díez, R., Matos, M., Rodrigues, L., Bronze, M.R., Rodríguez-Rojo, R., Cocero, M.J., Matias, A.A. (2019). Microwave and ultrasound pre-treatments to enhance anthocyanins extraction from different wine lees. *Food Chemistry*, 272, 258–266.

Sang, J., Sang, J., Maa, Q., Hou, X.-F., Li, C.-Q. (2017). Extraction optimization and identification of anthocyanins from *Nitraria tangutorun* Bobr. seed meal and establishment of a green analytical method of anthocyanins. *Food Chemistry*, 218, 386–395.

Santos, L.B., Assis, R.S., Barreto, J.A., Bezerra, M.A., Novaes, C.G., Lemos, V.A. (2022). Deep eutectic solvents in liquid-phase microextraction: contribution to green chemistry. *TrACac-Trends in Analytical Chemistry*, 146, 116478.

Shen, M., Liu, K., Liang, Y., Liu, G., Sang, J., Li, C. (2020). Extraction optimization and purification of anthocyanins from *Lycium ruthenicum* Murr. and evaluation of tyrosinase inhibitory activity of the anthocyanins. *Journal of Food Science*, 85(3), 696–706.

Steingass, C.B., Burkhardt, J., Bäumer, V., Kumar, K., Mibus-Schoppe, H., Zinkernagel, J., Esquivel, P., Jiménez, V.M., Schweiggert, R. (2023). Characterisation of acylated anthocyanins from red cabbage, purple sweet potato, and *Tradescantia pallida* leaves as natural food colourants by HPLC-DAD-ESI(+)-QTOF-MS/MS and ESI(+)-MSn analysis. *Food Chemistry*, 416, 135601.

Tena, N., Asuero, A.G. (2022). Up-to-date analysis of the extraction methods for anthocyanins: Principles of the techniques, optimization, technical progress, and industrial application. *Antioxidants*, 11, 286.

Umsza-Guez, M.A., Vázquez-Espinosa, M., Chinchilla, N., Aliaño-González, M.J., Oliveira de Souza, C., Ayena, K., Barbero, G.F., Palma, M., Carrera, C. (2023). Enhancing anthocyanin extraction from wine lees: A comprehensive ultrasound-assisted optimization study. *Antioxidants*, 12, 2074.

UzelYas, R.A. (2017). A practical method for isolation of phenolic compounds from black carrot utilizing pressurized water extraction with in-site particle generation in hot air assistance. *Journal of Supercritical Fluids*, 120, 320–327.

Wang, F., Zhang, S., Deng, G., Xu, K., Xu, H., Liu, J. (2022). Extracting total anthocyanin from purple sweet potato using an effective ultrasound-assisted compound enzymatic extraction technology. *Molecules*, 27, 4344.

Wang, Y., Luan, G., Zhou, W., Meng, J., Wang, H., Hua, N., Suo, Y. (2018). Subcritical water extraction, UPLC-Triple-TOF/MS analysis and antioxidant activity of anthocyanins from *Lycium ruthenicum* Murr. *Food Chemistry*, 249, 119–126.

Wang, Z., Yan, Y., Nisar, T., Zou, L., Yang, X., Niu, P., Sun, L., Guo, Y. (2018). Comparison and multivariate statistical analysis of anthocyanin composition in *Lycium ruthenicum* Murray from different regions to trace geographical origins: The case of China. *Food Chemistry*, 246, 233–241.

Xue, H., Zha, M., Tang, Y., Zhao, J., Du, X., Wang, Y. (2024). Research progress on the extraction and purification of anthocyanins and their interactions with proteins. *Molecules*, 29(12), 2815.

Yang, J., Kuang, M.-T., Yang, L., Huang, W., Hu, J.-M. (2023). Modern interpretation of the traditional application of Shihu – a comprehensive review on phytochemistry and pharmacology progress of *Dendrobium officinale*. *Journal of Ethnopharmacology*, 302, 115912.

Yang, Y., Kilmartin, P.A. (2025). Advancing anthocyanin extraction: Optimising solvent, preservation, and microwave techniques for enhanced recovery from merlot grape marc. *Food Chemistry*, 472, 142648.

Yi, J., Qiu, M., Zhu, Z., Dong, X., Decker, E.A., McClements, D.J. (2021). Robust and recyclable magnetic nano biocatalysts for extraction of anthocyanin from black rice. *Food Chemistry*, 364, 130447.

Yu, Y., Shiau, S., Pan, W., Yang, Y. (2024). Extraction of bioactive phenolics from various anthocyanin-rich plant materials and comparison of their heat stability. *Molecules*, 29, 5256.

Yuan, T., Wang, L., Chen, L., Zhong, J., Lin, Y., Wang, Y., Lin, C., Fan, H. (2024). Combinatorial preparation and structural characterization of anthocyanins and aglycones from purple-heart radish for evaluation of physicochemical stability and pancreatic lipase inhibitory activity. *Food Chemistry*, 446, 138832.

Yücetepe, M., Özaslan, Z.T., Karakus, M.S., Akalan, M., Karaaslan, A., Karaaslan, M., Başyiğit, B. (2024). Unveiling the multifaceted world of anthocyanins: Biosynthesis pathway, natural sources, extraction methods, copigmentation, encapsulation techniques, and future food applications. *Food Research International*, 187, 114437.

Yun, D., Wu, Y., Yong, H., Tang, C., Chen, D., Kan, J., Liu, J. (2024). Recent advances in purple sweet potato anthocyanins: Extraction, isolation, functional properties and applications in biopolymer-based smart packaging. *Foods*, 13, 3485.

Zhang, S., Lin, S., Zhang, J., Liu, W. (2024) Ultrasound-assisted natural deep eutectic solvent extraction of anthocyanin from *Vitis davidii* Foex. pomace: Optimization, identification, antioxidant activity and stability. *Heliyon*, 10, e33066.

Zhao, X., Zhang, S.-S., Zhang, X.-K., He, F., Duana, C.-Q. (2020). An effective method for the semi-preparative isolation of high-purity anthocyanin monomers from grape pomace. *Food Chemistry*, 310, 125830.

Zhou, Z., Yang, D. (2022). Economical and eco-friendly isolation of anthocyanins from grape pomace with higher efficiency. *Food Chemistry: X*, 15, 100419.

Zhu, S., Song, X., Huang, Y., Jiang, Y., Yao, J., Li, Z., Huang, Z., Dai, F. (2024) Acylated anthocyanins from *Dendrobium officinale* Kimura & Migo: Structural characteristics, antioxidant and hypoglycemic activities. *Food Chemistry*, 455, 139952.

Zhu, Z., Guan, Q., Kouba, M., Barba, F.J., Roohinejad, S., Cravotto, G., Yang, X., Li, S., He, J. (2017). HPLC-DAD-ESI-MS2 analytical profile of extracts obtained from purple sweet potato after green ultrasound-assisted extraction. *Food Chemistry*, 215, 391–400.

Physical and Chemical Stability of Acylated Anthocyanins

5

Fernando Pina, Victor de Freitas, and Joana Oliveira

5.1 INTRODUCTION

There is a huge variety of anthocyanins (ACs) spread in nature that differ in the number of hydroxyl and methoxyl moieties, the nature and number of bonded sugars to their structure, the aliphatic or aromatic carboxylates bonded to the sugar in the molecule, and the position of these bonds. The six most common anthocyanidins in vascular plants are cyanidin, peonidin, pelargonidin, malvidin, delphinidin, and petunidin, which differ in the number of hydroxyl and methoxyl moieties on their B-ring.

With a few exceptions, ACs are always glycosylated at C3. Besides the C3 position, other sugars can also be attached to any hydroxyls at C5, C7, C3′, C5′, and even C4′. The most common sugar is glucose, but rhamnose, xylose, galactose, arabinose, and fructose, and rutinose, sophorose, gentobiose, sambubiose, xylosyl rutinose, and glycosyl rutinose may also be present.[1]

The sugar residues may be further acylated with organic acids. Common acylating agents include phenolic acids such as *p*-hydroxybenzoic, caffeic, *p*-coumaric, ferulic, and sinapic acid, and a range of aliphatic acids such as acetic, malic, malonic, oxalic, and succinic acid. Cinnamic acids may themselves bear glycosidic sugars. Aromatic and aliphatic acylation may occur in the same molecule, leading to the formation of a polyacylated compound. Acyl substituents are usually bonded to the C6 sugar but may also occur in the C4 position.[1]

While identified by the flavylium cation, ACs are much more than this core structure molecule.[2] Flavylium cation of any AC, including the acylated form, can generate five species in an acidic medium, flavylium cation (AH^+), quinoidal base (A), hemiketal (B) *cis* and *trans* chalcones (Cc and Ct), which are reversibly interconnected by four chemical reactions – proton transfer, hydration, tautomerization, and isomerization – as exemplified in an acidic medium, Scheme 5.1. ACs are intrinsically unstable, and their behavior depends on the kinetic balance between reversible and irreversible (degradation) reactions. Within plant and food matrices and all along the digestive tract, ACs encounter a large range of pH conditions. In particular, the bulk of ACs ingested reaches the large intestine (neutral to moderately alkaline pH), where they are extensively degraded by a combination of chemical and microbial processes.[3,4] Extension to the basic medium is thus a requirement to rationalize the role played by ACs in these environments.

Temperature, pH, and light affect both reversible and irreversible reactions. Before discussing the particularities of acylated ACs, the *modus operandi* to account for the kinetics and thermodynamics of all types of ACs is described.

DOI: 10.1201/9781003519591-6

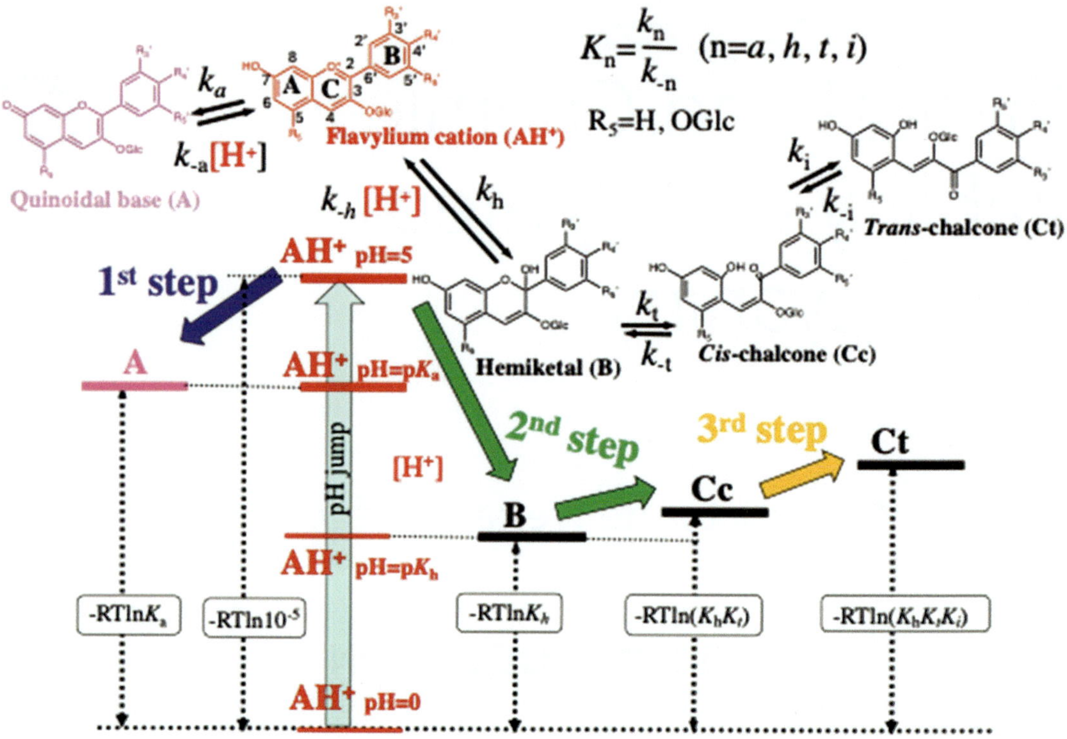

SCHEME 5.1 Energy-level diagram of ACs. In basic media, anionic quinoidal bases, *cis* and *trans* chalcones (anionic hemiketal is generally unstable) are formed as transient species or at the equilibrium.

5.2 BASIC INFORMATION TO QUANTIFY THE ANTHOCYANIN MULTI-EQUILIBRIUM INCLUDING DEGRADATION

The construction of a relative energy-level diagram[5,6] of the several species defined above allows for rationalizing the kinetics and the thermodynamics of any AC. The diagram is straightforwardly constructed, provided that the equilibrium constants are calculated. Section 5.2.2 presents the most reliable procedure to calculate the equilibrium constants.

According to Scheme 5.1, by increasing the proton concentration, the multi-equilibria converge to flavylium cation, which becomes the sole species in ACs generally for pH ≤ 1, justifying its use to define the system. For each pH value, the kinetics of the multi-equilibria are determined by the relative energy level of the flavylium cation. The kinetics can be conveniently studied by adding base to equilibrated solutions of the flavylium cation (direct pH jumps). Besides, acid can be added to solutions equilibrated at higher pH values back to pH ≤ 1 (reverse pH jumps). In basic medium, anionic quinoidal bases, and *cis* and *trans* chalcones (anionic hemiketal is generally unstable) are formed as transient species or at the equilibrium. During reverse, pH jumps from basic solutions, and within the mixing time of the stopped-flow apparatus, the anionic species are converted into their respective neutral species or flavylium cations, in the case of quinoidal bases.

5.2.1 Direct pH Jumps

In Scheme 5.1, the kinetic steps triggered upon a direct pH jump to pH 5 are presented. The first reaction corresponds to the proton transfer to water to produce the quinoidal bases. These reactions are by far the fastest of the multi-equilibrium occurring in submilliseconds.[7,8] This step is too fast to be followed by a stopped-flow equipment and requires temperature jumps[7] or flash photolysis in the case of the compounds exhibiting excited state proton transfer.[8]

$$k_{1st} = k_a + k_{-a}[H^+] \tag{5.1}$$

Nevertheless, the absorption of the flavylium cation in equilibrium with the quinoidal base (and the successive acid–base equilibria of the anionic quinoidal bases) can be collected by stopped flow, 10 ms after the direct pH jump, allowing titration to be carried out at representative wavelengths from which the respective acidity constants are calculated.

The second step in ACs in acidic medium is the hydration reaction followed by a faster tautomerization reaction, and consequently, the former is the controlling step. One important property of the quinoidal base discovered by Bouillard and Dubois[7] is the lack of reactivity of the quinoidal base, which is a kinetic product. Consequently, the system evolves toward the equilibrium exclusively from the AH^+ hydration, and therefore, the hydration rate decreases with the increase in pH.

$$k_{2nd} = \chi_{AH^+} k_h + \chi_B k_{-h}[H^+] = \frac{[H^+]}{[H^+]+K_a} k_h + \frac{1}{1+K_t} k_{-h}[H^+] \tag{5.2}$$

The kinetics of the second step is given by Eq. 5.2, where χ_{AH^+} and χ_B are, respectively, the mole fraction of AH^+ in its equilibrium with A and the mole fraction of B in its equilibrium with Cc. As shown below, for pH < 7, the third step is much slower than the first and second steps, allowing a pseudo-equilibrium involving all species except *trans*-chalcone, identified by the symbol (^), to be reached.

Finally, the system reaches the equilibrium through the *cis–trans* isomerization (third step) that occurs in several hours according to Eq. 5.3.

$$k_{3rd} = \chi_{Cc} k_i + \chi_{Ct} k_{-i} = \frac{K_h K_t}{[H^+]+K_h(1+K_t)} + \chi_{Ct} k_{-i} \tag{5.3}$$

Here, χ_{Cc} is the mole fraction of Cc at the pseudo-equilibrium and χ_{Ct} is the mole fraction of Ct equal to 1 if no Ct^- is in equilibrium with Ct.

In basic medium, the first reaction is the same as that in acid medium, and the system evolves to the equilibrium from the OH^- attack, accompanied by a parallel degradation reaction.[9] It should be emphasized that in basic medium, the *cis–trans* isomerization is very fast.[9]

5.2.2 Reverse pH Jumps

Reverse pH jumps are the most reliable procedure to calculate the equilibrium constants necessary to construct energy-level diagrams. Reverse pH jumps are based on the change in regime.[10] This occurs in very acidic medium when the hydration becomes faster than tautomerization.

The trace obtained upon a pH jump from equilibrated solutions, e.g., at pH 4.2, is shown in Figure 5.1a. The kinetics is bi-exponential, and there is an initial amplitude. The attribution of the three amplitudes is straightforwardly explained by means of Scheme 5.2. The initial amplitude is due to the proton transfer occurring during the mixing time of the stopped flow and regards the conversion of quinoidal bases in the flavylium cation, the first step. The bi-exponential trace can be decomposed into a faster trace and a slower one. As mentioned above, at pH 1, there is a change of regime, and the faster trace

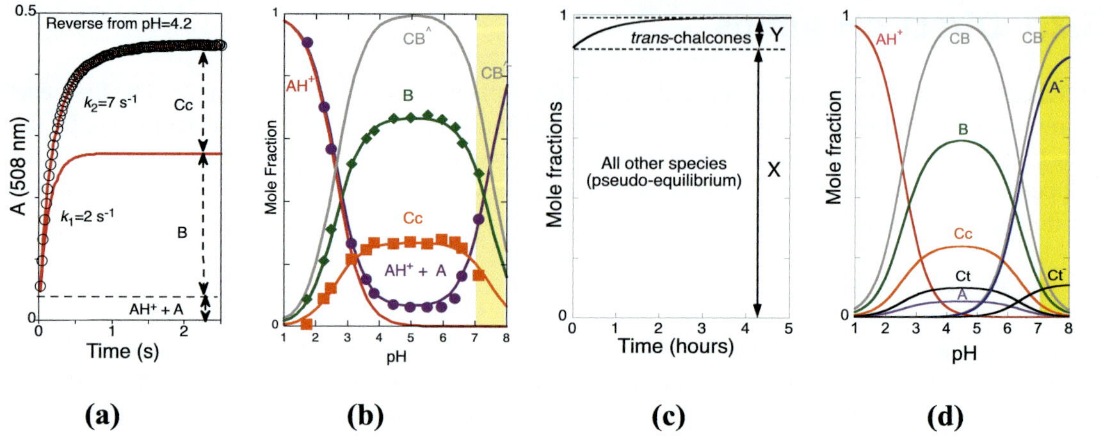

FIGURE 5.1 (a) Stopped-flow trace of malvidin 3-*O*-glucoside after a reverse pH jump from pH 4; (b) mole fraction distribution at the pseudo-equilibrium; (c) reverse pH jump as in (a) monitored by a standard spectrophotometer; (d) mole fraction distribution at the equilibrium.

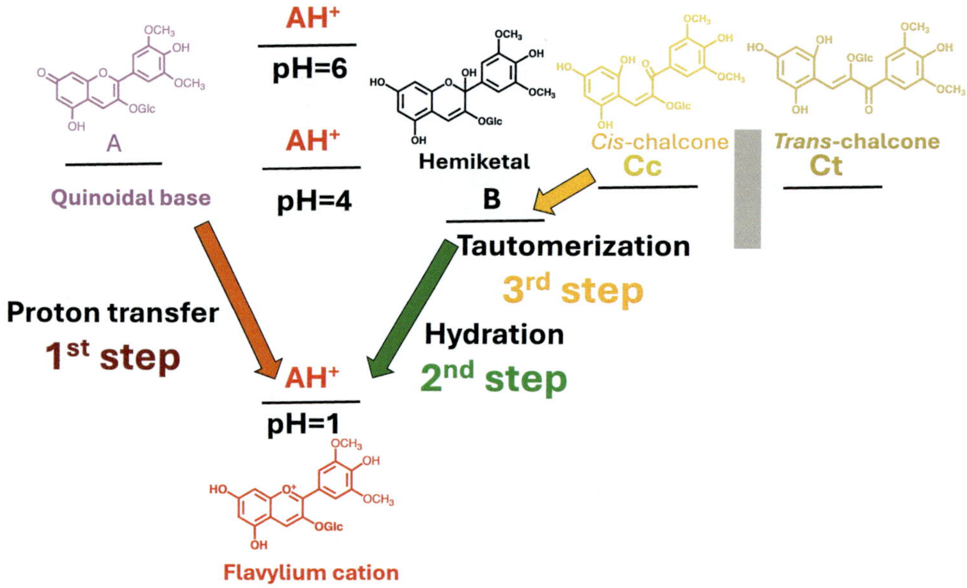

SCHEME 5.2 Kinetic steps that follow a reverse pH jump from equilibrated solutions for 4 < pH < 6 of malvidin 3-*O*-glucoside, plateau of Figure 5.1b.

corresponds to the conversion of the hemiketal into flavylium cation, while the slower trace corresponds to the formation of more flavylium cations from *cis*-chalcone *via* hemiketal. Normalization of these three amplitudes gives the mole fraction distribution of the pseudo-equilibrium species, allowing for the calculation of the respective equilibrium constants (Figure 5.1b).[11] The mole fraction of *trans*-chalcone can be obtained from a reverse pH jump monitored by a standard spectrophotometer (Figure 5.1c). The initial amplitude is the sum of all species except *trans*-chalcone, and the slow trace regards the conversion of this last species into more flavylium cations. Normalization of the traces allows for the calculation of the *trans*-chalcone mole fraction and correction of those of the pseudo-equilibrium reported in Figure 5.1b to the equilibrium (Figure 5.1d).

The stopped-flow experiments can be extended to the basic medium since all anionic species are transformed into the flavylium cation (anionic quinoidal bases) or into the respective neutral species. The traces are like those reported in Figure 5.1, but the amplitudes are different since they reflect the mole fraction distribution in the initial basic pH of the reverse pH jump.

5.2.2.1 The Kinetics Toward the Equilibrium and Degradation Rates

This is an important issue to evaluate the effect of the acylation on the behavior of ACs. The discussion will be made using simultaneously Figure 5.2 and Scheme 5.3.

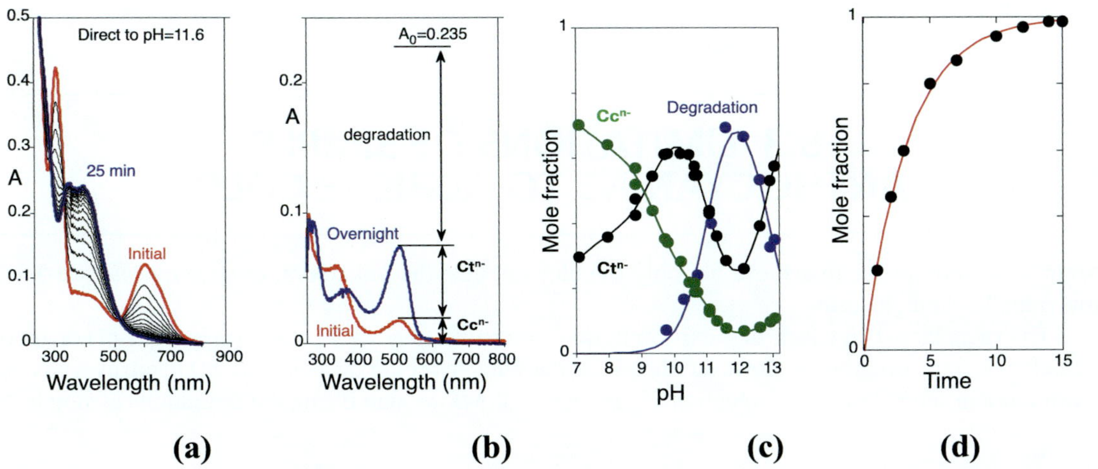

FIGURE 5.2 (a) Spectral variations after a direct pH jump of cyanidin 3,5-diglucoside to pH 11.6; (b) reverse pH jump to pH 1; (c) mole fraction distribution of the species, including degradation after 25 min; (d) representation of the mole fractions for an optimum pH as a function of time delay.

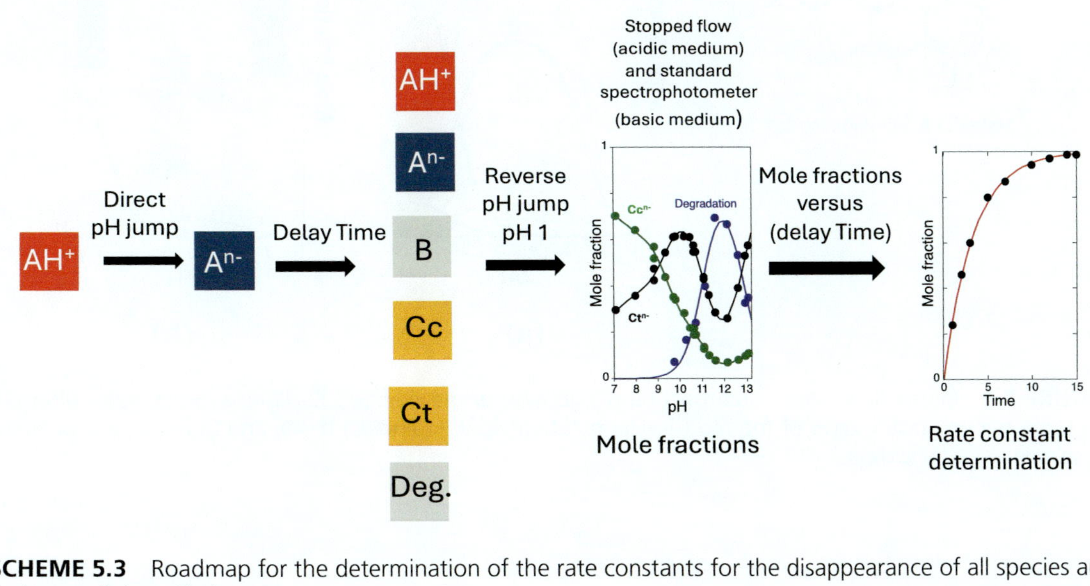

SCHEME 5.3 Roadmap for the determination of the rate constants for the disappearance of all species and their overall degradation rate.

Figure 5.2a shows the spectral variations after a direct pH jump of cyanidin 3,5-*O*-diglucoside to pH 11.6, but the procedure is the same in acylated ACs. After the direct pH jump, a reverse pH jump to pH 1 was carried out after 25 min, but can also be made for other time points (delay time). In basic media, only anionic quinoidal base and chalcones are observed, and consequently, the initial flavylium absorption after the reverse pH jump (monitored by a standard spectrophotometer) corresponds to the *cis*-chalcone and quinoidal base, if present at the initial basic pH. The appearance of more flavylium overnight is due to the conversion of *trans*-chalcone. The difference between the total flavylium absorption and the expected absorption if no degradation occurs leads to the degradation fraction (Figure 5.2b). Normalizing these amplitudes gives the respective mole fractions. Extension to a pH range allows for the mole fraction distribution after the selected delay time (Figure 5.2c). Finally, the rate constant for any species disappearance and for the degradation time is calculated as illustrated in Figure 5.2d, representing the respective mole fraction as a function of the delay time.

These are indispensable information that should be collected to quantify the kinetic processes, and particularly the degradation rates.

5.3 LIMITATIONS OF SIMPLE ANTHOCYANINS TO ACHIEVE COLOR

Simple ACs are unable to achieve not only the blue but also the red colors, as shown in Figure 5.3, for malvidin 3,5-*O*-diglucoside.

The vacuolar pH, at which ACs exist in flowers and fruits, may vary considerably, but a pH of approximately 5 is often observed.[12] However, the pH of the vacuoles could be as low as 2.0 in citrus fruits[13] and 7.7 or more in some flowers.[14] Inspection of Figure 5.3 allows for identifying the limitations of simple ACs

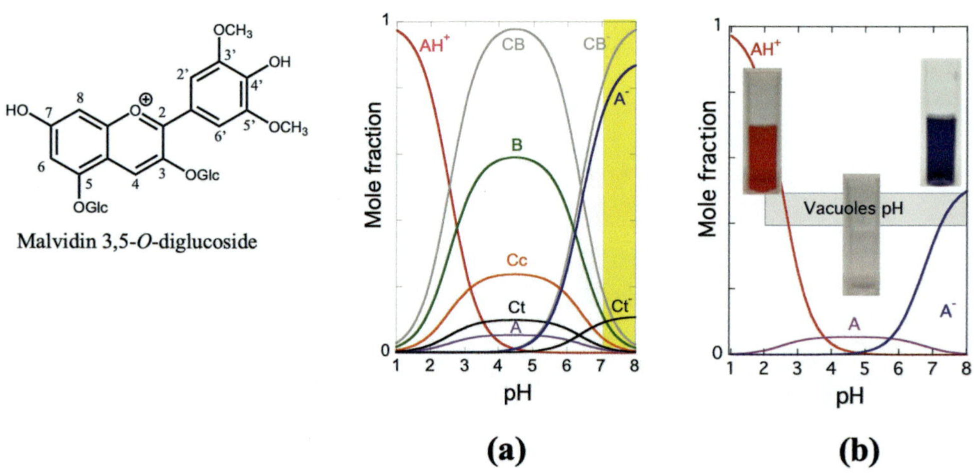

FIGURE 5.3 (a) pH-dependent mole fraction distribution of malvidin 3,5-*O*-diglucoside in water after 6 h; (b) mole fraction distribution of the red (flavylium cation), pink (quinoidal base), and blue (anionic quinoidal base) species after 2 days.

to give red to blue colors: (1) the pH domain of the flavylium cation is limited to very acidic vacuoles; (2) in moderately acidic vacuoles, neither red nor blue colors are expressed, and around the neutrality, the blue anionic quinoidal base is unstable.

Nature developed three main strategies to overcome this limitation: inter- and intramolecular co-pigmentation and metalloanthocyanins (Scheme 5.4). Intermolecular co-pigmentation is more adequate to increase the pH domain of the red flavylium cation, while intramolecular co-pigmentation and metalloanthocyanins are the main strategies used by nature to achieve the blue color, together with its stability.

While intermolecular co-pigmentation can be used to extend the pH domain of the flavylium cation, the blue color requires, in general, intramolecular co-pigmentation or the supramolecular structures of metalloantocyanins. In both cases, acylated ACs play a central role.

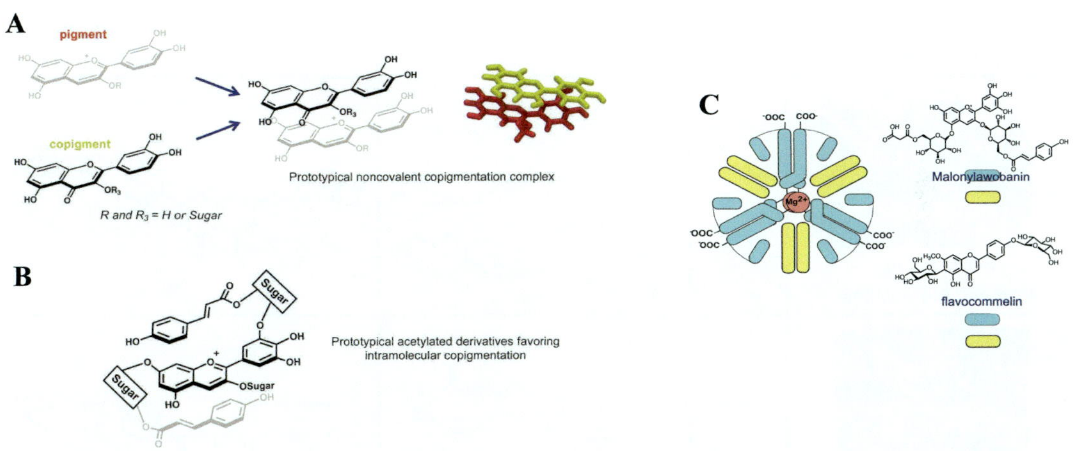

SCHEME 5.4 Inter- (A) and intramolecular (B) co-pigmentation after Trouillas et al.[15] and Yoshida et al.[16] (C) Metalloanthocyanins. Adapted with permission from ref.15. Copyright 2020, American Chemical Society.

5.3.1 Metalloanthocyanins

Metalloanthocyanins were comprehensively reviewed by Yoshida et al.[17] They are self-assembling supramolecular structures containing two planes where three ACs and three intercalating flavanones are linked to a metal (Schemes 5.4C and 5.5). The two plans are rotated 30º. All ACs contain a catechol unit that allows for the complexation of the respective anionic quinoidal base and one acylated unit that reinforces the bonding by π–π stacking. To our knowledge, no physical–chemical studies regarding their kinetics have been reported, but the stability of the supramolecules was noticed.

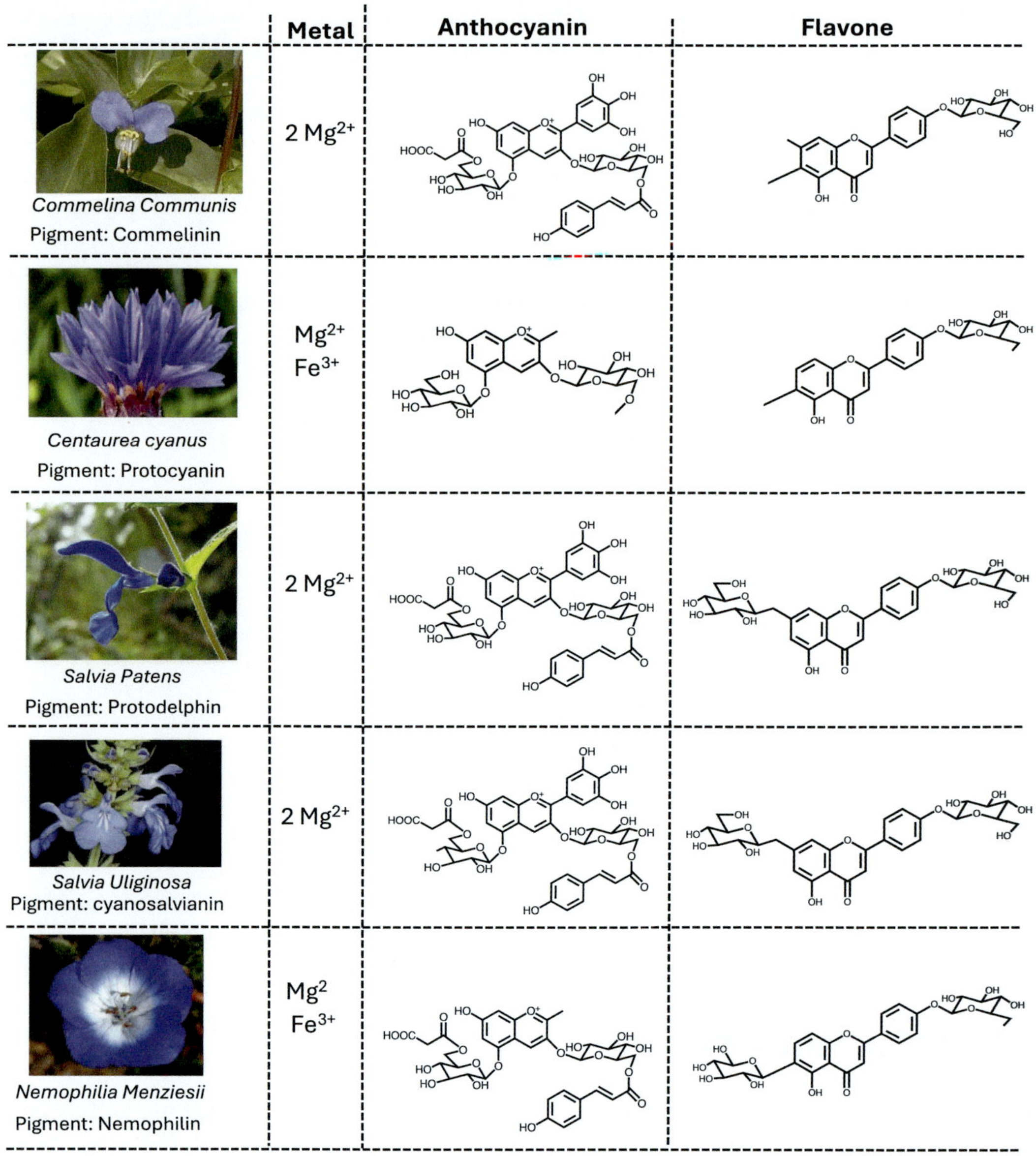

SCHEME 5.5 Adapted from Ref. 17.

5.4 ACYLATED ANTHOCYANINS: THE EFFECT OF THE NUMBER OF ACYLATED UNITS

Many studies on the stability of acylated ACs focus on extracts. While very important for applications in several domains such as food coloring and health effects, these works are not appropriate to the scope of this chapter, which requires purified compounds. In this section, we will discuss the impact of the number of acylate units on the physical–chemical properties of these pigments.

5.4.1 Monoacylated Anthocyanins

The comparison between the structures of malvidin 3-*O*-glucoside (Oenin) and malvidin 3-*O*-(6-*O*-coumaroyl) glucoside (OeninCoum) (Scheme 5.6) was reported using ^{1}H NMR and circular dichroism.[18] The CD signal regarding Oenin is similar to those previously reported for other simple ACs and was attributed to self-aggregation.[19–21] When temperature increases, the CD signal disappears, suggesting that the aggregation is disrupted (Figure 5.4). In the case of OeninCoum, the CD signal similar to the one observed in Oenin also disappears, but one signal is still observed at high temperature. It was interpreted that in OeninCoum, the self-aggregation also disappears with increasing temperature, but the signal regarding intramolecular co-pigmentation involving the acylated unit remains.

This interpretation was corroborated by variable temperature NMR experiments. The signals for H-6 and H-8 of the A-ring are shielded in OeninCoum compared to Oenin and move upfield in the NMR spectra, suggesting a folded structure such as the one shown in Figure 5.5.

Another example of the effect of one acylated unit was reported (Figure 5.6).[22] In Table 5.1, the rate and equilibrium constants in acidic medium are presented. The acylation (1) decreases the mole fraction

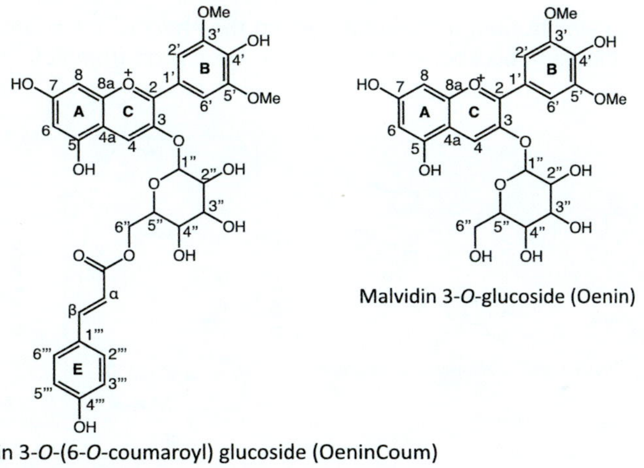

SCHEME 5.6 Malvidin 3-*O*-(6-*O*-coumaroyl) glucoside (OeninCoum) and malvidin 3-*O*-glucoside (Oenin).[18]

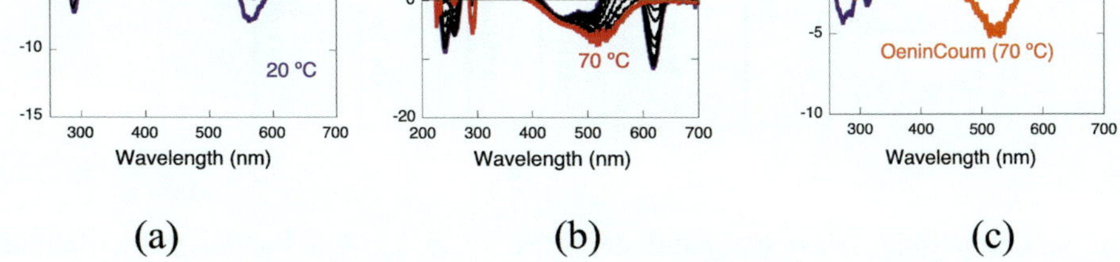

FIGURE 5.4 (a) Circular dichroism spectra of Oenin in water at 2.5 × 10^{-4} M, pH 1.0, as a function of temperature (20–80°C). (b) The same for OeninCoum in water at 2.5 × 10^{-4} M, pH 1.0, at 20–70°C. (c) Comparison of the CD spectra of OeninCoum and Oenin under the same conditions. Adapted with permission from Ref. 18. Copyright 2020, Elsevier.

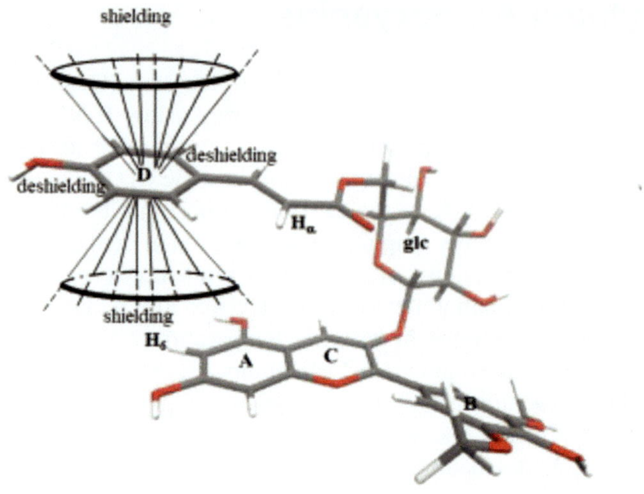

FIGURE 5.5 Proposed OeninCoum model representing the effect of the coumaroyl moiety on the ^{1}H NMR chemical shifts on the flavylium backbone. Adapted with permission from Ref. 18. Copyright 2020, Elsevier.

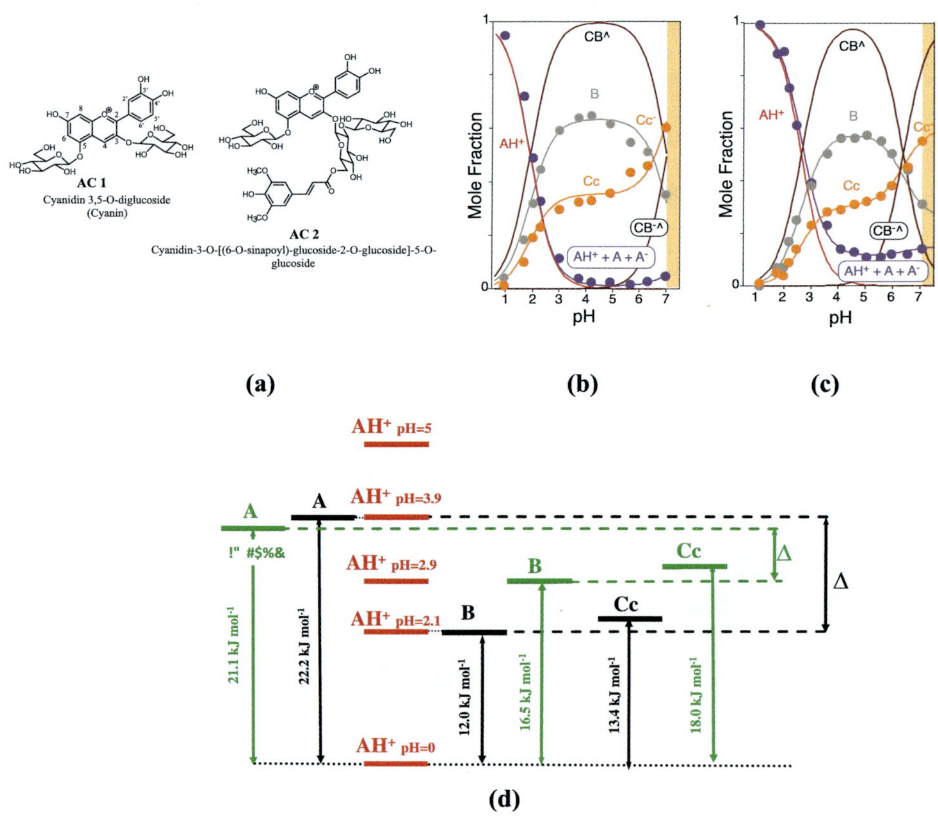

FIGURE 5.6 (a) Cyanin (AC1) and one acylated derivative (AC2). (b) Mole fraction distribution of AC1 at the pseudo-equilibrium in acidic medium calculated by reverse pH jumps. (c) The same for AC2. The mole fraction distribution of *trans*-chalcone is very small in both compounds. (d) Relative energy-level diagram of both compounds: black (AC1); green (AC2).

TABLE 5.1 Equilibrium and Rate Constants for the Second Kinetic Step in Acidic Medium[22(1)]

	pK_a	$k_h(s^{-1})$	$k_{-h}(s^{-1})$	$k_t(s^{1})$	$k_{-t}(s^{1})$	pK_h	K_t	$k_t^H(M^{-1}s^{-1})$	$pK^{\wedge a}$
AC1	3.9	0.22	27	0.1	0.2	2.1	0.57	10.3	1.9
AC2	3.7	0.25	188	0.06	0.1	2.9	0.54	5.4	2.6

(1) Estimated error 10%.

of the colorless hemiketal in acidic medium and increases that of the colored quinoidal base; (2) increases the pH domain of the flavylium cation; and (3) moderately decreases the rate of tautomerization and isomerization of the neutral and mono-anionic species.

Another way to visualize the influence of the acylated unit is the energy-level diagram of Figure 5.6d. Acylation reduces the energy level between the colorless hemiketal (major species) and quinoidal base (minor one), contributing to a higher relative concentration of the colored species.

5.4.2 Poly-Acylated Anthocyanins

The thermodynamics and kinetics of the ACs extracted from a morning glory variety, the flower of *Pharbitis (Ipomoea) nil* (a pelargonidin derivative), were reported by Dangles et al.[23] (Scheme 5.7). Four of these ACs were studied: pigments 1 with no acylated unit, pigments 1 and 2 with 1 acylated unit, and pigment 4 with two acylated units.

A similar study was conducted on the ACs extracted from the flower of *Ipomoea tricolor* (Scheme 5.8).[24] In this case, pigment 1 does not have any acylated unit, pigment 2 has one acylated unit, and pigment 3 has three acylated units.

The relative energy level of the colored quinoidal base and the colorless hemiketal is shown in Scheme 5.9. While the number of acylated units modestly changes the energy level of the quinoidal bases, there is a dramatic effect on the energy level of the hemiketal.

Inspection of Scheme 5.9 indicates that the number of acylated units is crucial for the stabilization of the quinoidal base. Indeed, three acylated units stabilize the quinoidal base of *Ipomea tricolor* and invert significantly the stability of the hemiketal/quinoidal base. This reduction in the stability of the quinoidal base with the increase in acyl groups was also observed for *Pharbitis (Ipomea) nil*. This fact has implications for the relative stability of the anionic quinoidal bases and the stabilization of the blue color.

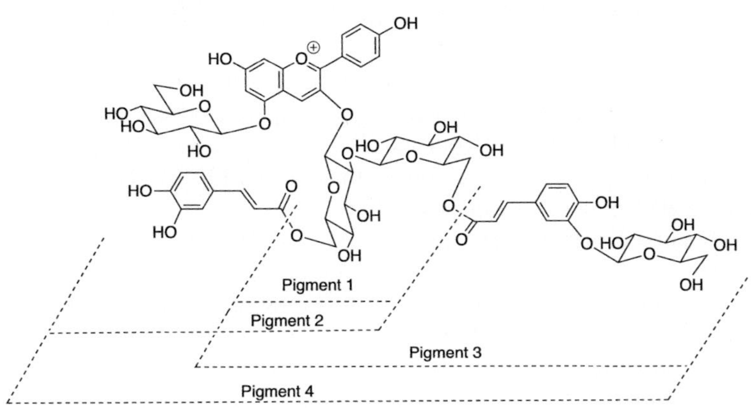

SCHEME 5.7 Anthocyanins extracted from *Pharbitis (Ipomoea) nil*.[22]

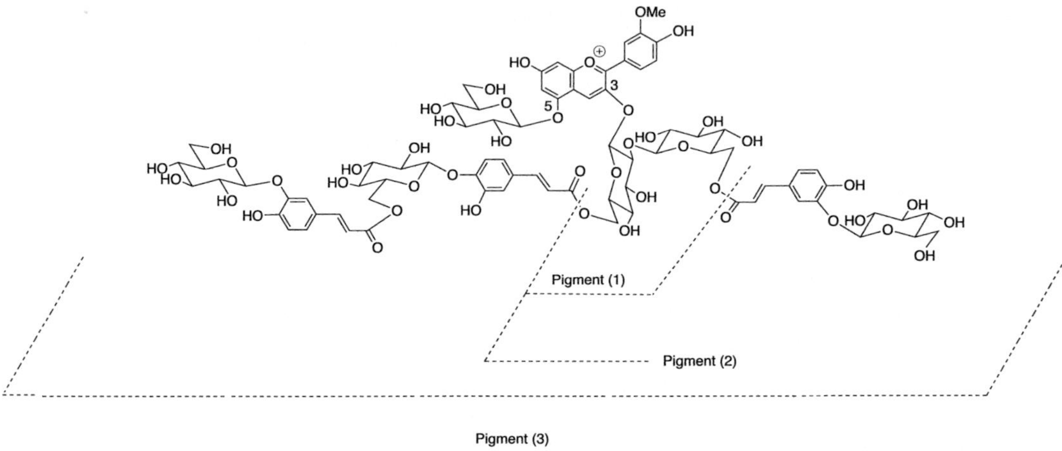

SCHEME 5.8 Anthocyanins extracted from *Ipomoea tricolor*.[23]

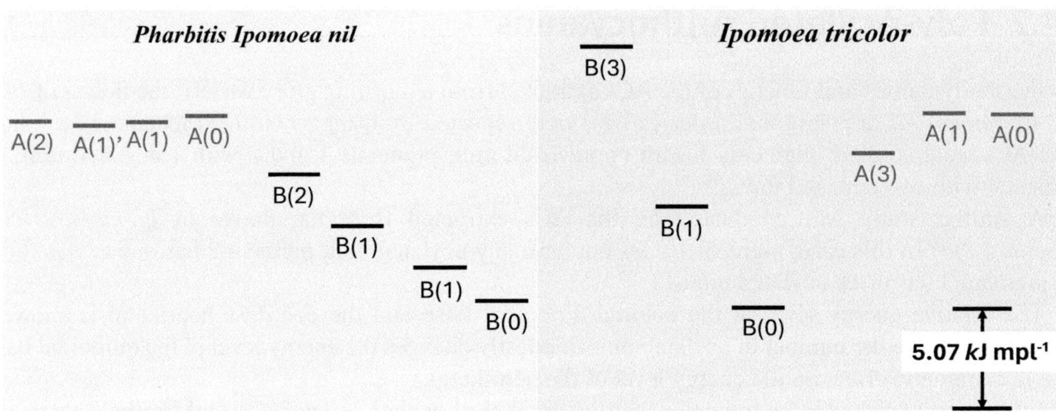

SCHEME 5.9 Relative energy-level diagram of the quinoidal bases and hemiketals as a function of the number of acylated units. In the case of *Ipomoea tricolor*, the equilibrium hemiketal/quinoidal base was considered. Numbers in brackets correspond to the number of acyl groups.

5.4.3 Polyacylated Anthocyanins in Basic Medium

The mole fraction distribution of heavenly blue AC (from *Ipomoea tricolor*) pigment 3 in Scheme 5.8 extended to the basic medium is shown in Scheme 5.10. At approximately pH 7, relatively small changes in pH could lead to large color changes from purple to blue. Vacuolar pH measurements of morning glory petals in the abaxial epidermis have shown that the purplish red buds and blue open flowers (Scheme 5.10b) were at pH 6.6 and 7.7, respectively.[14]

While the mole fraction distribution shown in Scheme 5.10a regards the situation *in vitro*, which could be slightly different from that *in vivo*, it explains the color difference between buds and petals. At neutral pHs, there is a mixture of pink-magenta with blue colors, explaining the color of the buds, while at moderately basic pH, blue is the dominant color. Moreover, there is an increase in the stability of the blue color.

In conclusion, the acylation of ACs dramatically extends the pH domain of the red color and stabilizes the blue color in neutral to moderately basic environments. The acylated sugars in positions 3′ and 5′ increase the stability of these molecules.

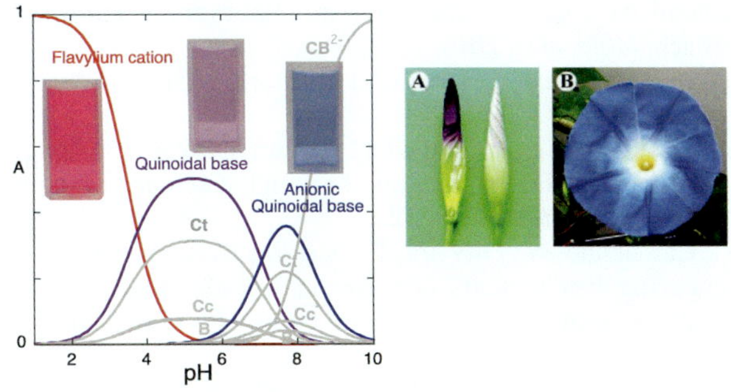

(a) (b)

SCHEME 5.10 (a) Mole fraction distribution of pigment 3 of the blue anthocyanin (*Ipomoea tricolor*) including the anionic forms and (b) photo of the buds and flowers.

5.5 CONCLUSIONS

Acylated ACs stand out for their enhanced physical–chemical stability, primarily due to their intramolecular co-pigmentation and potential for supramolecular interactions. These effectively protect the chromophore against environmental stressors like pH variation, light exposure, and oxidative conditions. This stability extends their pH domain; stabilizes vibrant colors, particularly reds and blues; and mitigates degradation under diverse conditions. These properties make them particularly valuable as natural dyes for textiles and colorants in food matrices, ranging from acidic beverages to neutral or mildly basic products. Their potential health benefits due to their antioxidant properties further amplify their appeal in functional food applications.[25–26] However, several challenges remain in fully harnessing acylated ACs. Current knowledge highlights gaps in understanding their behavior in complex food systems, including interactions with other components like proteins, lipids, and polysaccharides. The influence of processing conditions (e.g., high temperature or freeze–thaw cycles) on their stability and bioavailability is also insufficiently studied. Furthermore, the specific effects of acylation patterns, including the number, structure, and positioning of acyl groups, on kinetic and thermodynamic properties require deeper investigation.

Future research must integrate advanced analytical methods and experimental designs to address these gaps. These efforts will pave the way for optimizing the use of acylated ACs in food systems and potentially broader applications, ensuring their sustainable and effective deployment in industries.

REFERENCES

1. Andersen, O.M.M.; Markham, K.R., Eds. *Flavonoids: Chemistry, Biochemistry and Applications.* Taylor & Francis. 2006.
2. Pina, F.; Oliveira, J.; de Freitas, V. Anthocyanins and derivatives are more than flavylium cations. *Tetrahedron.* 2015, *71*, 3107–3114.
3. de Ferrars, R.M.; Czank, C.; Zhang, Q.; Botting, N.P.; Kroon, P.A.; Cassidy, A.; Kay, C.D. The pharmacokinetics of anthocyanins and their metabolites in humans. *Br. J. Pharmacol.* 2014, *171*, 3268–3282.
4. Dangles, O. Anthocyanins as natural food colorings: The chemistry behind and challenges still ahead. *J. Agric. Food Chem.* 2024, *72*, 12356–12372.
5. Pina, F.; Melo, M.J.; Maestri, M.; Ballardini, R.; Balzani, V. Photochromism of 4′-methoxyflavylium perchlorate: A "write-lock-read-unlock-erase" molecular switching system. *J. Am. Chem. Soc.* 1997, *119*, 5556–5561.

6. Basilio, N.; Pina, F. Chemistry and photochemistry of anthocyanins and related compounds: A thermodynamic and kinetic approach. *Molecules*. 2016, *21*, 25.
7. Brouillard, R.; Dubois, J.E. Mechanism of structural transformations of anthocyanins in acidic media. *J. Am. Chem. Soc.* 1977, *99*, 1359–1364.
8. Macanita, A.L.; Moreira, P.F.; Lima, J.C.; Quina, F.H.; Yihwa, C.; Vautier-Giongo, C. Proton transfer in anthocyanins and related flavylium salts: Determination of ground-state rate constants with nanosecond laser flash photolysis. *J. Phys. Chem. A*. 2002, *106*, 1248–1255.
9. Seco, A.; Pereira, A.R.; Camuenho, A.; Oliveira, J.; Dias, R.; Brás, N.F.; Basílio, N.; Parola, A.J.; Lima, J.C.; de Freitas, V. et al. Comparing the chemistry of malvidin-3-*O*-glucoside and malvidin-3,5-*O*-diglucoside networks: A holistic approach to the acidic and basic paradigms with implications in biological studies. *J. Agric. Food Chem.* 2024, *72*, 7497–7510.
10. Pina, F. Anthocyanins and related compounds: Detecting the change of regime between rate control by hydration or by tautomerization. *Dyes Pigm.* 2014, *102*, 308–314.
11. Mendoza, J.; Basilio, N.; de Freitas, V.; Pina, F. New procedure to calculate all equilibrium constants in flavylium compounds: Application to the copigmentation of anthocyanins. *ACS Omega*. 2019, *4*, 12058–12070.
12. Felle, H.H. pH regulation in anoxic plants. *Ann. Bot.* 2005, *96*, 519–532.
13. Echeverria, E.; Burns, J.; Felle, H. Compartmentation and cellular conditions controlling sucrose breakdown in mature acid lime fruits. *Phytochemistry.* 1992, *31*, 4091–4095.
14. Yoshida, K.; Kondo, T.; Okazaki, Y.; Katou, K. Cause of blue petal color. *Nature.* 1995, *373*, 291.
15. Trouillas, P.; Sancho-García, J.C.; De Freitas, V.; Gierschner, J.; Otyepka, M.; Dangles, O. Stabilizing and modulating color by copigmentation: Insights from theory and experiment. *Chem. Rev.* 2016, *116*, 4937–4982.
16. Kondo, T.; Oyama, K.; Yoshida, K. Chiral molecular recognition on formation of a metalloanthocyanin: A supramolecular metal complex pigment from blue flowers of salvia patens. *Angew. Chem. Int. Edit.* 2001, *40*, 894–897.
17. Yoshida, K.; Mori, M.; Kondo, T. Blue flower color development by anthocyanins: From chemical structure to cell physiology. *Nat. Prod. Rep.* 2009, *26*, 884–915.
18. Mendoza, J.; Oliveira, J.; Araújo, P.; Basílio, N.; Teixeira, N.; Brás, N.F.; Pina, F.; Yoshida, K.; de Freitas, V. The peculiarity of malvidin 3-*O*-(6-*O*-*p*-coumaroyl) glucoside aggregation. Intra and intermolecular interactions. *Dyes Pigm.* 2020, article number 108382.
19. Gavara, R.; Petrov, V.; Quintas, A.; Pina, F. Circular dichroism of anthocyanidin 3-glucoside self-aggregates. *Phytochem.* 2013, *88*, 92–98.
20. Hoshino, T. Anthocyanins self-aggregates. 7: Self-association of flavylium cations 3,5-diglucosides studied by circular dichroism and H-1-NMR. *Phytochem.* 1992, *31*, 647–653.
21. Hoshino, T.; Matsumoto, U.; Goto, T. Evidences of the self-association of anthocyanins. 1: Circular-dichroism of cyanin anhydrobase. *Tetrahedron Lett.* 1980, *21*, 1751.
22. Yang, P.; Basílio, N.; Liao, X.; Xu, Z.; Dangles, O.; Pina, F. Influence of acylation by hydroxycinnamic acids on the reversible and irreversible processes of anthocyanins in acidic to basic aqueous solution. *J. Agric. Food Chem.* 2024, *72*, 25955–25971.
23. Dangles, O.; Saito, N.; Brouillard, R. Kinetic and thermodynamic control of flavylium hydration in the pelargonidin cinnamic acid complexation – origin of the extraordinary flower color diversity of *Pharbitis nil*. *J. Am. Chem. Soc.* 1993, *115* (8), 3125–3132.
24. Mendoza, J.; Basilio, N.; Pina, F.; Kondo, T.; Yoshida, K. Rationalizing the color in heavenly blue anthocyanin: A complete kinetic and thermodynamic study. *J. Phys. Chem. B*. 2018, *122*, 4982–4992.
25. Matsumoto, H.; Nakamura, Y.; Hyramaya, M.; Yoshiki, Y.; Okubo, K. Antioxidant activity of black currant anthocyanins aglycones and their glycosides measured by chemiluminescence in a neutral pH region and human plasma. *J. Agric. Food Chem.* 2002, *50*, 5034–5037.
26. Liu, S.S.; Zhang, Y. Antioxidant properties and electrochemical activity of anthocyanins in mulberries. *J. Food Meas. Charact.* 2024, *18*, 3569–3576.

PART II

The Bioavailability of Acylated Anthocyanins

Bioaccessibility and Absorption of Acylated Anthocyanins

6

Margarida Teixeira, Hélder José Couto Oliveira, and Patrícia Correia

6.1 INTRODUCTION

More than ever, there is growing interest in exploring the health benefits associated with anthocyanin consumption, with substantial research evidence highlighting their strong antioxidant capacity and their potential metabolic, anti-inflammatory, anticancer, cardioprotective, and neuroprotective effects. Such evidence may provide valuable insights into the possible therapeutic applications of these phytochemicals [1, 2].

Despite numerous studies on the health benefits of dietary anthocyanins, their bioaccessibility and bioavailability remain important research areas. For dietary phytochemicals to exert their associated health benefits, they must be absorbed, enter the systemic circulation, and reach target tissues at sufficient concentrations to produce a biological effect. Thus, bioavailability is crucial for their efficacy [3, 4]. Bioavailability involves the assessment of five key stages: (1) liberation (from the food matrix), (2) absorption, (3) distribution, (4) metabolism, and (5) elimination, collectively referred to as LADME [5]. To provide a more precise understanding of the processes behind the availability of a phytonutrient, concepts such as bioaccessibility and bioactivity were introduced [6]. Bioaccessibility describes the amount of free anthocyanins released from the food matrix during digestion and hence being available for absorption in the gastrointestinal tract (GIT) [5]. Bioactivity is the ability of anthocyanins to produce a beneficial physiological effect by being absorbed and distributed to tissues where they trigger a health-promoting response [7].

Research evidence supports a quite low bioavailability of anthocyanins, typically less than 2%, with only trace amounts detected in the bloodstream and target tissues [3]. However, this low detection rate may not accurately reflect their bioavailability. This is largely due to the significant pre-systemic metabolism of anthocyanins, which leads to the formation of metabolites that may have greater bioactivity than the parent compounds. This highlights the importance of assessing bioavailability and metabolism for a more accurate evaluation of anthocyanin bioactivity [8]. Furthermore, the structure of anthocyanins plays a key role in their absorption, with pelargonidin-based anthocyanins exhibiting higher absorption rates than those with additional B-ring substituents [9]. As discussed in previous chapters, acylation enhances anthocyanin stability, depending on the type, structure, and attachment site of the acyl group. Aromatic acids stabilize anthocyanins through mechanisms such as co-pigmentation, while aliphatic acids may

DOI: 10.1201/9781003519591-8

provide stability primarily through steric hindrance. Such chemical properties confer anthocyanins higher stability under food processing and storage conditions, protecting them from degradation [10, 11]. Overall, the structural differences between nonacylated and acylated anthocyanins not only contribute to the higher stability of the latter but also affect their distinct absorption, bioavailability, and colonic degradation [12, 13].

In this chapter, the bioaccessibility of acylated anthocyanins during digestion will be explored, with particular emphasis on factors that influence their release from the food matrix. The mechanisms involved in the absorption and distribution of these compounds in the body will be discussed, providing insights into how acylation affects these processes compared to their nonacylated counterparts. A detailed analysis of the impact of acylation on anthocyanin bioaccessibility and absorption shall be presented, shedding light on the key challenges and considerations for optimizing their bioavailability for potential health/ technological applications.

6.2 FACTORS INFLUENCING THE RELEASE OF ANTHOCYANINS DURING DIGESTION

The bioaccessibility of acylated anthocyanins during digestion is influenced by several factors. Structural characteristics, such as glycosylation and acylation patterns, not only affect the stability of anthocyanins but also play a key role in their release from the food matrix in which they are embedded. Moreover, interactions with macronutrients, namely, proteins, fibers, and polysaccharides, can compromise their release. In addition to matrix-related factors, the conditions of the GIT are crucial in determining the release and absorption of anthocyanins. For instance, pH, digestive enzymes, bile acids, gut microbiota, and the mobility and permeability of the intestines all influence the release of anthocyanins during the digestion process [14].

6.2.1 Food Matrix and Interactions with Macronutrients

The acidic environment within plant vacuoles preserves anthocyanin stability. However, once ingested, anthocyanins are subjected to various physiological conditions that result in structural modifications. These changes depend largely on the parent structure of the anthocyanins. For instance, acylated anthocyanins exhibit different release dynamics compared to their nonacylated counterparts, which significantly affects their bioaccessibility. As a result, acylated and nonacylated forms display distinct absorption rates and bioavailability, with acylation playing a key role in determining their overall stability and efficacy in the body [12, 15].

The composition and structure of the food matrix influence the release and stability of anthocyanins during digestion. In more complex matrices, such as whole fruits and vegetables, anthocyanins may be less accessible for digestion than simpler forms such as juices or extracts. In the former, anthocyanins can bind to carbohydrates, proteins, and cell walls, as well as other phenolic compounds. The complexity of food matrices may offer a protective effect by shielding anthocyanins from degradation by digestive enzymes and fluids until they reach the small intestine [16]. Podsędek et al. (2014) investigated the impact of the food matrix on the digestibility of red cabbage anthocyanins, by comparing whole red cabbage with a liquid anthocyanin-rich extract. Anthocyanin recovery after digestion ranged from 17.96% to 112.50% for whole red cabbage, whereas the anthocyanin-rich extract showed a recovery rate of 2.97% to 30.88%, suggesting that anthocyanin stability during digestion was highly influenced by the food matrix [17]. Furthermore, anthocyanin bioaccessibility is influenced by its interactions with dietary components such as proteins, fibers, and polysaccharides, making it important to understand how anthocyanins interact with macronutrients in the GIT [18]. The synergistic or antagonistic effects among these components significantly affect the bioaccessibility of anthocyanins [5]. Furthermore, the same authors reported that acylated anthocyanins from red cabbage show substantial differences in anthocyanin absorption when compared to nonacylated forms. The results revealed that nonacylated anthocyanins were absorbed at rates four times higher than acylated forms [17]. This lower bioavailability of acylated anthocyanins suggests they remain longer in the GIT, where they may exert important health benefits.

In summary, the acylation of anthocyanins, which increases their structural complexity, influences their binding to the food matrix, often leading to stronger interactions with components such as proteins and cell walls. While this can impair their release and reduce their bioavailability, acylation also protects anthocyanins from degradation during digestion, helping preserve their bioactivity until they reach the gut. The interplay between the food matrix and anthocyanin binding is a critical determinant of anthocyanin release during digestion. Understanding these factors is essential for developing strategies to optimize the bioaccessibility of anthocyanins in dietary sources, formulate functional foods and nutraceutical ingredients, and maximize their potential health benefits [19, 20].

6.2.2 Enzymatic Activity – Digestive Enzymes and β-Glucosidases

The processes of food disintegration and nutrient release within the human digestive system are highly complex. Amylase, lipase, and protease are the three main classes of enzymes responsible for the breakdown of macronutrients, indirectly aiding in the release of anthocyanins from the food matrix during digestion [21].

Digestion begins in the oral cavity, where food is mechanically broken down by chewing and chemically processed by enzymes. Saliva, produced by the salivary glands, contains amylase, an enzyme that initiates the breakdown of starch into maltose. The oral cavity is also home to thousands of bacterial species that release their own enzymes to assist with digestion (Figure 6.1) [14, 22]. Along with enzymes from the oral epithelium and oral microbiota, glucosidases, particularly salivary β-glucosidase, contribute to the enzymatic hydrolysis of ingested anthocyanins by cleaving the glycosidic bonds that attach anthocyanins to sugar moieties, converting nonacylated anthocyanins into their aglycone forms [23]. Oliveira et al. (2019) studied the stability of purple sweet potato and grape skin anthocyanin-rich extracts (rich in acylated versus nonacylated anthocyanins, respectively) by incubating them in human saliva ex vivo for 10 min. Their findings showed no significant stability differences, regardless of the presence of matrix components from purple sweet potato or red wine [24]. In a different approach, Kamonpatana et al. (2012) demonstrated that monoglycosylated anthocyanins were more prone to degradation by saliva compared to their di- and tri-glycosylated forms, based on their study of anthocyanin-rich extracts from five different fruits [25]. These observations highlight that anthocyanins undergo biotransformation immediately after ingestion, marking the beginning of further structural modifications [26].

The stomach is the primary site for digestion. Although carbohydrate digestion can begin in the mouth, protein digestion starts only in the stomach and is activated by the acidic environment [27]. This acidic environment also triggers the secretion of pepsin, which is primarily produced by the pancreas and is involved in protein breakdown. It is widely known that anthocyanins can bind to proteins in the food matrix, and this proteolytic activity may facilitate their release, increasing their bioaccessibility [22]. In addition to pepsin, gastric lipase is present in the stomach and is responsible for the breakdown of fat. While anthocyanins are not directly bound to lipids, many anthocyanin-rich foods contain fats or may be consumed alongside lipid-rich foods, and the breakdown of these fats aids in the release of anthocyanins [14]. Podsędek et al. (2014) observed that the concentration of acylated anthocyanins increased 1.6–2.2 times following gastric digestion of red cabbage. Additionally, of the anthocyanins studied, nonacylated cyanidin 3-diglucoside-5-glucoside showed the lowest recovery during the digestion of whole red cabbage [17]. These findings indicate that the acidic environment of the stomach may enhance the release of anthocyanins from the solid food matrix. Furthermore, in an *in vitro* gastric model, nonacylated anthocyanins from grape skin (mainly C3 monoglucosides) showed a slight reduction in their amount compared to the acylated anthocyanins from

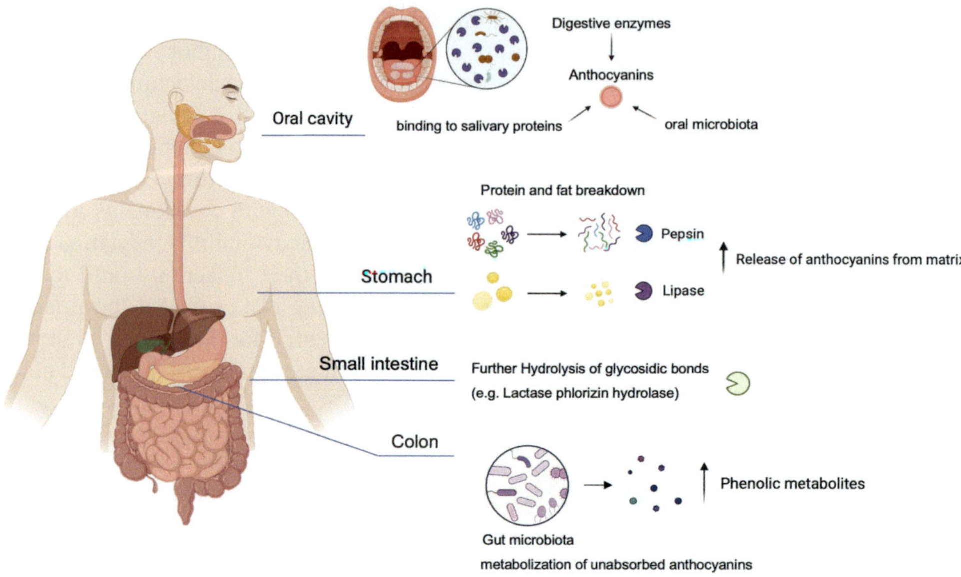

FIGURE 6.1 Key phases during digestion and metabolization of anthocyanins through the gastrointestinal system.

purple sweet potato (acylated mono- and diglucosides), both with the addition of food matrix components. However, this difference in degradation was not statistically significant [24].

In the small intestine, various enzymes, including proteases (such as trypsin and chymotrypsin), pancreatic lipase, and pancreatic amylase, along with coenzymes, bile, and inorganic salts, complete the breakdown of macronutrients. The remaining proteins, carbohydrates, and fats from the previous digestion steps are further hydrolyzed, with the macronutrient hydrolysis completed at this stage [28]. Hydrolysis of glycosidic bonds occurs beyond the oral cavity, mainly in the small intestine, where lactase phlorizin hydrolase (LPH), a β-glucosidase, can further hydrolyze anthocyanins, converting them into their aglycone forms (more easily absorbed) [29–31]. A digestion study conducted by Ribnicky et al. (2014) [32] used the TIM-1 system to compare the bioaccessibility of anthocyanins from blueberry juice and blueberry protein-rich flour (defatted soy flour, DSF), both containing the same amount of anthocyanins. The results indicated that anthocyanins bound to the protein matrix in DSF had a higher bioaccessibility (36%) than those in blueberry juice (26%); however, this difference was not statistically significant. Additionally, the total recovery of anthocyanins was greater for those DSF-bound anthocyanins (63%) than for those in the juice (36%). These findings suggest that the matrix may protect anthocyanins during digestion, allowing more anthocyanins to reach the colon. In the same study, anthocyanins with acylated sugar moieties showed the highest bioaccessibility regardless of whether a high-fat meal was consumed. This aligns with the stability of acylated anthocyanins under both acidic and neutral conditions. Moreover, several studies have shown that acylated anthocyanins from purple potato [33], purple sweet potato [34], and red cabbage [17, 35] may undergo hydrolysis, as their levels decrease in small intestine simulations without cellular absorption models. In a comparative study, simulated small intestinal conditions reduced acylated anthocyanins by 27%–43% in purple sweet potato (22%–31% with the food matrix). In contrast, nonacylated red wine anthocyanins showed a more substantial reduction, with a 49%–52% decrease without matrix components and a 30%–45% decrease when the matrix components were present [24].

In the colon, the gut microbiota produces glycosidases that can help in the breakdown of glycosylated anthocyanins. However, human β-glucosidases are less likely to break down anthocyanins with uncommon glycosides, such as the acylated forms, thus making them more resistant to degradation. As a result, the pharmacokinetics of anthocyanins with more complex *O*-glycoside substitutions may differ greatly from those of simpler monoglucosides [36].

6.2.3 pH Effect and Gastrointestinal Transit

The mouth, with a pH ranging from 5.6 to 7.9, serves as the initial site of food breakdown [37]. The brief oral digesting time, typically 5 s to 2 min, is insufficient to significantly impact their stability since anthocyanin degradation at neutral or basic pH is a time-dependent process [38]. Although anthocyanins are rather stable at these pH levels, enzymatic hydrolysis is the primary mechanism by which they are released from the food matrix, with chewing and saliva moistening playing important roles [12, 22].

Anthocyanins undergo substantial structural transformations as food transitions from the neutral pH environment of the mouth to the acidic conditions of the stomach. Anthocyanins transition from the unstable quinoidal base form to the more stable flavylium cation form that predominates at pH levels of 7 and 3, respectively. In the stomach, gastric secretions induce pH fluctuations ranging from 1.5 to 3.5 [3, 37, 38]. Pepsin and other enzymes that denature the food matrix and facilitate anthocyanin release are activated in this acidic environment, facilitating the release of anthocyanins and enhancing their stability [22]. However, compounds that are more tightly bound, especially those attached to proteins or cell wall constituents, may not be fully released. [39, 40]. As they pass through the upper GIT, acylated anthocyanins show notable resistance to degradation, which helps them retain their structure and bioactivity [24, 41]. With stomach transit times varying from 15 min to 4 h, some release may occur, particularly from simpler matrices like juices [22].

Anthocyanins that were not absorbed in earlier digestion stages mainly exist in their flavylium cation form but shift to hemiketal pseudobase and eventually quinoidal base structures upon reaching the alkaline conditions of the intestine, with pH values ranging from 6.7 to 7.4. This structural shift alters their stability, with greater impact on nonacylated anthocyanins. Free anthocyanins can undergo biotransformation into smaller molecules such as phenolic acids [37, 42]. Despite such structural changes, the pH of the small intestine promotes the release of anthocyanins from the food matrix, as pancreatic enzymes hydrolyze macronutrients [43]. Moreover, by interacting with lipids at both the bulk and interfacial levels, bile salts further promote anthocyanin solubilization, especially those bound inside lipid matrices. By enhancing lipolysis, this self-assembly mechanism improves the solubilization of anthocyanins [44]. Once more, acylated anthocyanins exhibit greater resistance to degradation under neutral to alkaline conditions, as their acyl groups provide protection, slow their breakdown, and enable them to maintain their native structure further along the GIT [35, 45]. Moreover, the stability of acylated anthocyanins varies depending on the type of phenolic acid present in their structure. For instance, red cabbage anthocyanins acylated with sinapic acid have shown higher susceptibility to degradation than those acylated with other hydroxycinnamic acids [35]. Most anthocyanin release occurs in the small intestine, where the breakdown of macronutrients within the food matrix promotes anthocyanin absorption [8]. The small intestine transit time is around 2–5 h, and this time interval is a key determinant for bioavailability [46]. Slower gastrointestinal transit allows for extended digestion and absorption, enhancing anthocyanin uptake, whereas faster transit can limit release and absorption, resulting in more anthocyanins reaching the colon [47]. Altogether, anthocyanin bioavailability is influenced by both gastric and intestinal motility, as a longer contact time generally results in enhanced absorption [42].

Anthocyanins that remain unabsorbed in the small intestine reach the colonic microbial environment and directly influence their metabolism. The colon has a slightly acidic pH in its ascending portion, between 5.5 and 7.5, which increases distally to pH values of 7.0–8.0 in the descending colon [46]. A key factor in lowering colonic pH is the production of short-chain fatty acids (SCFAs) due to microbial fermentation of dietary carbohydrates by gut microorganisms [48]. Briefly, structural changes in anthocyanins are believed to result primarily from spontaneous degradation due to gastrointestinal conditions or through metabolism by colonic microbiota [42]. The colonic transit time is highly variable owing to both inter- and intra-individual changes in the human microbiota, but it is usually around 24 h [46, 49]. A longer transit time allows greater exposure to microbial enzymes and consequently higher anthocyanin degradation and formation of metabolites [50].

6.2.4 Interaction with Gut Microbiota

A defining characteristic of the colonic environment is its rich microbiota, which contains over 10^{14} microorganisms that provide catalytic and hydrolytic action, thereby influencing both human endogenous and xenobiotic metabolism [8, 51, 52]. This intrinsic enzymatic activity, with β-glucosidase and esterase enzymes, aids in anthocyanin breakdown and consequently influences anthocyanin release, biotransformation, and bioavailability [53].

Although small intestine β-glucosidases are mainly responsible for the hydrolysis of anthocyanin glycosidic bonds, some glycosylated anthocyanins still reach the colon, where the gut microbiota contributes to the cleavage of these bonds [8]. This process is followed by the breakdown of the anthocyanin heterocycle, resulting in distinct molecular structures in which cleavage occurs on the B-ring or A-ring, leading to the formation of benzoic acids and phloroglucinol derivatives, respectively [51]. For instance, protocatechuic acid is the primary metabolite resulting from the ring fission of cyanidin-3-glucoside, as reported in several studies [54, 55]. Various studies have reported that even after simulated gastric and intestinal digestion, intact acylated anthocyanins from red cabbage (27%) [35] and purple tomato extracts (61%) [56] and from purple potato and carrot freeze-dried powders (71.8 ± 0.3% and 45 ± 1%, respectively) [57] were detected, indicating that they may reach the colon in their parent form. Acylated anthocyanins are metabolized by the colonic microbiota due to microbial esterase activity, as humans lack endogenous esterases [58]. Thus, acylated anthocyanins with higher stability can cross the upper GIT and reach the colon, where they are considerably metabolized by the gut microbiota [59]. Moreover, depending on the substitution pattern of the B-ring of the parent anthocyanin, the colonic microbiota hydrolyzes glycosides into aglycones and further breaks them down into simple phenolic acids [12, 42]. Several studies have corroborated that the human gut microbiota can degrade acylated anthocyanins. For example, acylated anthocyanins from purple potato [33] and red cabbage extract [17] were degraded when incubated with human fecal microflora in an *in vitro* gastrointestinal model. Moreover, pelargonidin sophorosides acylated with either hydroxycinnamic acids or malonic acid from a red radish anthocyanin extract were *in vitro* degraded into their respective hydroxycinnamic acids and 4-hydroxybenzoic acid [13]. Furthermore, acylated anthocyanins may be more easily broken down by gut microbiota than their nonacylated counterparts. In rats, cyanidin-3-(6″-malonyl)-glucoside appeared more susceptible to microbial breakdown than its nonacylated form. This conclusion is supported by a shift in the ratio of these two anthocyanins within the fecal contents, as a higher abundance of nonacylated anthocyanins was reported, as opposed to the original proportions found in red orange juice [60]. Altogether, these degradation products, along with the resultant aglycone structures, may be absorbed in the colon, potentially contributing to the bioactivity of dietary anthocyanins [8]. Several studies have shown that the resulting phenolic metabolites exhibit greater stability under neutral pH conditions than hydrolyzed aglycones and are more easily absorbed than intact anthocyanins [12].

Anthocyanins, beyond being metabolized by the gut microbiota, can modulate it and influence intestinal bacterial growth, thus resulting in a prebiotic effect [51]. Research conducted by Hidalgo et al. (2012) [61] investigated how several anthocyanins affect different intestinal bacterial populations. The findings revealed that all tested anthocyanin extracts promoted the growth of beneficial bacterial groups, specifically *Lactobacillus* spp. and *Bifidobacterium* spp., while reducing potentially harmful bacteria, such as *Clostridium histolyticum*. These beneficial strains possess β-glucosidase activity, and the glucose released from anthocyanin aglycones serves as an energy source for bacterial growth [62]. Likewise, *in vitro* incubation of purple sweet potato anthocyanin extract, rich in diacylated anthocyanins, with fecal samples from healthy human volunteers demonstrated probiotic effects. The extract promoted the growth of *Bifidobacterium* and *Lactobacillus/Enterococcus* spp., while suppressing the growth of *Prevotella*, *Clostridium histolyticum*, and *Bacteroides* without altering the overall bacterial count. This incubation also led to an increase in the total concentration of SCFAs produced [63].

6.3 ANTHOCYANIN GI ABSORPTION AND METABOLISM

Anthocyanins may cross cell membranes through four main mechanisms: paracellular, transcellular, carrier-mediated, and receptor-mediated transport. Absorption through these pathways depends on various physical properties of anthocyanins, including hydrophobicity, molecular weight, ionization constants, and pH stability [64].

6.3.1 Absorption in the Stomach

The stomach plays a significant role in the absorption of anthocyanins. Research indicates that anthocyanidin glycosides are absorbed in the stomach as they are detected in the plasma shortly after consumption [65, 66]. However, they are often considered unavailable for passive diffusion because they must be deglycosylated for absorption. The presence of anthocyanidin glycosides in plasma after intake suggests a specific transport mechanism for anthocyanins that does not rely on deglycosylation in the GIT [67]. It has been suggested that anthocyanins are transferred to gastric epithelial cells *via* bilitranslocase, an organic anion transporter found in both the gastric mucosa and the liver plasma membrane [51, 67]. Approximately 20%–25% of the gastric uptake of dietary anthocyanins is through bilitranslocase [68]. Different anthocyanin structures can act as substrates for bilitranslocase, with mono- and di-glucosides showing higher affinities than aglycones. However, acylation reduces this affinity. For instance, malvidin-3-*O*-(6-*O*-acetoyl)-β-glucoside (K1 = 58.3 μM) has a 41-fold lower affinity than malvidin glucoside (1.4 μM). Acylation with hydroxycinnamic acid, such as malvidin-3-*O*-(6-*O*-*p*-coumaroyl)-β-glucoside, showed no affinity, potentially due to steric hindrance [67]. Other transporters have also been investigated, including glucose transporters (GLUT1), sodium-coupled monocarboxylate transporters (SMCT1 and SMCT2), and organic anion transporter 2 (OAT2) [51]. An *in vitro* study using MKN-28 cells found that red wine anthocyanin transport was reduced in the presence of glucose, implying that GLUTs, especially GLUT1 and GLUT3, which are expressed in this cell line, play a role in anthocyanin gastric transport [69]. Furthermore, docking studies have shown that anthocyanins from blackberry juice and red wine can bind GLUTs through either their B-ring or glucose moiety [70]. While GLUT1 and GLUT3 transport both nonacylated and acylated anthocyanins, a higher number of glycosyl and acyl groups are associated with lower transport efficiencies, as demonstrated by purified mono- and diacylated purple sweet-fleshed potato peonidin-derived anthocyanins (7%–8%), compared to nonacylated malvidin-3-*O*-glucoside (10%) [70]. In another study by Ichiyanagi et al. (2006), shortly after oral administration of nasunin, an acylated anthocyanin from eggplant peel, in rodents, it appeared in its intact form in rat plasma within 15 min, with an absorption comparable to that of delphinidin-3-glucoside [71]. Despite their lower transport efficiency *in vitro*, rapid absorption of acylated anthocyanins in rodent studies reinforces the role of the stomach in their uptake. Another study by Harada et al. (2004) reported that after ingestion of an extract of the purple sweet potato tuber, only two of the eight acylated anthocyanins were rapidly absorbed in rats, peaking at 5 min after ingestion: cyanidin and peonidin derivatives, each with a caffeoyl–sophoroside–glucoside moiety. Conversely, the other six anthocyanins were all diacylated, suggesting that the gastric absorption of acylated anthocyanins is structure-dependent [72].

Although only native anthocyanins have been detected in the stomach, enzymes that contribute to anthocyanin conjugation, such as uridine 5′-diphosphate-glucuronosyltransferases (UGTs), sulfotransferases (SULTs), and catechol-*O*-methyltransferases (COMTs), are present in the gastric mucosa [42].

6.3.2 Absorption in the Small Intestine

The small intestine is a crucial site not only for anthocyanin glycoside uptake but also as the primary site for flavonoid absorption [8, 12]. Absorption varies across the duodenum, jejunum, and ileum and is also influenced by the structural characteristics of anthocyanins. The jejunum, which makes up nearly half of the length of the small intestine, is thought to be the optimal site for anthocyanin absorption. Conversely, absorption in the duodenum is generally low [73, 74].

At this level, anthocyanin absorption occurs *via* passive diffusion or through membrane transporters [61]. Anthocyanin aglycones are less hydrophilic than their glycosylated counterparts. Owing to their lipophilicity and small molecular size, aglycones are expected to cross the brush border of the small intestine *via* passive diffusion. However, this diffusion is only feasible if anthocyanins undergo enzymatic deglycosylation [8]. Although many flavonoid glycosides are hydrolyzed in the small intestine, anthocyanin glycosides can be absorbed directly by enterocytes, possibly involving GLUT2 and sodium glucose cotransporter 1 (SGLT1), the main mediator of intestinal glucose uptake [12, 75, 76]. GLUT2 is located on the apical surface of brush border cells and mobilizes to the membrane when high levels of glucose are present, thus facilitating glucose uptake. SGLT1 has been previously documented in the literature as a flavonoid transporter [77].

The structural features of anthocyanins influence their intestinal transport. Acylated anthocyanins, due to their larger molecular size and absence or a limited number of free sugar units (which impair their transporter binding), exhibit an overall lower transport efficiency than nonacylated structures [8, 12]. Additionally, *in vitro* studies with Caco-2 cells have shown that anthocyanins with more methoxy groups or fewer hydroxyl groups on the B-ring exhibit enhanced transport efficiency [78]. Another *in vitro* study reported the absorption of a purple carrot anthocyanin extract in a Caco-2 cell model, with acylated anthocyanins crossing the intestinal barrier, although they demonstrated lower permeability than nonacylated anthocyanins [79]. Conversely, another study using rat models revealed that the acylation of cyanidin-3-(6″-malonyl)-glucoside enhanced the intestinal absorption of anthocyanins, as a higher percentage of anthocyanins was absorbed from the intestinal lumen out of the perfused dose when compared to its nonacylated counterpart, cyanidin-3-glucoside [60]. In an *in vitro* study using a Caco-2 cell model, Han et al. (2020) reported that malvidin-3-*O*-(6-*O*-coumaroyl)-glucoside-5-*O*-glucoside exhibited an absorption rate similar to that of its nonacylated form [80]. In short, depending on their molecular size and the presence of free sugar moieties, acylated anthocyanins may exhibit distinct transport efficiencies when compared to their nonacylated counterparts; moreover, they tend to exhibit lower transport efficiencies.

6.3.3 Paracellular Transport

Paracellular transport consists of the passive movement of substances across the epithelium through the spaces between epithelial cells, occurring through diffusion or convection, driven by current transepithelial gradients. Unlike transcellular transporters, paracellular pathways are relatively unselective, mainly allowing molecules to pass based on their size and charge and are generally less tightly regulated [64, 81]. Tight junctions regulate paracellular transport by acting as the main barrier for the movement of ions and larger molecules across epithelial cells. Smaller, less complex anthocyanins are more likely to use this route than their larger, acylated counterparts. Furthermore, claudin proteins form paracellular channels that mediate selective transport across tight junctions [68, 81]. Since paracellular transport is concentration-dependent, it tends to be more prominent at higher concentrations of anthocyanins, where the capacity of transcellular transporters is exceeded [82]. Overall, while the paracellular transport of anthocyanins contributes to their absorption, it is often secondary to transcellular pathways involving specific transporters, which are far more well described in the literature.

6.3.4 Phase I and II Metabolism

Anthocyanins are subjected to phase I and II metabolic processes mainly in the liver but also in the intestinal epithelium and kidneys. Phase I metabolism comprises functionalization reactions through the modification or addition of a functional group *via* hydrolysis, oxidation, or reduction reactions. In phase II, compounds such as glucuronic acid, glutathione, and sulfuric acid are conjugated to the anthocyanidin or anthocyanin structure, with glucuronic acid being the preferred conjugate [83]. These reactions predominantly occur in the epithelial cells of the small intestine, colon, liver, and kidneys [8, 12]. Anthocyanins are frequently conjugated to glucuronic acid by UGTs, methyl groups by COMTs, and sulfates by SULTs [36]. Similar to nonconjugated anthocyanins, conjugated anthocyanins are thought to be transported *via* the same mechanism through the walls of the stomach and the small intestine [12]. Although glucuronidation of parent acylated anthocyanins has not been elucidated, it has been reported that these anthocyanins may be subjected to methylation in phase II reactions [60]. For instance, cyanidin-3-(6″-malonyl)-glucoside from red orange juice lyophilizate undergoes methylation during rat metabolism and is subsequently excreted in the urine, similar to its nonacylated form [60]. However, acyl groups must be removed before the conjugation reaction, as demonstrated by several clinical studies involving the consumption of differently processed red cabbage [84, 85] and grape juice [86] subjected to different food processing methods, both of which contain acylated and nonacylated anthocyanins. The results showed that only nonacylated anthocyanins were conjugated to methyl, glucuronide, or diglucuronide groups, suggesting that acyl groups may be removed before conjugation, allowing only nonacylated anthocyanins to undergo this process. Furthermore, phenolic metabolites may be conjugated during phase II metabolism [87]. In some cases, interconversion of anthocyanidins has been observed in the postprandial period, a process that is thought to be mediated by cytochrome P450, which removes methyl groups and adds hydroxyl groups (during phase I) and later adds methyl groups to hydroxyl groups, thus creating methoxy moieties (during phase II) [12].

6.3.5 Enterohepatic Circulation

During enterohepatic circulation (EHC), solutes are absorbed in the upper part of the GIT and transported into the liver *via* the portal vein. Xenobiotic conjugation occurs in the liver, and the resultant anthocyanin conjugates are combined into bile [36]. Instead of being released into systemic circulation, phase II conjugates are delivered sporadically *via* bile to the ileum, the final section of the small intestine. These metabolites can be deconjugated as they move along the GIT. They are either absorbed in the small intestine and colon and transported back to the liver, thus remaining in the EHC, or excreted and/or deposited in tissue reservoirs [12, 36]. This is confirmed by experimental data showing that anthocyanin conjugates remain in the urine for up to 24 h [88] and 48 h [87]. Additionally, not one but two peaks of anthocyanin concentration in the plasma have been reported [89]. Pharmacokinetic and pharmacodynamic studies indicate that EHC plays a vital role in enhancing the bioactivity of compounds by expanding their distribution and extending the half-life of their sulfated and glucuronide metabolites [90].

6.4 DISTRIBUTION IN THE BODY

6.4.1 Circulation of Anthocyanins and Metabolites

Following phase I and II reactions and the absorption of anthocyanins and their metabolites into the general circulation, they are transported and distributed to organs and tissues *via* plasma proteins such as albumin, with the unbound fraction being the first to undergo distribution [91]. Thus, the detection of anthocyanins in the plasma and serum is influenced by their binding to these proteins. Knowing both the concentration and type of anthocyanins and their metabolites in plasma is crucial for evaluating their potential physiological effects, as plasma delivers them to target sites where they may exert their activity [92]. Anthocyanins can be found as aglycones, glycosides, and conjugates, with concentrations typically in the pmol/L to nmol/L range [3]. Anthocyanin conjugates with glucuronide groups are typically the most prevalent form found in human plasma, peaking at around 30 min to 3 h after ingestion [93–96]. While anthocyanins can appear quickly in plasma due to gastric absorption (first peak), the majority of their absorption occurs in the small intestine (second peak) [89].

6.4.2 Tissue Distribution

Despite the low bioavailability of anthocyanins, their accumulation in tissues can still be detected [97]. Parent anthocyanins and their phase II metabolites have been identified in most tissues analyzed, with some exceptions [36]. In studies using rodent models, nonacylated anthocyanins were found in target tissues, including the kidneys [98–102], brain [100], heart, lungs, and prostate, after consuming rich sources of these compounds [99]. In terms of anthocyanin mean concentration, in the 13 studies compiled by Sandoval-Ramírez et al. (2018), the highest anthocyanin content was found in mice kidneys (2.17×10^5 pmol/g), followed by the mice liver (1.73×10^5 pmol/g), mice testes and lungs (1.49×10^5 pmol/g and 1.16×10^5 pmol/g, respectively), mice prostate (4.96×10^4 pmol/g), pig brain (6.08×10^3 pmol/g), and mice heart (3.6×10^3 pmol/g) [103]. Due to their polarity and growing evidence of GLUT-dependent active transport of anthocyanins in the stomach and small intestine, it is hypothesized that anthocyanins may cross the blood–brain barrier and enter the brain in their glycosylated form *via* the GLUT1 [74, 104]. Tissue uptake of acylated anthocyanins may differ significantly from that of their nonacylated counterparts. Following intraruminal administration of red cabbage lyophilizate in 5% (v/v) ethanol, both acylated and nonacylated anthocyanins and their metabolites have been detected in the plasma and urine of sheep, although only nonacylated anthocyanins have been identified in the cerebrospinal fluid [105]. This highlights the need for further research to better understand the mechanisms governing tissue uptake of acylated anthocyanins. Currently, research on human anthocyanin bioavailability mainly relies on blood, plasma, and urine measurements [3].

6.4.3 Excretion

In humans, anthocyanins and their metabolites are excreted through the urine, feces, and breath. Following the acute intake of $^{13}C_5$-cyanidin-3-*O*-glucoside, urine was found to be the primary excretion route during the first hour, whereas feces became the major route between 6 and 24 h. Peak elimination rate *via* breath was observed at 6 h [87, 106]. Notably, urinary concentrations of parent anthocyanins were approximately 30 times higher than their plasma concentrations [87]. Moreover, following the consumption of bilberry juice by healthy volunteers, the urinary concentrations of anthocyanin degradants and conjugates surpassed those of the parent compounds, with a 24-fold increase [107]. In healthy individuals, the urinary recovery of parent anthocyanins after a single meal is typically less than 1%, and acylation further reduces this recovery. The time to reach the maximum concentration (t_{max}) in urine ranges from 0.5 to 5 h. Furthermore, nonacylated anthocyanins have been identified in various forms in urine, predominantly as glucuronidated and methylated aglycones, glucuronidated and methylated glycosides, aglycone diglucuronides, anthocyanin (di)sulfates, and individual aglycones and glycosides [12]. Numerous clinical trials have shown that the detection of acylated anthocyanins in urine varies based on the source of anthocyanins and food processing methods used, suggesting a structure-dependent absorption. For instance, nonacylated anthocyanins have been identified in urine following the consumption of purple potato extract [108]. Conversely, another study successfully detected two acylated anthocyanins (of eight in total) after the ingestion of a beverage prepared from purple sweet potato extract [72]. These findings indicate that factors such as food matrix, preparation method, and structural characteristics of acylated anthocyanins significantly influence their absorption and excretion.

In a study by He et al. (2005), the fecal samples of rats fed with an anthocyanin-rich grape extract for 14 weeks revealed high concentrations of both diglycosylated and acylated anthocyanins, including *p*-coumaroyl-glucosides of petunidin, malvidin, and delphinidin. These types of anthocyanins exhibit higher gut stability than other nonacylated or monoglycosylated anthocyanins. Both acylation and diglycosylation lead to increased stability of these anthocyanins, which in turn results in higher concentration in feces [109]. Another study by Xu et al. (2021) revealed the presence of phenolic acids derived from A- and B-ring breakdown of rutinosides and glucosides of pelargonidin in feces, along with the disappearance of certain conjugates, which may be indicative of either the reabsorption of some metabolites through EHC or degradation in the colon by the gut microbiota [110]. At last, a study by Czank et al. (2013) detected significant amounts of volatile metabolites and auto-oxidation by-products derived from [^{13}C]-cyanidin-3-glucoside in exhaled breath after its oral intake. This respiratory elimination pathway accounted for 6.91 ± 1.59% of the ingested dose [87].

Overall, dietary anthocyanins may be excreted through urine, feces, and breath mainly as anthocyanin conjugates, phenolic acid metabolites, and volatile metabolites, respectively [3, 89].

6.5 FACTORS AFFECTING THE BIOAVAILABILITY OF (ACYLATED) ANTHOCYANINS

The bioavailability of anthocyanins is influenced by various factors, which can be broadly categorized into food-, host-, and phenolic-specific factors (Figure 6.2) [111].

Food processing techniques, including homogenization, lyophilization, and heat treatment, can significantly affect anthocyanin stability, absorption efficiency, and health benefits. Likewise, cooking procedures play a crucial role in determining the bioactive content and stability [111, 112]. For instance, fermentation influences anthocyanin bioavailability, as reported by a study on the recovery of acylated anthocyanins in human plasma after the consumption of fresh red cabbage, which was 10% higher than that observed after fermented red cabbage consumption [85]. As described earlier in this chapter, anthocyanins may need to be released from other components of the food matrix, such as polysaccharides, proteins, fibers, or other dietary ingredients consumed concomitantly [42]. The interactions between anthocyanins and the food matrix directly influence their bioaccessibility, thereby affecting their bioavailability [113]. For instance, using an *in vitro* gastrointestinal model, the co-digestion of crushed red cabbage and carotenoid-rich vegetables, including cherry tomatoes, baby spinach, and carrots, was observed. Anthocyanin bioaccessibility was increased in the mixtures of cherry tomatoes and baby spinach, enhancing anthocyanin recovery by 10% to 15%. A different outcome was observed in the co-digestion with carrot, in which anthocyanin bioaccessibility was not affected [114]. Moreover, high glucose content in foods can impair anthocyanin uptake by competing for the same transport mechanisms [73, 113]. Furthermore, the quantity of anthocyanins consumed plays a critical role in determining bioavailability. This was evidenced by Carkeet et al. (2008), who reported that higher doses of strawberry anthocyanins in healthy individuals resulted in increased levels of anthocyanins detected in the urine [115]. Additionally, external factors, including agronomic practices, cultivar type, and ripening stage, also affect anthocyanin bioavailability [111].

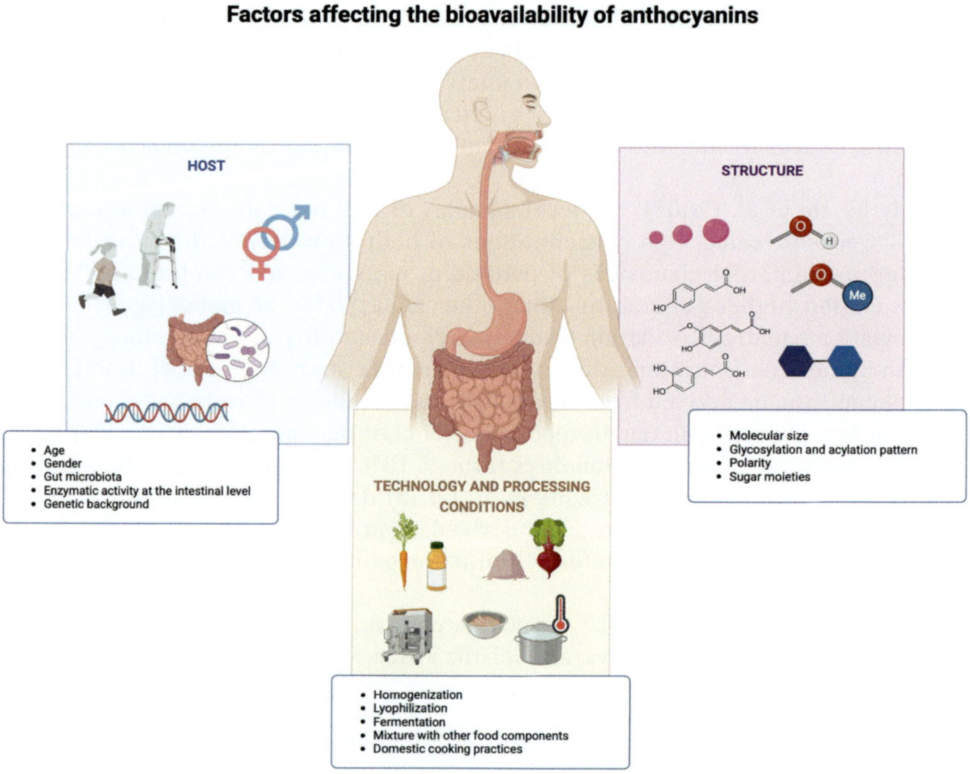

FIGURE 6.2 Variables (structural factors, host-related characteristics, and food processing methods) affecting the bioavailability of anthocyanins.

Host factors, such as genetic background, age, gender, and variations in gut microbiota and enzymatic activity at the intestinal level, account for significant interindividual differences in anthocyanin absorption, making the process of anthocyanin uptake widely diverse among individuals [5, 111].

Structural differences among anthocyanins also influence their bioavailability. In particular, acylated anthocyanins are more stable due to co-pigmentation effects but exhibit lower absorption than their nonacylated counterparts, which are more bioavailable but less stable [14, 51]. Interestingly, acylation does not seem to impair absorption when anthocyanins possess a free sugar unit necessary for gastrointestinal transport, regardless of its position [112]. Anthocyanidin polarity also influences its cellular uptake. Highly polar anthocyanidins, such as cyanidin, demonstrate greater uptake by NCI-N87 cells than less polar anthocyanidins, such as malvidin and petunidin. This disparity may also be influenced by anthocyanin molecular size and weight, with facilitated transport mechanisms likely playing a role [116]. Furthermore, pH variations, with a gradual increase along the small intestine, significantly impact anthocyanin stability and absorption, with lower pH values promoting uptake through, for instance, proton co-transporters [73].

Intestinal transport mechanisms, including passive and facilitated diffusion and active transport, are critical for the bioavailability of anthocyanins and other bioactive compounds. Many bioactive molecules lack the physicochemical properties required for passive diffusion and rely on transmembrane transporters for absorption [5]. However, efflux transporters such as P-glycoprotein and ATP-binding cassette (ABC) transporters can impair anthocyanin absorption while stimulating their excretion, thus reducing their distribution to tissues and overall bioavailability [5, 51].

6.6 CHALLENGES AND RESEARCH GAPS

Although substantial progress has been made in understanding the bioaccessibility and absorption of acylated anthocyanins, key gaps still limit the full understanding of their gastrointestinal behavior and potential health benefits.

The structural diversity of acylated anthocyanins results in an array of distinct stabilities, absorption profiles, and interactions within biological systems. However, further research is needed to clarify how these variations in acyl groups and acylation sites impact their health effects, gut microbiota interactions, and metabolic fate [12]. The identity of their digestive metabolites remains unclear [117].

The food matrix significantly influences the release and absorption of acylated anthocyanins; however, the specific interactions with dietary components during digestion remain underexplored. Understanding these interactions is crucial for the development of dietary formulations with enhanced bioavailability. Knowledge of the effects of different macronutrients, such as glucose, starch, and proteins, on the transport efficiency of acylated anthocyanins is still limited [24]. In fact, research on anthocyanin–protein interactions has been underreported, with gaps in understanding their impact on bioavailability, health benefits, and digestion dynamics, particularly regarding complex stability and its influence on protein digestion [20]. Moreover, further research is needed to understand how food processing methods and complex meal compositions affect the bioaccessibility and absorption of acylated anthocyanins to better reflect daily dietary scenarios [118]. To bridge this gap, research must consider the combined effects of the food matrix, processing, meal composition, and digestion dynamics on the bioefficacy of acylated anthocyanins.

The metabolic fate of acylated anthocyanins is not yet fully understood. The role of gut microbiota and individual variability adds further complexity. The identities of digestive metabolites, including phenolic acids and phloroglucinaldehyde, are unclear, with many expected metabolites remaining undetected in chyme [117]. A deeper understanding of anthocyanin metabolism is required, emphasizing the role of enterohepatic recycling and phase I and II metabolic reactions, along with the diverse range of metabolic intermediates beyond the parent compounds and their breakdown products, to better elucidate their bioavailability and health effects [3, 118]. Moreover, the impact of gut bacteria on anthocyanin metabolism and the enzymes responsible for interindividual variability in anthocyanin bioavailability remain poorly understood. There are some inconsistent findings regarding the impact of anthocyanins on the gut microbiota, highlighting the need for further research. Overall et al. (2017) showed that different anthocyanin profiles from distinct sources affect the gut microbiota composition in different ways [119].

Emerging evidence suggests that the stomach plays a more significant role in anthocyanin absorption than previously recognized, with several *in vivo* studies indicating that gastric absorption has likely been underestimated [51]. To investigate this further, studying anthocyanin transport in MKN-28 or NCI-N87 cells lines, commonly used to model the stomach wall, may offer valuable insights into the exact mechanism by which anthocyanins are absorbed in the stomach [42]. Further research is needed to understand how gastric absorption contributes to the formation of bioactive metabolites and their potential health benefits, crucial for a more comprehensive understanding of anthocyanin transepithelial absorption and metabolism.

Although evidence collected from *in vitro* or animal models provide valuable information and represent a valid tool to explore this subject at a more initial stage, they do not fully capture the complexity of human digestion and metabolism. Larger-scale human clinical trials are essential to confirm these findings and establish the bioavailability, physiological relevance, and health benefits of acylated anthocyanins [12]. Although the short-term benefits of anthocyanin-rich diets have been studied, the long-term health effects remain imprecise. Longitudinal research is needed to assess their roles in chronic or degenerative disease prevention, gut microbiota modulation, and systemic inflammation [12, 120]. Variations in experimental conditions, analytical methods, and reporting standards make it difficult to compare the findings across studies. Establishing standardized protocols for evaluating bioaccessibility, absorption, and metabolism is crucial for generating comparable, consistent, and reliable data [12].

6.7 CONCLUDING REMARKS

This chapter provides critical insights into the bioaccessibility and absorption of acylated anthocyanins, emphasizing the complex interplay among structural diversity, food matrix interactions, metabolic pathways, and tissue distribution. Despite significant progress, key challenges remain, including the need for standardized bioavailability assessment methods, more comprehensive human clinical trials, and a deeper understanding of the long-term health effects.

The inherently low bioavailability of anthocyanins, which is further reduced by acylation, highlights the importance of future studies focused on phenolic metabolites arising from their degradation. Advancing research in these areas is essential to optimize the health benefits of acylated anthocyanins. A clearer understanding of their absorption mechanisms and metabolic fate will pave the way for more effective dietary strategies and therapeutic applications, maximizing their potential to support human health.

REFERENCES

1. Lakshmikanthan, M., et al., A comprehensive review on anthocyanin-rich foods: Insights into extraction, medicinal potential, and sustainable applications. *Journal of Agriculture and Food Research*, 2024. **17**: p. 101245.
2. Teixeira, M., et al., Anthocyanin-rich edible flowers, current understanding of a potential new trend in dietary patterns. *Trends in Food Science & Technology*, 2023. **138**: p. 708–725.
3. Lila, M.A., et al., Unraveling anthocyanin bioavailability for human health. *Annual Review of Food Science and Technology*, 2016. **7**: p. 375–393.
4. D'Archivio, M., et al., Bioavailability of the polyphenols: Status and controversies. *International Journal of Molecular Sciences*, 2010. **11**(4): p. 1321–1342.
5. Rein, M.J., et al., Bioavailability of bioactive food compounds: A challenging journey to bioefficacy. *British Journal of Clinical Pharmacology*, 2013. **75**(3): p. 588–602.
6. Nicolescu, A., et al., Bioaccessibility and bioactive potential of different phytochemical classes from nutraceuticals and functional foods. *Frontiers in Nutrition*, 2023. **10**: p. 1184535.
7. Dima, C., et al., Bioavailability and bioaccessibility of food bioactive compounds; overview and assessment by in vitro methods. *Comprehensive Reviews in Food Science and Food Safety*, 2020. **19**(6): p. 2862–2884.
8. He, J. and M.M. Giusti, Anthocyanins: Natural colorants with health-promoting properties. *Annual Review of Food Science and Technology*, 2010. **1**: p. 163–187.
9. Mattioli, R., et al., Anthocyanins: A comprehensive review of their chemical properties and health effects on cardiovascular and neurodegenerative diseases. *Molecules*, 2020. **25**(17).
10. Gençdağ, E., et al., Copigmentation and stabilization of anthocyanins using organic molecules and encapsulation techniques. *Current Plant Biology*, 2022. **29**: p. 100238.
11. Luo, X., et al., Acylation of anthocyanins and their applications in the food industry: Mechanisms and recent research advances. *Foods*, 2022. **11**(14).
12. Jokioja, J., B. Yang, and K.M. Linderborg, Acylated anthocyanins: A review on their bioavailability and effects on postprandial carbohydrate metabolism and inflammation. *Comprehensive Reviews in Food Science and Food Safety*, 2021. **20**(6): p. 5570–5615.
13. Fleschhut, J., et al., Stability and biotransformation of various dietary anthocyanins in vitro. *European Journal of Nutrition*, 2006. **45**(1): p. 7–18.
14. Ayvaz, H., et al., Anthocyanins: Metabolic digestion, bioavailability, therapeutic effects, current pharmaceutical/industrial use, and innovation potential. *Antioxidants*, 2022. **12**(1): p. 48.

15. Enaru, B., et al., Anthocyanins: Factors affecting their stability and degradation. *Antioxidants*, 2021. **10**(12).
16. Saura-Calixto, F., Concept and health-related properties of nonextractable polyphenols: The missing dietary polyphenols. *Journal of Agricultural and Food Chemistry*, 2012. **60**(45): p. 11195–11200.
17. Podsędek, A., et al., Matrix effects on the stability and antioxidant activity of red cabbage anthocyanins under simulated gastrointestinal digestion. *BioMed Research International*, 2014. **2014**(1): p. 365738.
18. Dangles, O. and J.A. Fenger, The chemical reactivity of anthocyanins and its consequences in food science and nutrition. *Molecules*, 2018. **23**(8).
19. Boulet, J.-C., et al., Focus on the relationships between the cell wall composition in the extraction of anthocyanins and tannins from grape berries. *Food Chemistry*, 2023. **406**: p. 135023.
20. Wu, H., G. Oliveira, and M.A. Lila, Protein-binding approaches for improving bioaccessibility and bioavailability of anthocyanins. *Comprehensive Reviews in Food Science and Food Safety*, 2022. **22**(1): p. 333–354.
21. Garvey, S.M., et al., Fungal digestive enzymes promote macronutrient hydrolysis in the INFOGEST static in vitro simulation of digestion. *Food Chemistry*, 2022. **386**: p. 132777.
22. Sensoy, I., A review on the food digestion in the digestive tract and the used in vitro models. *Current Research in Food Science*, 2021. **4**: p. 308–319.
23. Mallery, S.R., et al., Effects of human oral mucosal tissue, saliva, and oral microflora on intraoral metabolism and bioactivation of black raspberry anthocyanins. *Cancer Prevention Research*, 2011. **4**(8): p. 1209–1221.
24. Oliveira, H., et al., Comparison of the in vitro gastrointestinal bioavailability of acylated and non-acylated anthocyanins: Purple-fleshed sweet potato vs red wine. *Food Chemistry*, 2019. **276**: p. 410–418.
25. Kamonpatana, K., et al., Susceptibility of anthocyanins to ex vivo degradation in human saliva. *Food Chemistry*, 2012. **135**(2): p. 738–747.
26. Kamonpatana, K., et al., Anthocyanin structure determines susceptibility to microbial degradation and bioavailability to the buccal mucosa. *Journal of Agricultural and Food Chemistry*, 2014. **62**(29): p. 6903–6910.
27. Sullivan, R.J. and D.A. Cooley, *Digestion and Nutrition*. Facts on File, Incorporated, 2009.
28. Gallego, M., et al., Compositional, structural design and nutritional aspects of texture-modified foods for the elderly. *Trends in Food Science & Technology*, 2022. **119**: p. 152–163.
29. Day, A.J., et al., Absorption of quercetin-3-glucoside and quercetin-4′-glucoside in the rat small intestine: The role of lactase phlorizin hydrolase and the sodium-dependent glucose transporter. *Biochemical Pharmacology*, 2003. **65**(7): p. 1199–1206.
30. Sesink, A.L., et al., Intestinal uptake of quercetin-3-glucoside in rats involves hydrolysis by lactase phlorizin hydrolase. *The Journal of Nutrition*, 2003. **133**(3): p. 773–776.
31. Hassimotto, N.M.A., M.I. Genovese, and F.M. Lajolo, Absorption and metabolism of cyanidin-3-glucoside and cyanidin-3-rutinoside extracted from wild mulberry (Morus nigra L.) in rats. *Nutrition Research*, 2008. **28**(3): p. 198–207.
32. Ribnicky, D.M., et al., Effects of a high fat meal matrix and protein complexation on the bioaccessibility of blueberry anthocyanins using the TNO gastrointestinal model (TIM-1). *Food Chemistry*, 2014. **142**: p. 349–357.
33. Kubow, S., et al., Effects of simulated human gastrointestinal digestion of two purple-fleshed potato cultivars on anthocyanin composition and cytotoxicity in colonic cancer and non-tumorigenic cells. *Nutrients*, 2017. **9**(9): p. 953.
34. Kubow, S., et al., Biotransformation of anthocyanins from two purple-fleshed sweet potato accessions in a dynamic gastrointestinal system. *Food Chemistry*, 2016. **192**: p. 171–177.
35. McDougall, G.J., et al., Anthocyanins from red cabbage – stability to simulated gastrointestinal digestion. *Phytochemistry*, 2007. **68**(9): p. 1285–1294.
36. Kalt, W., Anthocyanins and their C6-C3-C6 metabolites in humans and animals. *Molecules*, 2019. **24**(22): p. 4024.
37. Braga, A.R.C., et al., Bioavailability of anthocyanins: Gaps in knowledge, challenges and future research. *Journal of Food Composition and Analysis*, 2018. **68**: p. 31–40.
38. Pineda-Vadillo, C., et al., The food matrix affects the anthocyanin profile of fortified egg and dairy matrices during processing and in vitro digestion. *Food Chemistry*, 2017. **214**: p. 486–496.
39. Arruda, H.S., et al., Anthocyanins recovered from agri-food by-products using innovative processes: Trends, challenges, and perspectives for their application in food systems. *Molecules*, 2021. **26**(9): p. 2632.
40. Kumar, M., et al., Optimization of the use of cellulolytic enzyme preparation for the extraction of health promoting anthocyanins from black carrot using response surface methodology. *LWT*, 2022. **163**: p. 113528.
41. Vidana Gamage, G.C., Y.Y. Lim, and W.S. Choo, Sources and relative stabilities of acylated and nonacylated anthocyanins in beverage systems. *Journal of Food Science and Technology*, 2022. **59**(3): p. 831–845.
42. Fernandes, I., V. de Freitas, and N. Mateus, Anthocyanins and human health: How gastric absorption may influence acute human physiology. *Nutrition and Aging*, 2014. **2**: p. 1–14.

43. Mat, D.J.L., et al., In vitro digestion of foods using pH-stat and the INFOGEST protocol: Impact of matrix structure on digestion kinetics of macronutrients, proteins and lipids. *Food Research International*, 2016. **88**: p. 226–233.
44. Macierzanka, A., et al., Bile salts in digestion and transport of lipids. *Advances in Colloid and Interface Science*, 2019. **274**: p. 102045.
45. Zhao, C.-L., et al., Stability-increasing effects of anthocyanin glycosyl acylation. *Food Chemistry*, 2017. **214**: p. 119–128.
46. Lozoya-Agullo, I., et al., Preclinical models for colonic absorption, application to controlled release formulation development. *European Journal of Pharmaceutics and Biopharmaceutics*, 2018. **130**: p. 247–259.
47. Kimura, T. and K. Higaki, Gastrointestinal transit and drug absorption. *Biological and Pharmaceutical Bulletin*, 2002. **25**(2): p. 149–164.
48. Walker, A.W., et al., pH and peptide supply can radically alter bacterial populations and short-chain fatty acid ratios within microbial communities from the human colon. *Applied and Environmental Microbiology*, 2005. **71**(7): p. 3692–3700.
49. Minnebo, Y., et al., Gut microbiota response to in vitro transit time variation is mediated by microbial growth rates, nutrient use efficiency and adaptation to in vivo transit time. *Microbiome*, 2023. **11**(1): p. 240.
50. Müller, M., E.E. Canfora, and E.E. Blaak, Gastrointestinal transit time, glucose homeostasis and metabolic health: Modulation by dietary fibers. *Nutrients*, 2018. **10**(3).
51. Fernandes, I., et al., Multiple-approach studies to assess anthocyanin bioavailability. *Phytochemistry Reviews*, 2015. **14**(6): p. 899–919.
52. Thursby, E. and N. Juge, Introduction to the human gut microbiota. *Biochemical Journal*, 2017. **474**(11): p. 1823–1836.
53. Luo, B., et al., Bioactive phytochemicals and their potential roles in modulating gut microbiota. *Journal of Agriculture and Food Research*, 2023. **12**: p. 100583.
54. Aura, A.M., et al., In vitro metabolism of anthocyanins by human gut microflora. *European Journal of Nutrition*, 2005. **44**(3): p. 133–142.
55. Vitaglione, P., et al., Protocatechuic acid is the major human metabolite of cyanidin-glucosides. *The Journal of Nutrition*, 2007. **137**(9): p. 2043–2408.
56. Li, H., et al., Bioaccessibility, in vitro antioxidant activities and in vivo anti-inflammatory activities of a purple tomato (Solanum lycopersicum L.). *Food Chemistry*, 2014. **159**: p. 353–360.
57. Zhang, H., et al., Bioaccessibility, bioavailability, and anti-inflammatory effects of anthocyanins from purple root vegetables using mono- and co-culture cell models. *Molecular Nutrition & Food Research*, 2017. **61**(10): p. 1600928.
58. Plumb, G.W., et al., Metabolism of chlorogenic acid by human plasma, liver, intestine and gut microflora. *Journal of the Science of Food and Agriculture*, 1999. **79**(3): p. 390–392.
59. Chen, K., et al., Anthocyanins as promising molecules affecting energy homeostasis, inflammation, and gut microbiota in type 2 diabetes with special reference to impact of acylation. *Journal of Agricultural and Food Chemistry*, 2023. **71**(2): p. 1002–1017.
60. Felgines, C., et al., Absorption and metabolism of red orange juice anthocyanins in rats. *British Journal of Nutrition*, 2006. **95**(5): p. 898–904.
61. Hidalgo, M., et al., Metabolism of anthocyanins by human gut microflora and their influence on gut bacterial growth. *Journal of Agricultural and Food Chemistry*, 2012. **60**(15): p. 3882–3890.
62. Modrackova, N., et al., Bifidobacterium β-glucosidase activity and fermentation of dietary plant glucosides is species and strain specific. *Microorganisms*, 2020. **8**(6).
63. Zhang, X., et al., The modulatory effect of anthocyanins from purple sweet potato on human intestinal microbiota in vitro. *Journal of Agricultural and Food Chemistry*, 2016. **64**(12): p. 2582–2590.
64. Yun, Y., Y.W. Cho, and K. Park, Nanoparticles for oral delivery: Targeted nanoparticles with peptidic ligands for oral protein delivery. *Advanced Drug Delivery Reviews*, 2013. **65**(6): p. 822–832.
65. Passamonti, S., et al., The stomach as a site for anthocyanins absorption from food. *FEBS Letters*, 2003. **544**(1–3): p. 210–213.
66. Talavéra, S., et al., Anthocyanins are efficiently absorbed from the stomach in anesthetized rats. *The Journal of Nutrition*, 2003. **133**(12): p. 4178–4182.
67. Passamonti, S., U. Vrhovsek, and F. Mattivi, The interaction of anthocyanins with bilitranslocase. *Biochemical and Biophysical Research Communications*, 2002. **296**(3): p. 631–636.
68. Speciale, A., et al., An overview on the cellular mechanisms of anthocyanins in maintaining intestinal integrity and function. *Fitoterapia*, 2024. **175**: p. 105953.
69. Oliveira, H., et al., Experimental and theoretical data on the mechanism by which red wine anthocyanins are transported through a human MKN-28 gastric cell model. *Journal of Agricultural and Food Chemistry*, 2015. **63**(35): p. 7685–7692.

70. Oliveira, H., et al., GLUT1 and GLUT3 involvement in anthocyanin gastric transport-nanobased targeted approach. *Scientific Reports*, 2019. **9**(1): p. 789.
71. Ichiyanagi, T., et al., Gastrointestinal uptake of nasunin, acylated anthocyanin in eggplant. *Journal of Agricultural and Food Chemistry*, 2006. **54**(15): p. 5306–5312.
72. Harada, K., et al., Absorption of acylated anthocyanins in rats and humans after ingesting an extract of *Ipomoea batatas* purple sweet potato tuber. *Bioscience, Biotechnology, and Biochemistry*, 2004. **68**(7): p. 1500–1507.
73. Cahyana, Y., et al., Factors affecting cellular uptake of anthocyanins: The role of pH, glucose and anthocyanin structure. *Nutrients*, 2022. **14**(22): p. 4807.
74. Manolescu, B.N., et al., Dietary anthocyanins and stroke: A review of pharmacokinetic and pharmacodynamic studies. *Nutrients*, 2019. **11**(7): p. 1479.
75. Kishnani, P.S., Glycogen storage diseases. In *Encyclopedia of the Neurological Sciences* (Second Edition), M.J. Aminoff and R.B. Daroff, Editors. Oxford: Academic Press, 2014. p. 454–459.
76. Rieg, T. and V. Vallon, Development of SGLT1 and SGLT2 inhibitors. *Diabetologia*, 2018. **61**(10): p. 2079–2086.
77. Faria, A., et al., Absorption of anthocyanins through intestinal epithelial cells – putative involvement of GLUT2. *Molecular Nutrition & Food Research*, 2009. **53**(11): p. 1430–1437.
78. Yi, W., et al., Absorption of anthocyanins from blueberry extracts by caco-2 human intestinal cell monolayers. *Journal of Agricultural and Food Chemistry*, 2006. **54**(15): p. 5651–5658.
79. Olejnik, A., et al., Purple carrot anthocyanins suppress lipopolysaccharide-induced inflammation in the co-culture of intestinal Caco-2 and macrophage RAW264.7 cells. *Food & Function*, 2016. **7**(1): p. 557–564.
80. Han, F., et al., In vitro gastrointestinal absorption of red wine anthocyanins – impact of structural complexity and phase II metabolization. *Food Chemistry*, 2020. **317**: p. 126398.
81. Yu, A.S.L., Paracellular transport as a strategy for energy conservation by multicellular organisms? *Tissue Barriers*, 2017. **5**(2): p. e1301852.
82. Goff, J.P., Invited review: Mineral absorption mechanisms, mineral interactions that affect acid-base and antioxidant status, and diet considerations to improve mineral status. *Journal of Dairy Science*, 2018. **101**(4): p. 2763–2813.
83. Tillement, J.P. and D. Tremblay, 5.02 – clinical pharmacokinetic criteria for drug research. In *Comprehensive Medicinal Chemistry II*, J.B. Taylor and D.J. Triggle, Editors. Oxford: Elsevier, 2007. p. 11–30.
84. Charron, C.S., et al., Effect of dose size on bioavailability of acylated and nonacylated anthocyanins from red cabbage (*Brassica oleracea* L. var. *capitata*). *Journal of Agricultural and Food Chemistry*, 2007. **55**(13): p. 5354–5362.
85. Wiczkowski, W., D. Szawara-Nowak, and J. Romaszko, The impact of red cabbage fermentation on bioavailability of anthocyanins and antioxidant capacity of human plasma. *Food Chemistry*, 2016. **190**: p. 730–740.
86. Stalmach, A., et al., Identification of (poly)phenolic compounds in concord grape juice and their metabolites in human plasma and urine after juice consumption. *Journal of Agricultural and Food Chemistry*, 2011. **59**(17): p. 9512–9522.
87. Czank, C., et al., Human metabolism and elimination of the anthocyanin, cyanidin-3-glucoside: A 13C-tracer study123. *The American Journal of Clinical Nutrition*, 2013. **97**(5): p. 995–1003.
88. de Ferrars, R.M., et al., Methods for isolating, identifying, and quantifying anthocyanin metabolites in clinical samples. *Analytical Chemistry*, 2014. **86**(20): p. 10052–10058.
89. Fang, J., Bioavailability of anthocyanins. *Drug Metabolism Reviews*, 2014. **46**(4): p. 508–520.
90. Zhou, X., et al., Enterohepatic circulation of glucuronide metabolites of drugs in dog. *Pharmacology Research & Perspectives*, 2019. **7**(4): p. e00502.
91. Ionescu, C. and M.R. Caira, Drug metabolism in context. In *Drug Metabolism: Current Concepts*, C. Ionescu and M.R. Caira, Editors. Dordrecht, The Netherlands: Springer, 2005. p. 1–40.
92. Murkovic, M., U. Adam, and W. Pfannhauser, Analysis of anthocyane glycosides in human serum. *Fresenius' Journal of Analytical Chemistry*, 2000. **366**(4): p. 379–381.
93. Banaszewski, K., et al., A pilot study to investigate bioavailability of strawberry anthocyanins and characterize postprandial plasma polyphenols absorption patterns by Q-TOF LC/MS in humans. *Journal of Berry Research*, 2013. **3**: p. 113–126.
94. Felgines, C., et al., Influence of glucose on cyanidin 3-glucoside absorption in rats. *Molecular Nutrition & Food Research*, 2008. **52**(8): p. 959–964.
95. Kuntz, S., et al., Uptake and bioavailability of anthocyanins and phenolic acids from grape/blueberry juice and smoothie in vitro and in vivo. *British Journal of Nutrition*, 2015. **113**(7): p. 1044–1055.
96. Mullen, W., et al., Bioavailability of pelargonidin-3-O-glucoside and its metabolites in humans following the ingestion of strawberries with and without cream. *Journal of Agricultural and Food Chemistry*, 2008. **56**(3): p. 713–719.

97. Gui, H., et al., Current knowledge of anthocyanin metabolism in the digestive tract: Absorption, distribution, degradation, and interconversion. *Critical Reviews in Food Science and Nutrition*, 2023. **63**(22): p. 5953–5966.
98. Felgines, C., et al., Radiolabelled cyanidin 3-*O*-glucoside is poorly absorbed in the mouse. *British Journal of Nutrition*, 2010. **103**(12): p. 1738–1745.
99. Marczylo, T.H., et al., Pharmacokinetics and metabolism of the putative cancer chemopreventive agent cyanidin-3-glucoside in mice. *Cancer Chemotherapy and Pharmacology*, 2009. **64**(6): p. 1261–1268.
100. Talavéra, S., et al., Anthocyanin metabolism in rats and their distribution to digestive area, kidney, and brain. *Journal of Agricultural and Food Chemistry*, 2005. **53**(10): p. 3902–3908.
101. Tsuda, T., F. Horio, and T. Osawa, Absorption and metabolism of cyanidin 3-O-β-D-glucoside in rats. *FEBS Letters*, 1999. **449**(2–3): p. 179–182.
102. Vanzo, A., et al., Exceptionally fast uptake and metabolism of cyanidin 3-glucoside by rat kidneys and liver. *Journal of Natural Products*, 2011. **74**(5): p. 1049–1054.
103. Sandoval-Ramírez, B.A., et al., Anthocyanin tissue bioavailability in animals: Possible implications for human health. A systematic review. *Journal of Agricultural and Food Chemistry*, 2018. **66**(44): p. 11531–11543.
104. Li, P., et al., Protective effects of anthocyanins on neurodegenerative diseases. *Trends in Food Science & Technology*, 2021. **117**: p. 205–217.
105. Platosz, N., et al., The blood–cerebrospinal fluid barrier is selective for red cabbage anthocyanins and their metabolites. *Journal of Agricultural and Food Chemistry*, 2020. **68**(31): p. 8274–8285.
106. De Ferrars, R.M., et al., The pharmacokinetics of anthocyanins and their metabolites in humans. *British Journal of Pharmacology*, 2014. **171**(13): p. 3268–3282.
107. Kalt, W., et al., Anthocyanin metabolites are abundant and persistent in human urine. *Journal of Agricultural and Food Chemistry*, 2014. **62**(18): p. 3926–3934.
108. Jokioja, J., et al., Phenolic metabolites in the urine and plasma of healthy men after acute intake of purple potato extract rich in methoxysubstituted monoacylated anthocyanins. *Molecular Nutrition & Food Research*, 2021. **65**(9): p. 2000898.
109. He, J., B.A. Magnuson, and M.M. Giusti, Analysis of anthocyanins in rat intestinal contents – impact of anthocyanin chemical structure on fecal excretion. *Journal of Agricultural and Food Chemistry*, 2005. **53**(8): p. 2859–2866.
110. Xu, Y., et al., Bioavailability, absorption, and metabolism of pelargonidin-based anthocyanins using sprague–dawley rats and caco-2 cell monolayers. *Journal of Agricultural and Food Chemistry*, 2021. **69**(28): p. 7841–7850.
111. Cosme, P., et al., Plant phenolics: Bioavailability as a key determinant of their potential health-promoting applications. *Antioxidants*, 2020. **9**(12): p. 1263.
112. He, J., et al., Dietary polyglycosylated anthocyanins, the smart option? A comprehensive review on their health benefits and technological applications. *Comprehensive Reviews in Food Science and Food Safety*, 2022. **21**(4): p. 3096–3128.
113. Eker, M.E., et al., A review of factors affecting anthocyanin bioavailability: Possible implications for the inter-individual variability. *Foods*, 2019. **9**(1): p. 2.
114. Phan, M.A.T., M.P. Bucknall, and J. Arcot, Co-ingestion of red cabbage with cherry tomato enhances digestive bioaccessibility of anthocyanins but decreases carotenoid bioaccessibility after simulated in vitro gastro-intestinal digestion. *Food Chemistry*, 2019. **298**: p. 125040.
115. Carkeet, C., B.A. Clevidence, and J.A. Novotny, Anthocyanin excretion by humans increases linearly with increasing strawberry dose. *The Journal of Nutrition*, 2008. **138**(5): p. 897–902.
116. Sigurdson, G.T., et al., Aglycone structures and glycosylations affect anthocyanin transport and uptake in human gastric epithelial (NCI-N87) cells. *Journal of Food Composition and Analysis*, 2018. **65**: p. 33–39.
117. Victoria-Campos, C.I., et al., Gastrointestinal metabolism and bioaccessibility of selected anthocyanins isolated from commonly consumed fruits. *Food Chemistry*, 2022. **383**: p. 132451.
118. Bohn, T., et al., Mind the gap – deficits in our knowledge of aspects impacting the bioavailability of phytochemicals and their metabolites – a position paper focusing on carotenoids and polyphenols. *Molecular Nutrition & Food Research*, 2015. **59**(7): p. 1307–1323.
119. Overall, J., et al., Metabolic effects of berries with structurally diverse anthocyanins. *International Journal of Molecular Sciences*, 2017. **18**(2): p. 422.
120. Smeriglio, A., et al., Chemistry, pharmacology and health benefits of anthocyanins. *Phytotherapy Research*, 2016. **30**(8): p. 1265–1286.

Metabolization and Gut Microbiota Role

7

Ana Rita Monteiro and Iva Fernandes

7.1 INTRODUCTION

The well-documented biological activities of anthocyanins are closely related to their bioavailability. To reach systemic circulation, anthocyanins undergo absorption and metabolism processes within the human body, ultimately shaping their fate and potential effects (1). Once ingested, anthocyanins, like any other (poly)phenol, begin to undergo chemical transformations through the action of digestive enzymes. Subsequently, recognized as xenobiotics by the human body, they undergo extensive phase I and II metabolism. Phase I reactions include oxidation, reduction, and hydrolysis. In contrast, phase II metabolism involves conjugation reactions with, e.g., methyl, glucuronic acid, and sulfate groups, primarily in the enterocytes of the small intestine and hepatocytes of the liver (2, 3). The large intestine is also responsible for the postprandial metabolism of the anthocyanins by rapidly hydrolyzing anthocyanin glycosides. Beyond being metabolized by the gut microbiota, anthocyanins also modulate its composition and activity, highlighting a bidirectional interaction with potential implications for human health (4, 5). The structure of anthocyanins plays a crucial role in their absorption and metabolic transformation, with studies showing that pelargonidin-based anthocyanins are absorbed more efficiently than those with other backbones. Their metabolism also exhibits a strong structure-dependency, as pelargonidin-based compounds are predominantly converted into metabolites, whereas a greater proportion of anthocyanins with different backbones remain intact after absorption (6). In the study conducted by Felgines and colleagues, eight male Wistar rats were fed a diet enriched with freeze-dried strawberries to evaluate the urinary excretion and metabolism of strawberry anthocyanins. Additionally, *in situ* rat models (12 male Wistar rats) were used to investigate the gastric and intestinal absorption of pelargonidin-3-glucoside. The main findings, compared to previous results obtained with blackberry anthocyanins, revealed that intact pelargonidin glycosides were recovered in urine along with glucuronidated metabolites. Notably, the amount of pelargonidin glucuronide excreted was similar to that of pelargonidin-3-glucoside (6). Similarly, in other rat models, only trace amounts of glucuronide metabolites were detected in urine following cyanidin-3-glucoside administration (7). In human studies, the same research group demonstrated that pelargonidin glucuronides were the predominant strawberry anthocyanin metabolites found in urine. The urinary excretion of pelargonidin glucuronide was approximately 25 times higher than that of pelargonidin-3-glucoside, whereas the excretion of anthocyanidin glucuronides after blackberry consumption was only five times higher than that of cyanidin-3-glucoside (8, 9). Considering the anthocyanin backbone, pelargonidin possesses only a single hydroxyl group on the B-ring, preventing methylation and potentially making it more susceptible to glucuronidation

DOI: 10.1201/9781003519591-9

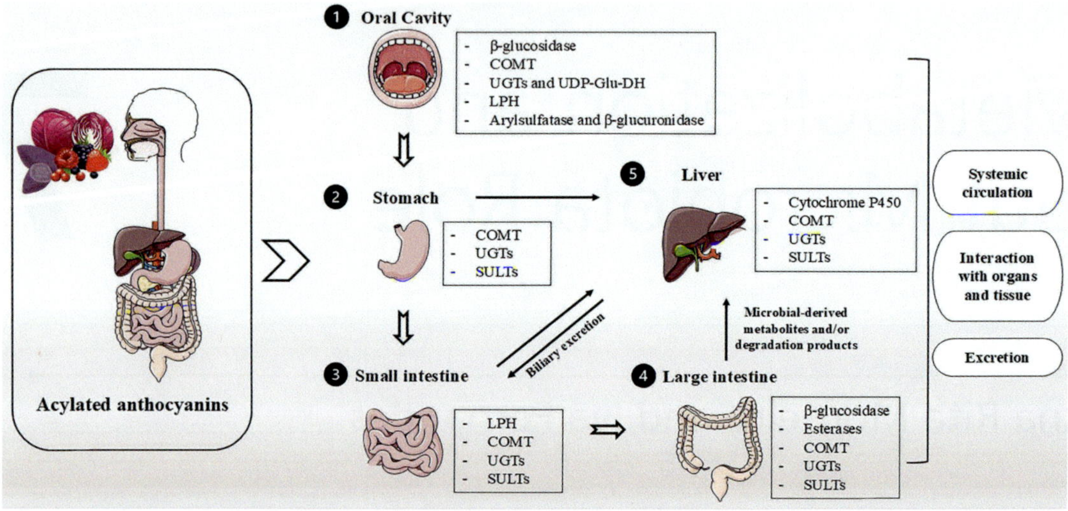

FIGURE 7.1 Schematic representation of the main metabolic sites of anthocyanins in humans and the enzymes involved (from human and microbiota origins).

Abbreviations: COMT, catechol-O-methyltransferase; UGTs, UDP-glucuronosyl transferases; UDP-Glu-DH, UDP-glucose-dehydrogenase; LPH, lactase phlorizin hydrolase; SULTs, sulfotransferases.

than other anthocyanins, such as cyanidin-3-glucoside (6). The key conclusion from these studies is that pelargonidin-3-glucoside is rapidly absorbed in both the stomach and small intestine, similar to cyanidin-3-glucoside (6). However, pelargonidin-3-glucoside undergoes glucuronidation to a significantly greater extent than cyanidin-3-glucoside, highlighting the influence of the aglycone structure on anthocyanin metabolism (6). Notably, the acylation of anthocyanins leads to a significant decline in their bioavailability (10, 11), further highlighting the impact of structural modifications on their metabolic fate. Figure 7.1 schematically represents the main metabolic sites of anthocyanins in humans.

7.1.1 Oral Cavity

The oral cavity is the first site where anthocyanins begin their structural transformation and metabolic modification journey. Chewing anthocyanin-rich foods with the teeth releases these compounds from the vacuolar structures of plant cells, increasing their bioaccessibility. This mechanical breakdown mixes the food bulk with saliva, which is essential in the initial phase of anthocyanin metabolism. The enzymatic hydrolysis occurring in the oral cavity is primarily driven by β-glucosidase in saliva, the oral epithelium, and the oral microbiota. β-Glucosidase catalyzes the cleavage of glycosidic bonds, releasing aglycone forms of anthocyanins, which exhibit distinct chemical properties and enhanced reactivity (2, 4). Although further studies are needed to fully understand the role of the oral cavity in anthocyanin metabolism, existing evidence suggests that this process is structure-dependent and mediated by oral microbiota (4). Even considering the short residence time of the food matrix in the mouth, studies conducted by Kamonpatana and colleagues revealed that the metabolization of anthocyanins is affected by the type of aglycone and glycoside moiety (12, 13). Specifically, *ex vivo* studies in human saliva demonstrated that nonacylated anthocyanin glycosides derived from delphinidin and petunidin degraded more readily than those of cyanidin, malvidin, and peonidin. This suggests that the structure of the anthocyanin backbone plays a critical role in their stability and susceptibility to enzymatic action. Furthermore, the type of sugar attached to the anthocyanin molecule also influences hydrolysis rates, with monosaccharides being more easily cleaved than di- and trisaccharides, as observed when incubating fruit and berry extracts rich in nonacylated anthocyanins in human saliva. The presence and position of hydroxyl (–OH) and methoxy ($-OCH_3$) substituents in the B-ring of aglycones further determine their stability, impacting the rate and extent of their degradation in the oral cavity (12, 13). In a study by Mallery and colleagues, ten clinically healthy, nonsmoking volunteers (aged 19–61 years) participated in a rinse study, and these same individuals – along with one additional participant – provided saliva samples for the functional β-glucosidase activity assay. Their findings implicated β-glucosidase from the oral microbiota in the transformation of anthocyanins from freeze-dried black raspberries, which are particularly rich in cyanidin-3-rutinoside, cyanidin-3-xylosylrutinoside, cyanidin-3-glucoside, and cyanidin-3-sambubioside (14). After the oral digestion of anthocyanins, residual anthocyanins, protocatechuic acid, cyanidin-3-glucoside, and glucuronidated anthocyanins can be detected in saliva. Further remarkable findings were reported by Mallery and colleagues, showing that protocatechuic acid, derived from cyanidin rather than cyanidin-3-glucoside, was readily detected in saliva, and its glucuronide form appeared after absorption by the oral epithelium. Immunohistochemistry analyses of 15 donor tissues further identified key enzymes and transporters, including catechol-*O*-methyltransferase (COMT), UDP-glucuronosyl transferases (UGTs), UDP-glucose-dehydrogenase (UDP-Glu-DH), breast cancer resistance protein (BCRP), β-glucosidase, lactase phlorizin hydrolase (LPH), and sodium-dependent glucose co-transporter 1 (SGLT1) at the end of the salivary ducts and stratified squamous surface epithelia (14). These findings suggest that glucuronidation, methylation, or conjugation of anthocyanins with glutathione might occur in the mucosa. Moreover, the detection of arylsulfatase and β-glucuronidase in human saliva, two key enzymes responsible for deglucuronidation in phase II metabolism of anthocyanins, further highlights the complex and dynamic metabolic processes potential of the oral cavity (14). As a result of enzymatic hydrolysis, anthocyanin aglycones become more lipophilic and, consequently, more easily absorbed. However, due to their instability, they can be rapidly broken down into protocatechuic acid. In oral cells, UGTs induce the glucuronidation of aglycones, anthocyanins, and protocatechuic acid, with some of these glucuronidated products returning to the oral cavity *via* BCRP efflux. These compounds are then subjected to deglucuronidation by arylsulfatase and β-glucuronidase in saliva, generating metabolites available for subsequent circulation (15). The oral and gut microbiomes are naturally compartmentalized by the oral–gut barrier, which, along with physical distance and chemical defenses like gastric acid and bile, prevents microbial crossover. However, if this barrier is compromised, microorganisms may migrate between these regions, enabling cross-communication (16). Thus, oral microbiota could reach the intestinal mucosa in case of barrier disruption or other malfunctions. At the same time, gut-to-oral microbial transmission could occur through inter- and intrapersonal mechanisms (15, 16). This potential overlap suggests that oral and gut microbial profiles may share similarities, leading to analogous anthocyanin metabolism in both the oral cavity and the intestinal tract. Notably, these microbial populations metabolize monoglycosylated anthocyanins and deglycosylate more complex diglycosylated and acylated anthocyanins, a process that will be explored in the following sections (15).

7.1.2 Stomach

Anthocyanins, including their acylated forms, exhibit remarkable stability in the stomach's acidic environment (pH 1.5–4), with some structural variations influencing their resistance to gastric conditions. Despite the stomach not being classically described as a major site of absorption, several studies indicate that anthocyanin uptake can occur here, mediated by specific transporters whose efficiency depends on the compound's structure (2, 15). Moreover, the time-dependent absorption process suggests that slower gastric emptying could enhance anthocyanin absorption and improve their bioavailability. This highlights the stomach's underappreciated role in digestion and the early modulation of anthocyanin metabolism and bioavailability (2, 15). The stomach possesses conjugative enzymes that contribute to the metabolism of anthocyanins, since they are detected in the blood circulation and urine in intact, methylated, glucuronidated, and sulfoconjugated forms (17). The organic anion carrier bilitranslocase is primarily responsible for transport efficiency, and its affinity with anthocyanin depends on the structure of the anthocyanin's glycoside portion (15). Besides acylated anthocyanins, mono- and diglycosylated forms are good ligands for bilitranslocase (18).

In a study conducted by Harada and colleagues, the absorbability of acylated anthocyanins from purple sweet potato extract was evaluated in humans (six healthy volunteers) and 36 male Sprague–Dawley rats. Only two of the eight acylated anthocyanins (cyanidin and peonidin derivatives of caffeoyl-sophoroside-glucoside) were detected after ingestion, thus highlighting structure-dependent absorption (19). Considering stability, several studies emphasize the stability of acylated anthocyanins in simulated *in vitro* gastric conditions, which are further released from the food matrix (2). The remarkable stability of acylated anthocyanins in the acidic gastric environment suggests they may withstand substantial degradation, with only limited metabolism occurring through phase II enzymes like methylation, glucuronidation, and sulfoconjugation. This stability likely enables acylated anthocyanins to reach the intestine in a more intact form than their nonacylated counterparts, potentially influencing their bioavailability and interaction with the gut microbiota. However, further studies are needed to understand the extent and implications of their gastric metabolism.

7.1.3 Small Intestine

The small intestine plays a critical role not only in the absorption but also in the metabolism of acylated anthocyanins. Their stability and hydrophobicity, determined by their structural characteristics, influence their capacity to permeate the intestinal barrier (2, 15). Acylated anthocyanins have been shown to cross the intestinal epithelium, as demonstrated in studies using purple carrot anthocyanins and Caco-2 cells (2). However, their transport efficiency was lower than that of nonacylated anthocyanins (2).

In the study by Novotny and colleagues, which investigated the kinetics of anthocyanin metabolism, 12 volunteers participated in a crossover study, consuming purple carrots containing five distinct anthocyanins: cyanidin-3-(xylose-glucose-galactoside), cyanidin-3-(xylose-galactoside), cyanidin-3-(xylose-sinapoyl-glucose-galactoside), cyanidin-3-(xylose-feruloyl-glucose-galactoside), and cyanidin-3-(xylose-coumaroyl-glucose-galactoside), in three different treatments, each separated by a 3-week washout period. The treatments included 250 g of raw purple carrots, 250 g of cooked purple carrots, and 500 g of cooked purple carrots. The results demonstrated that absorption, gastrointestinal transit, and plasma elimination are strongly influenced by anthocyanin structure. Acylated compounds, such as cyanidin-3-(xylose-sinapoyl-glucose-galactoside) and cyanidin-3-(xylose-feruloyl-glucose-galactoside), showed lower absorption efficiencies compared to their nonacylated counterparts, like cyanidin-3-(xylose-glucose-galactoside) and cyanidin-3-(xylose-galactoside) (18). Moreover, the study also revealed that acylated anthocyanins from purple carrots exhibited a shorter mathematically modeled half-life of absorption to plasma in the upper gastrointestinal tract compared to nonacylated ones, suggesting a more limited and shorter absorption site for acylated anthocyanins (18). This finding underscores the influence of anthocyanin structure on absorption dynamics and supports the notion that acylation significantly impacts these compounds' bioavailability and metabolic fate. Interestingly, in the study conducted by Felgines and colleagues, the acylation of cyanidin-3-(6″-malonyl)-glucoside favored intestinal absorption over its nonacylated counterpart, cyanidin-3-glucoside, in rat models – 12 male Wistar rats were fed a control diet supplemented with red orange juice lyophilizate (200 g/kg diet) for 12 days, followed by *in situ* experiments (10). Han and colleagues investigated the transport efficiency of different malvidin derivatives using human gastric (MKN-28) and intestinal (Caco-2) cell models. Specifically, malvidin-3-*O*-glucoside, malvidin-3,5-*O*-diglucoside, and malvidin-3-*O*-(6-*O*-coumaroyl)-glucoside-5-*O*-glucoside, extracted from *Vitis davidii* red wine, were used to assess the influence of glucoside, diglucoside, and coumaroylated glucoside moieties on absorption (20). Conversely, their findings revealed that malvidin-3-*O*-(6-*O*-coumaroyl)-glucoside-5-*O*-glucoside was absorbed at similar rates to its nonacylated counterpart in Caco-2 cells, suggesting that the coumaroylation did not hinder its intestinal transport (20).

LPH from enterocytes has been suggested to hydrolyze anthocyanins, influencing their stability and subsequent absorption (2, 15). Several studies emphasize the susceptibility of acylated anthocyanins from purple potato, purple sweet potato, and red cabbage to hydrolysis, as their concentrations decreased in a simulated small intestine model without cellular absorption (2). McDougall and colleagues further investigated the stability of nonacylated, mono-, and diacylated anthocyanins – acylated with *p*-coumaric, caffeic, ferulic, and sinapic acids – from red cabbage during simulated gastrointestinal digestion. Their findings indicated that anthocyanins acylated with sinapic acid were more prone to degradation than those acylated with other hydroxycinnamic acids, highlighting the differential stability imparted by the type of acylation (21). Once inside the enterocytes, anthocyanin aglycones undergo extensive phase II metabolism, primarily producing glucuronidated and sulfated conjugates. Anthocyanins can be absorbed in their intact glycosylated forms *via* glucose transporters or hydrolyzed into aglycones to facilitate their cell entry. Under alkaline conditions, some anthocyanins may also degrade into smaller phenolic acids. Both anthocyanins and their aglycones are subjected to phase II modifications within the enterocytes, which play a crucial role in determining their bioavailability and subsequent biological activity (15).

7.1.4 Large Intestine

Acylated anthocyanins exhibit lower transport efficiency and higher digestion resistance than their nonacylated counterparts. This increased stability results in a larger proportion of these compounds reaching the colon, where they can suffer the action of microbial fermentation. This makes the colon a crucial site for the postprandial metabolism of anthocyanins, as unabsorbed forms undergo extensive biotransformation by gut microbiota (22). Unabsorbed anthocyanins reach the colon and are exposed to microbial and endogenous β-glucosidase. The catabolic processes undertaken by the gut microbiota include hydrolysis, hydroxylation, hydrogenation, decarboxylation, and dihydroxylation. In humans, there are no endogenous esterases to release the phenolic acids from the acylated anthocyanins; conversely, the esterases present in the intestinal microbiota are capable of doing so (22).

Bacterial metabolism of anthocyanins in the gut involves a series of transformations: the cleavage of glycosidic linkages, the breakdown of the anthocyanidin heterocycle (C-ring), and the degradation into phloroglucinol derivatives (A-ring) and benzoic acids (B-ring). Additionally, microbial activity promotes *O*-demethylation, forming simple phenolics and their sulfated, glucuronidated, and methylated derivatives. Other phenolic metabolites arising from this microbial catabolism include compounds such as catechol, pyrogallol, hydroxybenzoic acids, propionic acids, gallic acid, and *p*-coumaric acid, highlighting the diversity and complexity of anthocyanin biotransformation in the colon (2, 22). Human gut microbiota plays a crucial role in the catabolism of acylated anthocyanins.

In an *in vitro* study conducted by Fleschhut and colleagues, pure anthocyanins – such as cyanidin-3-glucoside, malvidin-3-glucoside, and peonidin-3-glucoside – and an aqueous extract of red radish were incubated with a human fecal suspension to investigate their biotransformation by gut microbiota. The results showed that glycosylated and acylated anthocyanins were rapidly degraded, with pelargonidin sophorosides acylated with hydroxycinnamic and/or malonic acids breaking down into 4-hydroxybenzoic acid and hydroxycinnamic acids (23). The proposed formation pathway is schematically illustrated in Figure 7.2.

Similarly, Stalmach and colleagues incubated 2 ml of concentrated Concord grape juice, rich in both nonacylated and acylated anthocyanins (*O*-acetyl glycosides, *O*-*p*-coumaroyl-*O*-diglucosides, and *O*-*p*-coumaroylglucosides of delphinidin, cyanidin, petunidin, peonidin, and malvidin) – representing 46% of the total phenolic content of the juice – with human feces, resulting in the production of a variety of phenolic acids, such as phenylacetic acids, phenyllactic acids, and hydroxybenzoic acids, alongside simple phenolic degradants like catechol, resorcinol, pyrogallol, and phloroglucinol (24). Acylated anthocyanins appear to be more readily degraded by gut microbiota compared to their nonacylated counterparts. In a study involving obese mice, a 12-week chronic intake of a Concord grape supplement rich in acylated anthocyanins (the freeze-dried whole berry powder was standardized based on total anthocyanin content and incorporated into the high-fat diet (HFD) at a concentration of 400 μg/g of total anthocyanins) led to a ten-fold increase in total anthocyanin content in feces compared to the levels observed during a late test week when gut microbiota was knocked down using antibiotics. This increase was notably higher than

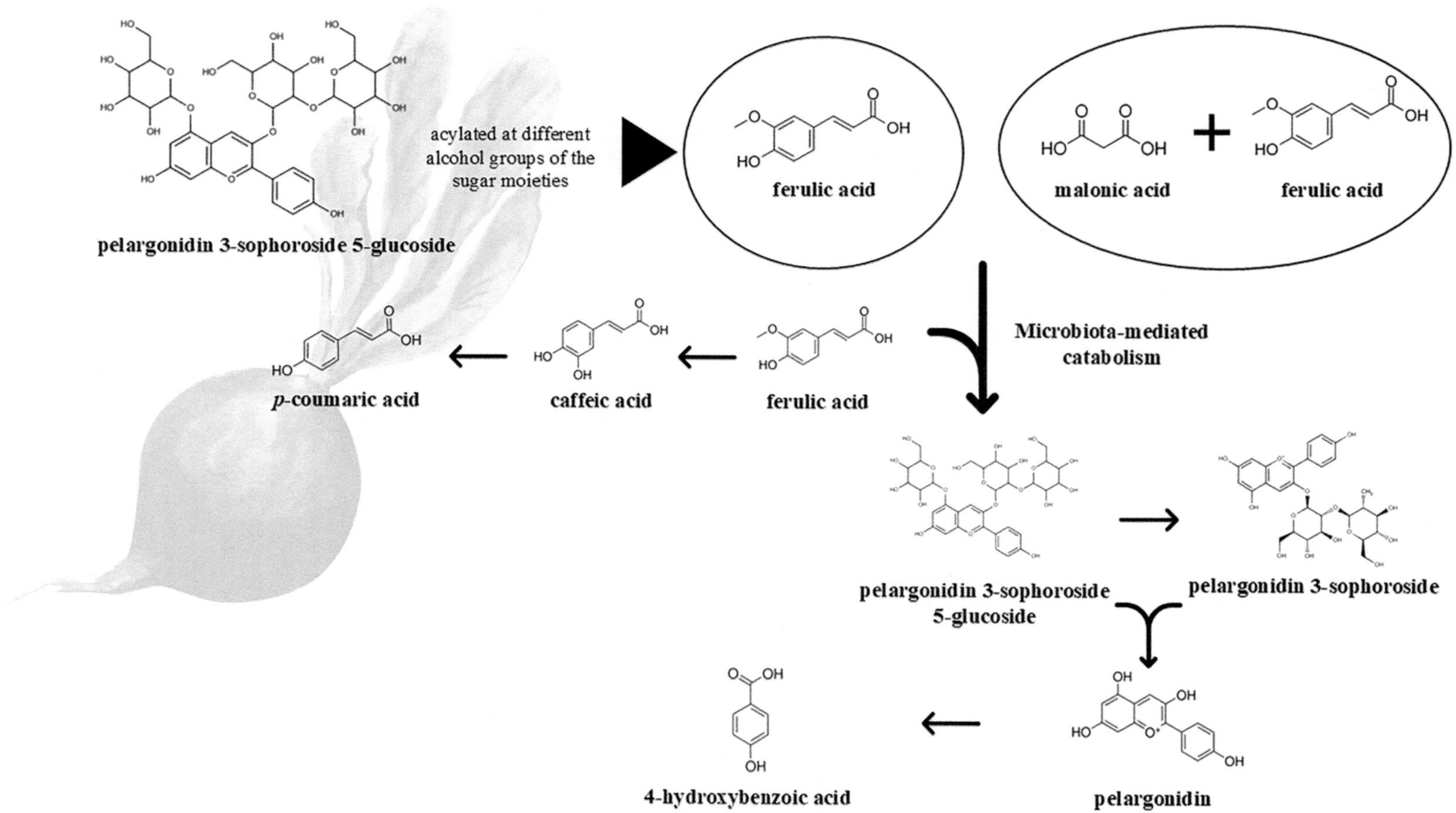

FIGURE 7.2 Schematic representation of a proposed pathway for the catabolism of acylated anthocyanins from red radish, as described by Fleschhut and colleagues.

that observed following the intake of berry supplements rich in nonacylated anthocyanins (25). Moreover, cyanidin-3-(6″-malonyl)-glucoside was found to be more readily available for microbial degradation in rats compared to its nonacylated counterpart, as the ratio of these two anthocyanins in cecal contents shifted to favor the nonacylated form when compared to their original ratio in red orange juice (10). Beyond their initial breakdown, some phenolic acids produced by microbial activity can undergo further transformation. For instance, caftaric acid, chlorogenic acid, and caffeic acid esters were converted into 3-(3′-hydroxyphenyl) propionic acid and benzoic acid. These microbial metabolites may either be reabsorbed for further metabolism or excreted *via* urine (15).

7.1.5 Liver

Once reabsorbed, anthocyanin-derived microbial metabolites enter the portal circulation and reach the liver, the primary organ responsible for extensive metabolic transformation through phase I and II metabolism. Phase I reactions, including demethylation and hydrolysis, prepare these compounds for subsequent conjugation. In contrast, phase II metabolism – glucuronidation, sulfation, and methylation – enhances their solubility and facilitates their systemic distribution or excretion (15, 22). Enzymes like cytochrome P450 1A (CYP1A), UGTs, COMT, and sulfotransferases (SULTs) are highly expressed in the liver and small intestine, playing key roles in these processes. Glucuronidation and sulfation, catalyzed by UGTs and SULTs, respectively, improve hydrophilicity, promoting the excretion of anthocyanin metabolites *via* urine and bile. In contrast, COMT methylation increases the metabolites' lipophilicity, enhancing their potential for tissue accumulation and prolonging their biological activity. Through enterohepatic circulation, some of these conjugated metabolites can be secreted back into the intestine *via* bile, where they may undergo further microbial transformation and reabsorption, creating a dynamic cycle that impacts their bioavailability and physiological effects (15).

Enterohepatic circulation has been suggested to occur due to the persistence of anthocyanin conjugates in human urine up to 24 h and even 48 h, alongside a second rise in anthocyanin concentration in plasma (2). In the study by Kalt and colleagues, after the intake of 250 mL of blueberry juice, parent anthocyanins (cyanidin, delphinidin, malvidin, pelargonidin, peonidin, and petunidin) accounted for only 4%, while anthocyanin metabolites comprised 96% of the total urinary anthocyanins over the following 24 h. These metabolites were dominated by aglycones, particularly aglycone glucuronides, highlighting the role of phase II metabolism. The diversity of anthocyanin metabolites included positional isomers of anthocyanin conjugates and chalcones, suggesting complex metabolic transformations. The persistence of these metabolites indicated active enterohepatic recycling, leading to a prolonged residence time and enhanced systemic exposure. Notably, the prevalence of metabolites based on pelargonidin, which is absent in blueberry juice, suggests ongoing dehydroxylation and demethylation of other anthocyanins, likely driven by xenobiotic metabolism and colonic bacterial activity (26). The liver plays a pivotal role in the secondary metabolism of anthocyanins, transforming them into simpler phenolic acids and driving their extensive phase II conjugation. Studies with human liver microsomes have demonstrated this capacity, showing that after gastrointestinal digestion, anthocyanin degradation products such as protocatechuic acid and 4-hydroxybenzoic acid undergo further biotransformation (27). Woodward and colleagues observed that pelargonidin was primarily converted into 4-hydroxybenzoic acid, with about 65% of this metabolite subsequently conjugated into two distinct glucuronides. Similarly, cyanidin was metabolized into protocatechuic acid, with 43% transformed into three glucuronide conjugates. This hepatic metabolism modifies the chemical structure of anthocyanins and plays a crucial role in determining their bioavailability and biological activity, underscoring the liver's central function in anthocyanin biotransformation (27).

7.2 MICROBE-DERIVED METABOLITES: A KEY ROUTE FOR MODULATING BIOAVAILABILITY

The metabolic transformation of anthocyanins into phenolic derivatives with enhanced biological activity has become a key focus in understanding their health-promoting effects. While parent anthocyanins exhibit limited bioavailability and rapid clearance, their microbe-derived metabolites often demonstrate greater stability, absorption, and biological efficacy. Acylated anthocyanins, which resist early-phase digestion and reach the colon largely intact, play a crucial role in this process as they undergo extensive microbial fermentation (2, 15). As previously mentioned, in the colon, anthocyanins are exposed to the enzymatic action of gut microbiota, including β-glucosidases and esterases, leading to the cleavage of glycosidic and acylated linkages. This results in the release of aglycones and the subsequent breakdown of the anthocyanidin structure into a diverse array of phenolic acids and simple phenolic compounds. Metabolites such as protocatechuic acid, vanillic acid, syringic acid, and ferulic acid emerge from this biotransformation and have been shown to possess superior antioxidant, anti-inflammatory, and metabolic regulatory properties compared to their parent anthocyanins (2, 22, 28, 29). Czank and colleagues examined blood, urine, breath, and fecal samples of eight male participants who consumed 500 mg of isotopically labeled cyanidin-3-glucoside to evaluate the absorption, distribution, metabolism, and elimination of anthocyanin-rich foods (30). This study highlights the human system's extensive metabolism and retention of anthocyanin-derived compounds. The phenolic metabolites detected in blood were predominantly conjugated *via* glucuronidation and sulfation, demonstrating their active participation in phase II metabolism. Moreover, the presence of methylated and glucuronide anthocyanins in urine further supports the hepatic transformation and subsequent excretion of these bioactive compounds. The fecal profile offered additional insights into the microbial transformation of anthocyanins. The disappearance of certain conjugates alongside the emergence of phenolic acids derived from the breakdown of the A- and B-ring suggests ongoing microbial metabolism in the colon, with the possibility of enterohepatic recirculation or direct excretion (30). This pattern aligns with findings from De Ferrars and colleagues, where distinct phenolic acids were detected selectively in urine or plasma, implying further tissue-specific metabolism, biliary elimination, or microbial catabolism followed by reabsorption (31). Furthermore, the study by Xu and colleagues in male Sprague–Dawley rats on oral administration of pelargonidin-3-rutinoside (isolated from strawberries) revealed a predominance of demethylated pelargonidin-3-rutinoside in plasma, likely resulting from gut microbial activity (32). This observation underlines the crucial role of gut microbiota in modulating the bioavailability and systemic presence of anthocyanin derivatives, emphasizing the transformative impact of colonic fermentation on these bioactive compounds. Together, these findings underscore that the intricate interplay between microbial metabolism and hepatic transformation positions microbe-derived phenolic metabolites as pivotal players in the health benefits attributed to anthocyanin-rich foods. Their enhanced bioactivity suggests that the true impact of anthocyanins may lie not only in their parent structures but also in the diverse and potent metabolites they produce through gut–liver axis interactions.

The limited number of studies investigating the degradants of acylated anthocyanins underscores a significant research gap in understanding their biological impact (2). This is exemplified by the study conducted by Chen and colleagues, which investigated and compared the impact of acylated anthocyanins from purple potatoes and nonacylated anthocyanins from bilberries on cecal and fecal metabolites, as well as their modulatory effects on gut microbiota in diabetic Zucker rats (33). The quantification of anthocyanins in freeze-dried extracts was performed using high-performance liquid chromatography (HPLC), with cyanidin-3-glucoside as the reference standard. The bilberry extract had an anthocyanin concentration of 423.9 mg/g, predominantly composed of nonacylated glucosides, galactosides, and arabinosides of delphinidin, petunidin, cyanidin, peonidin, and malvidin. In contrast, the purple potato extract contained

248.7 mg/g of anthocyanins, which were almost entirely acylated. The most abundant compound in this extract was petunidin coumaroyl-rutinoside-glucoside, followed by peonidin coumaroyl-rutinoside-glucoside and petunidin caffeoyl-rutinoside-glucoside. While all bilberry anthocyanins were nonacylated, 98.97% of those in purple potato were acylated. To assess their biological effects, Zucker diabetic fatty rats fed a HFD were gavaged daily with either nonacylated anthocyanins from bilberries or acylated anthocyanins from purple potatoes at doses of 25 and 50 mg/kg body weight for 8 weeks. The study provided valuable insights into the distinct metabolic and microbial responses elicited by acylated anthocyanins. The study demonstrated that acylated anthocyanins influenced gut microbiota composition and metabolic output differently (with more pronounced regulatory and beneficial effects on diabetes) from their nonacylated counterparts, suggesting a unique route of bioavailability and potential health benefits (33). These findings emphasize the need for further investigation to elucidate the specific mechanisms through which acylated anthocyanins and their microbe-derived metabolites contribute to health outcomes, especially given their apparent resilience to upper gastrointestinal digestion and subsequent transformation in the colon.

7.3 CONCLUDING REMARKS AND FUTURE PERSPECTIVES

The acylation of anthocyanins plays a crucial role in enhancing their resistance to gastrointestinal degradation, allowing them to reach the intestine in a more intact form. This structural modification influences not only their bioavailability but also their interactions with gut microbiota, leading to distinct metabolic transformations. Emerging evidence suggests that acylated and nonacylated anthocyanins may give rise to different phenolic metabolites, which could have unique biological activities. However, further studies are needed to elucidate the specific metabolic pathways and health effects of acylated anthocyanins. Future research should focus on identifying and characterizing these metabolites, and on understanding their bioactivity and systemic impact. An in-depth understanding of these differences will be fundamental for optimizing dietary strategies and developing functional foods enriched with anthocyanins to maximize their health benefits.

ACKNOWLEDGMENTS

This work was financially supported by the FCT project FERMEN.TO – FERmented foods to struggle MEtabolic syNdrome: An inTegrated *in vitro* dynamic approach (2023.00164.RESTART). This work received support and help from FCT/MCTES (LA/P/0008/2020 DOI 10.54499/LA/P/0008/2020, UIDP/50006/2020 DOI 10.54499/UIDP/50006/2020, and UIDB/50006/2020 DOI 10.54499/UIDB/50006/2020), through national funds. Ana Rita Monteiro is grateful for her Ph.D. grant from FCT (2024.03649.BDANA). Iva Fernandes acknowledges her assistant professor contract CEECINST/00064/2021.

REFERENCES

1. Li F, Sun Q, Chen L, Zhang R, Zhang Z. Unlocking the health potential of anthocyanins: a structural insight into their varied biological effects. *Crit Rev Food Sci Nutr.* 2024:1–21.
2. Jokioja J, Yang B, Linderborg KM. Acylated anthocyanins: a review on their bioavailability and effects on postprandial carbohydrate metabolism and inflammation. *Compr Rev Food Sci Food Saf.* 2021;20(6):5570–615.

3. Cardona F, Andrés-Lacueva C, Tulipani S, Tinahones FJ, Queipo-Ortuño MI. Benefits of polyphenols on gut microbiota and implications in human health. *J Nutr Biochem.* 2013;24(8):1415–22.
4. Tian L, Tan Y, Chen G, Wang G, Sun J, Ou S, et al. Metabolism of anthocyanins and consequent effects on the gut microbiota. *Crit Rev Food Sci Nutr.* 2019;59(6):982–91.
5. Faria A, Fernandes I, Norberto S, Mateus N, Calhau C. Interplay between anthocyanins and gut microbiota. *J Agric Food Chem.* 2014;62(29):6898–902.
6. Felgines C, Texier O, Besson C, Lyan B, Lamaison JL, Scalbert A. Strawberry pelargonidin glycosides are excreted in urine as intact glycosides and glucuronidated pelargonidin derivatives in rats. *Br J Nutr.* 2007;98(6):1126–31.
7. Talavéra S, Felgines C, Texier O, Besson C, Gil-Izquierdo A, Lamaison JL, et al. Anthocyanin metabolism in rats and their distribution to digestive area, kidney, and brain. *J Agric Food Chem.* 2005;53(10):3902–8.
8. Felgines C, Talavéra S, Gonthier MP, Texier O, Scalbert A, Lamaison JL, et al. Strawberry anthocyanins are recovered in urine as glucuro- and sulfoconjugates in humans. *J Nutr.* 2003;133(5):1296–301.
9. Felgines C, Talavera S, Texier O, Gil-Izquierdo A, Lamaison JL, Remesy C. Blackberry anthocyanins are mainly recovered from urine as methylated and glucuronidated conjugates in humans. *J Agric Food Chem.* 2005;53(20):7721–7.
10. Felgines C, Talavéra S, Texier O, Besson C, Fogliano V, Lamaison J-L, et al. Absorption and metabolism of red orange juice anthocyanins in rats. *Br J Nutr.* 2006;95(5):898–904.
11. Charron CS, Clevidence BA, Britz SJ, Novotny JA. Effect of dose size on bioavailability of acylated and nonacylated anthocyanins from red cabbage (*Brassica oleracea* L. var. *capitata*). *J Agric Food Chem.* 2007;55(13):5354–62.
12. Kamonpatana K, Giusti MM, Chitchumroonchokchai C, Morenocruz M, Riedl KM, Kumar P, et al. Susceptibility of anthocyanins to ex vivo degradation in human saliva. *Food Chem.* 2012;135(2):738–47.
13. Kamonpatana K, Failla ML, Kumar PS, Giusti MM. Anthocyanin structure determines susceptibility to microbial degradation and bioavailability to the buccal mucosa. *J Agric Food Chem.* 2014;62(29):6903–10.
14. Mallery SR, Budendorf DE, Larsen MP, Pei P, Tong M, Holpuch AS, et al. Effects of human oral mucosal tissue, saliva, and oral microflora on intraoral metabolism and bioactivation of black raspberry anthocyanins. *Cancer Prev Res.* 2011;4(8):1209–21.
15. Gui H, Sun L, Liu R, Si X, Li D, Wang Y, et al. Current knowledge of anthocyanin metabolism in the digestive tract: absorption, distribution, degradation, and interconversion. *Crit Rev Food Sci Nutr.* 2023;63(22):5953–66.
16. Park SY, Hwang BO, Lim M, Ok SH, Lee SK, Chun KS, et al. Oral–gut microbiome axis in gastrointestinal disease and cancer. *Cancers (Basel).* 2021;13(7):1–20.
17. Fernandes I, Faria A, Calhau C, de Freitas V, Mateus N. Bioavailability of anthocyanins and derivatives. *J Funct Foods.* 2014;7(1):54–66.
18. Novotny JA, Clevidence BA, Kurilich AC. Anthocyanin kinetics are dependent on anthocyanin structure. *Br J Nutr.* 2012;107(4):504–9.
19. Harada K, Kano M, Takayanagi T, Yamakawa O, Ishikawa F. Absorption of acylated anthocyanins in rats and humans after ingesting an extract of *Ipomoea batatas* purple sweet potato tuber. *Biosci Biotechnol Biochem.* 2004;68(7):1500–7.
20. Han F, Oliveira H, Brás NF, Fernandes I, Cruz L, De Freitas V, et al. In vitro gastrointestinal absorption of red wine anthocyanins – impact of structural complexity and phase II metabolization. *Food Chem.* 2020;317(August 2019):126398.
21. McDougall GJ, Fyffe S, Dobson P, Stewart D. Anthocyanins from red cabbage – stability to simulated gastrointestinal digestion. *Phytochemistry.* 2007;68(9):1285–94.
22. Chen K, Kortesniemi MK, Linderborg KM, Yang B. Anthocyanins as promising molecules affecting energy homeostasis, inflammation, and gut microbiota in type 2 diabetes with special reference to impact of acylation. *J Agric Food Chem.* 2023;71(2):1002–17.
23. Fleschhut J, Kratzer F, Rechkemmer G, Kulling SE. Stability and biotransformation of various dietary anthocyanins in vitro. *Eur J Nutr.* 2006;45(1):7–18.
24. Stalmach A, Edwards CA, Wightman JD, Crozier A. Colonic catabolism of dietary phenolic and polyphenolic compounds from concord grape juice. *Food Funct.* 2013;4(1):52–62.
25. Overall J, Bonney SA, Wilson M, Beermann A, Grace MH, Esposito D, et al. Metabolic effects of berries with structurally diverse anthocyanins. *Int J Mol Sci.* 2017;18(2).
26. Kalt W, Liu Y, McDonald JE, Vinqvist-Tymchuk MR, Fillmore SAE. Anthocyanin metabolites are abundant and persistent in human urine. *J Agric Food Chem.* 2014;62(18):3926–34.
27. Woodward GM, Needs PW, Kay CD. Anthocyanin-derived phenolic acids form glucuronides following simulated gastrointestinal digestion and microsomal glucuronidation. *Mol Nutr Food Res.* 2011;55(3):378–86.
28. Chen Y, Li Q, Zhao T, Zhang Z, Mao G, Feng W, et al. Biotransformation and metabolism of three mulberry anthocyanin monomers by rat gut microflora. *Food Chem.* 2017;237:887–94.

29. Eran Nagar E, Okun Z, Shpigelman A. Digestive fate of polyphenols: updated view of the influence of chemical structure and the presence of cell wall material. *Curr Opin Food Sci*. 2020;31:38–46.
30. Czank C, Cassidy A, Zhang Q, Morrison DJ, Preston T, Kroon PA, et al. Human metabolism and elimination of the anthocyanin, cyanidin-3-glucoside: a 13C-tracer study. *Am J Clin Nutr*. 2013;97(5):995–1003.
31. De Ferrars RM, Czank C, Zhang Q, Botting NP, Kroon PA, Cassidy A, et al. The pharmacokinetics of anthocyanins and their metabolites in humans. *Br J Pharmacol*. 2014;171(13):3268–82.
32. Xu Y, Li Y, Xie J, Xie L, Mo J, Chen W. Bioavailability, absorption, and metabolism of pelargonidin-based anthocyanins using Sprague–Dawley rats and Caco-2 cell monolayers. *J Agric Food Chem*. 2021;69(28):7841–50.
33. Chen K, Wei X, Kortesniemi M, Pariyani R, Zhang Y, Yang B. Effects of acylated and nonacylated anthocyanins extracts on gut metabolites and microbiota in diabetic Zucker rats: a metabolomic and metagenomic study. *Food Res Int*. 2022;153(January):110978.

Acylated Anthocyanins Evidence from *In Vivo* Studies

8

Ana Faria

8.1 INTRODUCTION

Anthocyanins are a subclass of phenolic compounds widely distributed in fruits and flowers, particularly those with vibrant pigmentation. Structurally, they are flavonoids characterized by a core flavonoid backbone with sugar moieties attached, which enhances their hydrophilicity and distinguishes them from other polyphenolic compounds (1, 2). Anthocyanins exhibit considerable structural diversity, ranging from simpler monoglycosylated forms to more complex polyglycosylated derivatives, which are predominantly found in flowers. In polyglycosylated anthocyanins, sugar residues can be attached at different positions on the flavonoid nucleus or sequentially at the same site (3). A subset of anthocyanins, known as acylated anthocyanins, contains additional acyl groups attached to sugar moieties, primarily at the 6−OH position (and less frequently at the 4−OH position). The most common acylating agents include *p*-hydroxybenzoic, *p*-coumaric, caffeic, ferulic, sinapic, malonic, acetic, and malic acids (4). Acylation confers increased chemical stability, making these anthocyanins more resistant to pH and temperature variations. This enhanced stability has implications for their potential use in the food industry and technological applications, such as natural colorants.

Acylated anthocyanins are present in a wide range of plant-based foods, including fruits, vegetables, berries, tubers, and edible flowers. Certain vegetables, such as radish, purple sweet potato, and red cabbage, contain a higher proportion of acylated anthocyanins compared to fruits (5).

The health benefits associated with anthocyanin-rich diets have been extensively documented, with most studies focusing on nonacylated, monoglycosylated anthocyanins. However, the specific biological effects of acylated anthocyanins remain less well characterized, likely due to the limited availability of targeted research. To date, few human studies have investigated the distinct physiological effects of acylated anthocyanins, as these compounds naturally co-occur with nonacylated anthocyanins and other phytochemicals in food matrices.

This chapter aims to summarize current knowledge on the biological effects of acylated anthocyanins, with a particular focus on findings from *in vivo* studies, including human trials.

DOI: 10.1201/9781003519591-10

8.2 BIOAVAILABILITY: KEY FACTORS AND CHALLENGES

Bioavailability, defined as the extent to which an ingested anthocyanin reaches its target tissue, remains a complex and not yet fully understood subject, despite significant advancements in recent years. Several factors, both biological and non-biological, influence anthocyanin bioavailability.

Non-biological factors such as dosage, chemical structure, source (pure compound or mixture), food or beverage matrix, processing methods, and analytical techniques all play crucial roles in determining anthocyanin bioaccessibility – its availability for absorption (6). On the biological side, individual factors like age, gender, body composition, microbiota, and metabolic capacity can significantly impact how anthocyanins are distributed to target tissues (7, 8). Additionally, characteristics of the gastrointestinal (GI) tract – including pH, concurrent food intake, digestive enzymes, bile acids, motility, permeability, and microbiota – strongly influence anthocyanin stability and absorption (9).

The chemical structure of anthocyanins plays a decisive role in their absorption efficiency. The presence of free hydroxyl groups and fewer methoxy ($-OCH_3$) groups tends to reduce their bioavailability (10, 11). Studies have shown that glycosylation patterns also affect absorption rates, with certain sugar residues enhancing uptake (7, 10–15) and the acylation pattern with a huge influence on anthocyanin availability.

The GI passage is crucial for anthocyanin absorption due to their inherent chemical instability, which affects their bioactivity. These compounds are highly sensitive to pH, becoming unstable at higher pH levels (16). Common analytical methodologies rely on detecting anthocyanins in their cationic form under acidic conditions, but incomplete conversion to this form can lead to underestimation (17). Moreover, anthocyanins undergo metabolism and biotransformation, meaning their original cationic form may not be fully recoverable, further contributing to discrepancies in quantification. The uncertainty regarding whether anthocyanins' biological effects stem from their cationic form, hemiacetal form, or metabolites complicates the interpretation of bioavailability data.

Study design plays a critical role in evaluating anthocyanin bioavailability. Variables such as food source, matrix, intake duration, and dosing strategy can significantly alter results.

The food matrix itself is another key factor. A more lipophilic environment can facilitate flavonoid solubilization and absorption, while the presence of ethanol has been shown to enhance anthocyanin transport across the intestinal epithelium (18).

Anthocyanins undergo extensive metabolism within the human body (19–21), yet many bioavailability studies fail to account for or quantify their metabolites, leading to underestimation. The complexity of these metabolites, influenced by GI conditions, xenobiotic metabolism, and microbiota activity, makes their detection challenging. Individual variability in xenobiotic transport and enzymatic biotransformation further complicates the study of anthocyanin metabolism and its biological effects (22). Additionally, anthocyanin absorption may be affected by the concurrent intake of other compounds. Since cells lack specific transport mechanisms for phenolics, anthocyanins must compete with other molecules such as organic cations and glucose for cellular uptake (18, 23, 24).

Although microbiota does not directly impact anthocyanin bioavailability, it plays a crucial role in their biological effects (25, 26).

8.3 BIOAVAILABILITY OF ACYLATED ANTHOCYANINS

For anthocyanins to exert biological effects, they must be absorbed into circulation. However, their bioavailability is generally reported as low, and differences in absorption between acylated and nonacylated anthocyanins have not been extensively explored.

The first dietary intervention study assessing acylated anthocyanin bioavailability investigated the absorption of purple sweet potato extract. Two acylated anthocyanins were detected in the plasma and urine of volunteers after purple sweet potato beverage consumption, albeit at a very low recovery rate (~0.02%) (27). Similarly, studies with purple carrots, tested in varying amounts and processing conditions (raw versus cooked), demonstrated lower recovery of acylated anthocyanins, with larger doses further reducing absorption, suggesting a saturable transport mechanism (28). Urinary excretion of an acylated anthocyanin was also reported following the consumption of a purple-fleshed sweet potato beverage (29). However, these studies did not assess anthocyanin metabolites but only native ones.

Charron et al. examined the bioavailability of red cabbage anthocyanins, comparing acylated and nonacylated forms. Of the 30 acylated anthocyanins ingested, eight were detected in urine alongside four metabolites, which were mostly glucuronidated or methylated metabolites (30). Nonetheless, nonacylated anthocyanins exhibited higher recovery rates than acylated anthocyanins. The increasing focus on anthocyanin metabolites produced in the human body has been facilitated by technological advancements, yet their low bioavailability presents challenges in detecting and quantifying metabolites above the analytical threshold.

Studies investigating the impact of acylation and plant matrix composition on anthocyanin bioavailability indicate that nonacylated anthocyanins are approximately four times more bioavailable than acylated forms, with plant matrix effects appearing negligible (31). A crossover study utilizing purple carrot as an anthocyanin source confirmed reduced absorption efficiency of acylated anthocyanins, which exhibited a shorter gastrointestinal half-life compared to nonacylated anthocyanins (10). Additionally, nonacylated anthocyanins displayed slower elimination rates (10), although metabolite profiling was not determined.

The influence of fermentation on anthocyanin bioavailability has also been investigated. Red cabbage fermentation reduced acylated anthocyanin content by approximately 15%, particularly affecting sinapic acid-acylated forms (32). Of the 20 anthocyanins ingested, 18 were acylated cyanidin derivatives. Overall, 18 native compounds and 12 metabolites were detected in plasma and/or urine. Bioavailability from fresh red cabbage was 10% higher than that from its fermented counterpart (32).

A more recent study examined metabolite production following the consumption of a purple potato extract rich in methoxy-substituted monoacylated anthocyanins. Unlike previous studies on anthocyanins' bioavailability, no parent anthocyanins were detected in plasma or urine (33). Instead, 28 phenolic metabolites were identified in urine and 14 in plasma, suggesting extensive microbial degradation of acylated anthocyanins (34). Acylation appears to reduce anthocyanin bioavailability, likely due to steric hindrance from glycosyl and acyl moieties, which may interfere with gastric absorption mechanisms (9, 35). However, reduced absorption may enhance microbial metabolism, highlighting a potential role for gut microbiota in anthocyanin biotransformation (36).

8.4 BIOLOGICAL EFFECTS OF ACYLATED ANTHOCYANINS

Despite numerous studies on anthocyanins' bioavailability, limited research has focused on the biological effects of acylated anthocyanin-rich foods. Consumption of red cabbage products has been associated with increased plasma antioxidant capacity, with fresh red cabbage eliciting a greater effect than fermented cabbage (32). Additionally, purple potato consumption has been linked to improved postprandial glycemic response compared to yellow potato consumption, with reductions in glucose and insulin area under the curve after an anthocyanin-rich meal (37). Some inflammatory markers, such as fibroblast growth factor-19 and interleukin (IL)-8, were also modulated by intervention (37). However, these findings are based on acute studies, and the chronic effect of acylated anthocyanin consumption remains largely unexplored in human trials.

One study assessed the impact of an 8-week intervention with a purple sweet potato beverage on hepatic biomarkers in men with borderline hepatitis, reporting a significant reduction in γ-glutamyl transferase (GGT) levels (38).

While human studies provide valuable insights and a more relevant model, they have inherent limitations. Research on acylated anthocyanins relies exclusively on whole-food interventions rather than on isolated compounds, limiting the identification and isolation of specific effects and underlying cellular mechanisms.

8.5 EVIDENCE FROM PRECLINICAL STUDIES

Animal models serve as a complementary approach due to their physiological similarities to humans and their utility for mechanistic approaches. These models allow for the administration of concentrated anthocyanin-rich extracts, facilitating metabolite identification and quantification.

The first bioavailability study was performed in rats and demonstrated direct absorption of acylated anthocyanins from purple sweet potato, with an associated increase in plasma antioxidant capacity (39). Igarashi et al. (2000) reported that acylated anthocyanins from red cabbage mitigated oxidative stress *in vivo*, which aligns with this observation (40). The absorption and metabolism of 6-*O*-caffeoylsophorose from red vinegar fermented with purple sweet potato were also studied in a rat model, making use of accurate analytical methods than the studies before, and found that this anthocyanin was intact in blood as well as its phase II metabolites and degradation products (41). Similarly, cyanidin 3-*O*-(6-*O*-feruloyl-β-D-glucopyranosyl)-(1 → 6)-[β-D-xylopyranosyl-(1 → 2)]-β-D-galactopyranoside from purple carrots was absorbed in the rat jejunum *via* the organic anion transport polypeptide and was found intact in systemic circulation (42). Acylated anthocyanins from purple yam were also recovered in rat blood and urine following oral administration (43). These findings reinforce the hypothesis that acylated anthocyanins can be absorbed and may exert biological effects.

Since evidence has shown intact acylated anthocyanins in blood and that they are able to reach target tissues, the ability of anthocyanins to cross the blood–cerebrospinal fluid (CSF) barrier was assessed in sheep. Only nonacylated anthocyanins were detected in CSF, whereas acylated anthocyanins, despite their presence in plasma and urine, were absent (44). The higher molecular weight and steric hindrance of acylated anthocyanins can hinder interaction with transporters necessary for central nervous system (CNS) uptake.

Acylated anthocyanins have been investigated in various disease contexts. In a renal disease model, highly acylated anthocyanins from purple sweet potato reduced serum uric acid levels and attenuated kidney damage in a hyperuricemic mice model, likely through anti-inflammatory mechanisms (45).

Several metabolic outcomes with acylated anthocyanin intervention have also been observed in animal studies. In a diabetic rat model, supplementation with purple potato extract lowered glycemia, improved glucose tolerance, and reduced glycated hemoglobin. Additionally, oxidative stress markers, such as malondialdehyde, decreased, while antioxidant enzyme activity increased (46). Similar antioxidant effects were reported in an aging mouse model, where red cabbage anthocyanin supplementation enhanced superoxide dismutase activity and decreased malondialdehyde (47).

The metabolic effects of acylated anthocyanins were further demonstrated in a rat model of obesity and diabetes. An 8-week intervention improved glycemic control, insulin sensitivity, and lipid profiles (48). Notably, acylated anthocyanins modulated hepatic gene expression related to glycolysis and gluconeogenesis in a manner distinct from nonacylated anthocyanins (49), reinforcing the need for further investigation.

The impact of anthocyanin-rich food consumption on metabolic risk factors has been examined in a mouse model. Berries rich in acylated anthocyanins exhibited lower bioactivity at physiological concentrations compared to sources predominantly containing nonacylated anthocyanins (50).

Given the steric hindrance and molecular size limitations affecting acylated anthocyanin absorption, and the emerging role of gut microbiota in their metabolism, further studies should explore the effects of acylated anthocyanins on microbiota composition and function.

8.6 INTERACTION OF ACYLATED ANTHOCYANINS WITH GUT MICROBIOTA

Gut microbiota represents the most extensive and functionally intricate ecosystem, predominantly residing within the GI tract (51). The human body harbors a vast and diverse consortium of microorganisms, including bacteria, archaea, viruses, fungi, and other eukaryotes, which colonize various anatomical niches and engage in complex symbiotic interactions with the host (52–54). Among these microbial communities, the majority of gut bacteria belong to the phyla Bacillota, Bacteroidota, Actinomycetota, Pseudomonadota, and Verrucromicrobia (55).

Recent technological advancements have facilitated in-depth research into the dynamic composition and functional roles of gut microbiota. Although the precise mechanisms governing host–microbiome interactions remain an active area of research, accumulating evidence implicates microbial dysbiosis – defined as an imbalance in microbial composition and function – in the pathogenesis of numerous diseases. These range from localized GI disorders to systemic pathologies, including autoimmune, hepatic, respiratory, cardiovascular, oncologic, metabolic, neurologic, and psychiatric conditions (56, 57).

Dysbiosis is characterized by an overgrowth of pathobionts (commensal microbes with pathogenic potential under specific conditions), a depletion of beneficial bacterial taxa, and/or a decline in microbial diversity (58, 59). Such perturbations can profoundly affect host physiology, contributing to metabolic disorders such as obesity and diabetes, as well as neurodegenerative and neuropsychiatric diseases. Environmental factors, particularly dietary patterns, play a pivotal role in the onset and perpetuation of dysbiosis. For instance, high-fat diets (HFDs) are well-documented to induce significant alterations in gut microbial composition, reducing bacterial diversity and fostering dysbiosis.

Among the various determinants influencing gut microbiome composition, diet is one of the most potent modulators. Dietary modifications can rapidly and profoundly reshape microbial communities (60). HFDs and high-sucrose diets have been demonstrated to detrimentally impact gut microbiota by decreasing bacterial diversity and promoting dysbiotic states (61, 62). In addition to HFDs, exogenous environmental factors, such as antibiotic exposure, pollutants (63, 64), and food additives – including dietary emulsifiers (65, 66) and noncaloric artificial sweeteners (67, 68) – have been implicated in gut microbiota perturbations similar to those observed in HFD-induced obesity models. Conversely, targeted dietary interventions aimed at modulating the gut microbiota, such as prebiotic supplementation, represent a promising avenue for mitigating or reversing dysbiosis.

Prebiotics were traditionally defined as nondigestible carbohydrates that selectively stimulate the growth of beneficial gut bacteria (69). However, this definition has evolved to encompass "a substrate that is selectively utilized by host microorganisms, conferring a health benefit" (70). This broader conceptualization allows for the inclusion of noncarbohydrate compounds with microbiota-modulating properties. Among these, polyphenols – specifically anthocyanins – have emerged as potential prebiotic candidates due to their capacity to influence microbial composition and function.

Anthocyanins, particularly polyglycosylated acylated anthocyanins, are particularly relevant in this context owing to their low bioavailability. As stated before, a substantial proportion of these compounds remains unabsorbed in the upper GI tract, and thereby reach the colon, where they interact with gut microbiota. This interaction may facilitate the selective proliferation of beneficial bacterial taxa while concurrently enabling microbial metabolism of anthocyanins into bioactive metabolites. Such prebiotic-like effects underscore the potential role of anthocyanins in dietary strategies aimed at preserving microbial homeostasis.

Once in the colon, anthocyanins serve as substrates for microbial fermentation, which can enhance the abundance of beneficial bacterial species and generate metabolites with putative health-promoting effects (36, 71).

Despite the limited number of studies investigating this topic, evidence suggests that acylated anthocyanin supplementation in a diabetic rat model influences gut microbiota composition (72). Specifically, an increased abundance of *Peptostreptococcaceae sp.* and a decreased abundance of *Parabacteroides* spp. have been observed. Additionally, anthocyanin supplementation has been associated with elevated fecal short-chain fatty acid levels, reduced colonic *Ruminococcus torques* and *Lachnospiraceae bacterium 4_1_37FAA* abundances, and enhanced oxidative phosphorylation, collectively supporting the role of acylated anthocyanins in promoting health through microbiota modulation (72). In this context, acylated anthocyanin extracts modulated gut metabolism and microbiota in diabetic models, exhibiting greater regulatory and beneficial effects on diabetes than their nonacylated counterparts (72).

In aged mice, supplementation with red cabbage resulted in a significant decrease in *Bifidobacterium* abundance and a slight increase in *Lachnospiraceae* and *Clostridia* abundances (47). These microbiota shifts were associated with gut–brain axis modulation, as evidenced by reduced hippocampal neuronal damage, attenuation of brain-derived neurotrophic factor degeneration, decreased inflammatory cytokines in the brain, and mitigation of cognitive impairment (47).

These interactions highlight the intricate relationship among diet, gut microbiota, and host health, reinforcing the importance of dietary polyphenols as modulators of microbial ecology and potential therapeutic agents for microbiota-associated disorders. Nevertheless, the full extent of the effects and therapeutic potential of acylated anthocyanins remains largely unexplored. There is a significant gap in human studies investigating microbiota-mediated health benefits, despite the promising potential of these compounds due to their low bioavailability (Figure 8.1). Furthermore, beyond their direct

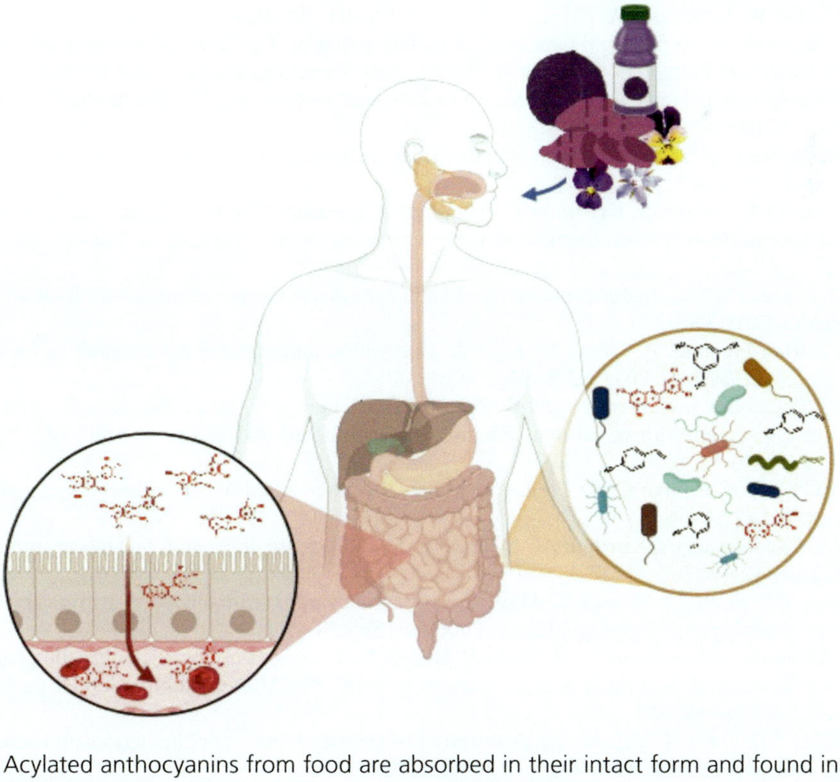

FIGURE 8.1 Acylated anthocyanins from food are absorbed in their intact form and found in blood. Unabsorbed anthocyanins reach the colon and are subject to colonic microbiota metabolization, giving rise to small metabolites and modulation microbiota composition. Created in https://BioRender.com.

microbiota-modulating effects, the microbial metabolites derived from anthocyanins remain underexplored. Low-molecular-weight metabolites produced predominantly through colonic microbiota activity are particularly relevant, as they may be bioavailable and access to target tissues.

8.7 CONCLUSION

While studies on acylated anthocyanins have been limited to a few sources, such as purple potatoes and red cabbage, expanding research to include alternative sources such as edible flowers could provide valuable insights. Preclinical models remain a crucial tool for unraveling the mechanistic pathways underlying the health effects of acylated anthocyanins and for evaluating their therapeutic potential. Despite their reported low bioavailability, the extensive health benefits associated with anthocyanin consumption suggest that their effects may be mediated by microbiota interactions and their metabolic byproducts. Future research should focus on elucidating these mechanisms and exploring their implications for human health.

REFERENCES

1. Lee MJ, Park JS, Choi DS, Jung MY. Characterization and quantitation of anthocyanins in purple-fleshed sweet potatoes cultivated in Korea by HPLC-DAD and HPLC-ESI-QTOF-MS/MS. *J Agric Food Chem.* 2013;61(12):3148–58.
2. Teixeira M, Tao W, Fernandes A, Faria A, Ferreira IMPLVO, He JR, et al. Anthocyanin-rich edible flowers, current understanding of a potential new trend in dietary patterns. *Trends Food Sci Tech.* 2023;138:708–25.
3. He J, Ye S, Correia P, Fernandes I, Zhang R, Wu M, et al. Dietary polyglycosylated anthocyanins, the smart option? A comprehensive review on their health benefits and technological applications. *Compr Rev Food Sci Food Saf.* 2022;21(4):3096–128.
4. Giusti MM, Wrolstad RE. Acylated anthocyanins from edible sources and their applications in food systems. *Biochem Eng J.* 2003;14(3):217–25.
5. Lakshmikanthan M, Muthu S, Krishnan K, Altemimi AB, Haider NN, Govindan L, et al. A comprehensive review on anthocyanin-rich foods: Insights into extraction, medicinal potential, and sustainable applications. *J Agr Food Res.* 2024;17.
6. Dangles O, Fenger JA. The chemical reactivity of anthocyanins and its consequences in food science and nutrition. *Molecules.* 2018;23(8).
7. Lila MA, Burton-Freeman B, Grace M, Kalt W. Unraveling anthocyanin bioavailability for human health. *Annu Rev Food Sci Technol.* 2016;7:375–93.
8. Marques C, Fernandes I, Norberto S, Sá C, Teixeira D, de Freitas V, et al. Pharmacokinetics of blackberry anthocyanins consumed with or without ethanol: A randomized and crossover trial. *Mol Nutr Food Res.* 2016;60(11):2319–30.
9. Fernandes I, Faria A, Calhau C, de Freitas V, Mateus N. Bioavailability of anthocyanins and derivatives. *J Funct Foods.* 2014;7:54–66.
10. Novotny JA, Clevidence BA, Kurilich AC. Anthocyanin kinetics are dependent on anthocyanin structure. *Brit J Nutr.* 2012;107(4):504–9.
11. Yi WG, Akoh CC, Fischer J, Krewer G. Absorption of anthocyanins from blueberry extracts by Caco-2 human intestinal cell monolayers. *J Agr Food Chem.* 2006;54(15):5651–8.
12. Baron G, Altomare A, Regazzoni L, Redaelli V, Grandi S, Riva A, et al. Pharmacokinetic profile of bilberry anthocyanins in rats and the role of glucose transporters: LC-MS/MS and computational studies. *J Pharmaceut Biomed.* 2017;144:112–21.
13. Mazza G, Kay CD, Cottrell T, Holub BJ. Absorption of anthocyanins from blueberries and serum antioxidant status in human subjects. *J Agr Food Chem.* 2002;50(26):7731–7.
14. Talavéra S, Felgines C, Texier O, Besson C, Manach C, Lamaison JL, et al. Anthocyanins are efficiently absorbed from the small intestine in rats. *J Nutr.* 2004;134(9):2275–9.

PART III

The Bioactivity of Acylated Anthocyanins

Antioxidant Properties of Acylated Anthocyanins

9

Gamze Güçlü, Ozlem Kilic-Buyukkurt, Serkan Selli, and Hasim Kelebek

9.1 INTRODUCTION

Anthocyanins are important water-soluble pigments responsible for the orange-red to blue-violet colors of fruits, vegetables, some cereals, and plants (Giusti and Wrolstad, 2003; Kelebek et al., 2006; Martín et al., 2017). Common sources of anthocyanins include vegetables and fruits like red cabbage, purple-fleshed sweet potato, eggplants, beets, elderberries, blueberries, black chokeberry, strawberries, and blackberries, as well as cereals like black beans, purple corn, black rice, and black lentils (Nieto, 2020; Tanguler et al., 2021). The term anthocyanin is derived from the Greek words *anthos*, meaning flower, and *kyanos*, meaning dark blue. Anthocyanins are compounds belonging to the flavonoid group, a subgroup of phenolic compounds (Bąkowska-Barczak, 2005). However, they differ from other flavonoids because the flavan ring in their structure forms oxygenated derivatives of a 2-phenylbenzopyrylium (flavylium) cation, or, in other words, due to their capacity to form flavylium cations (Miguel, 2011). Anthocyanins are also glycosylated products of anthocyanidins (Figure 9.1) having two benzoyl rings (A and B) in their structure separated by an oxygen-containing six-membered heterocyclic ring (C) (Zhao et al., 2017).

Anthocyanins are glycosidic compounds composed of both sugar and non-sugar components. The aglycone part, which includes non-sugar molecules, consists of phenolic compounds known as anthocyanidins. An anthocyanin molecule is generally formed by the attachment of an OH group to the C_3 position of the C-ring and sometimes the C_5, C_7 positions of the A-ring and $C_{3'}$, $C_{5'}$ positions of the B-ring in an anthocyanidin, resulting in the formation of a glycoside structure (Bąkowska-Barczak, 2005; Miguel, 2011). Anthocyanidins are most commonly found as *O*-glycosides (Fenger et al., 2021a), and they are typically composed of 3-monosides and 3,5-diglycosides, although sometimes 3,7-diglycosides or 3-triglycosides can also occur (Bąkowska-Barczak, 2005). Anthocyanidins are generally bonded to sugars such as glucose, rhamnose, xylose, galactose, arabinose, and rutinose through the hydroxyl group on the third carbon atom (C_3) (Liu et al., 2020). These sugars are responsible for the stability and solubility of aglycones (Nieto, 2020). Approximately 17 anthocyanidins are found in nature, and by binding sugars to them, anthocyanins of various colors can be formed. However, only six are of significant importance: pelargonidin, cyanidin, delphinidin, peonidin, petunidin, and malvidin, and by their modification, a wide variety of anthocyanins can be produced (Miguel, 2011; Zhao et al., 2017). Anthocyanins vary according

DOI: 10.1201/9781003519591-12

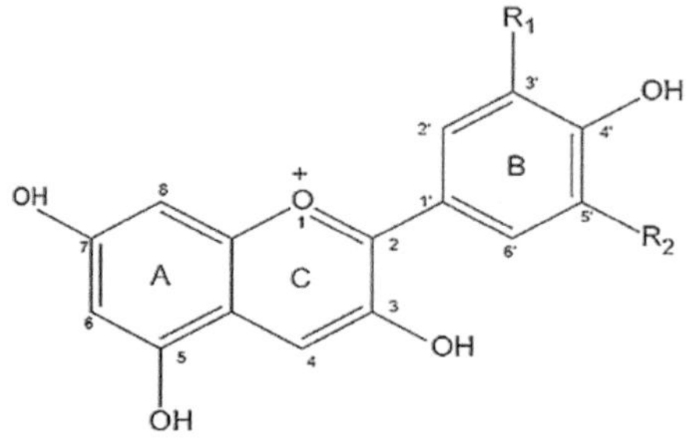

	R_1	R_2
Pelargonidin	H	H
Cyanidin	OH	H
Delphinidin	OH	OH
Peonidin	OCH_3	H
Petunidin	OCH_3	OH
Malvidin	OCH_3	OCH_3

FIGURE 9.1 Structure of anthocyanidin and chemical structures of the six common anthocyanidins.

to the number and position of hydroxyl $(-OH)$ and methoxyl $(-OCH_3)$ groups on the basic anthocyanidin skeleton.

Anthocyanins can be acylated by the attachment of phenolic acids, such as caffeic and sinapic acids, and aliphatic acids, such as acetic and malic acids, to their sugar moieties (Miguel, 2011). Acylation enhances the stability of anthocyanins compared to their nonacylated counterparts, with the degree of stability depending on the type, number, and binding site of the acyl groups (Jokioja et al., 2021). It improves anthocyanin stability through intermolecular and/or intramolecular co-pigmentation and self-association reactions. As a result, acylated anthocyanins exhibit greater color and pigment stability, and studies have demonstrated their potential as natural alternatives to synthetic dyes in various food products (Malien-Aubert et al., 2001; He et al., 2010; La and Giusti, 2022).

Anthocyanins are extremely reactive substances and thus susceptible to degradation reactions. Many factors influence anthocyanin stability, including the pH, the presence of oxygen, light exposure, temperature, and the environment (Xu et al., 2019; Sasmaz et al., 2024). Regarding this issue, it has been reported that in pomegranate (*Punica granatum* cv. Caner) juice obtained through different extraction methods (using halved pomegranate, arils, and macerated arils), maceration of the arils led to a decrease in total anthocyanin content, possibly due to factors such as light exposure, oxygen, temperature, and storage time (Budiene et al., 2021). Additionally, the type, position, and number of sugars attached to the aglycone structure also have an effect (Martín et al., 2017). As the number of hydroxyl groups in the structure increases, the blueness intensifies, while an increase in methoxy groups enhances the redness (He et al., 2010). The color of anthocyanin also changes as an indicator depending on the pH of the environment (Miguel, 2011). At low pHs, anthocyanins appear purple-red, and the color is more stable in acidic conditions (Jokioja et al., 2021), while at higher pHs, they shift to green-blue. In general, increasing the number of hydroxyl groups in a flavylium cation structure reduces its stability. However, the increase in the degree of methylation and the glycosylation of free hydroxyl groups positively affects the stability of anthocyanins (Liu et al., 2020).

Anthocyanins are potential health bioactives that exhibit various biological properties (Budiene et al., 2021; Selli et al., 2023). Studies on anthocyanins indicate that these compounds may reduce the risk of various diseases, including diabetes, cardiovascular disease, and cancer, due to their bioactive properties, such as antioxidant and anti-inflammatory effects (Zafra-Stone et al., 2007; Hu et al., 2016; Wu et al., 2018). However, for anthocyanin compounds to be bioavailable, they must be absorbed from the intestine, added to the circulation, transported, and distributed to the damaged area (Niki, 2010; Oliveira et al., 2020). Anthocyanins demonstrate antioxidant activity primarily because of the flavylium cation present in their structure. These compounds are recognized for their ability to prevent or inhibit oxidation and to scavenge free radicals due to these antioxidant properties. The antioxidant action of anthocyanins occurs through two main mechanisms: donating hydrogen atoms or facilitating electron transfer. Therefore, the antioxidant activity of anthocyanins depends on the number and position of the hydroxyl groups and the conjugated double bonds in their structures (Martín et al., 2017).

This chapter aims to provide a comprehensive understanding of anthocyanins by exploring their structure, natural sources, and the characteristics of acylated anthocyanins. Additionally, it delves into the antioxidant mechanisms associated with these compounds and examines the structure–activity relationships that influence their bioactivity. This exploration intends to offer a detailed perspective on the functional properties of acylated anthocyanins.

9.2 MECHANISMS OF ANTIOXIDANT ACTION

Oxidation is a chain reaction that occurs in the presence of oxygen and can occur naturally during the processing and storage of foods. By-products resulting from oxidation cause deterioration in the quality of food. These oxidation by-products are known to be potentially harmful to humans in terms of causing various inflammations and cancer formation (López-Pedrouso et al., 2020). On the one hand, the redox system of living organisms plays a crucial role in maintaining a healthy equilibrium for human life. Some free radicals, including nitric oxide (NO), superoxide (O^{2}) radical anion, and related reactive oxygen species (ROS) and/or reactive nitrogen species (RNS), are indispensable for the survival of living organisms as they facilitate signaling processes within cells; in certain instances, the equilibrium of this redox system may be disrupted, leading to the potential occurrence of oxidative stress (Miguel, 2011; López-Pedrouso et al., 2020). On the other hand, oxidative stress is also a harmful process that can seriously alter structures such as cell membranes and DNA. When generated in excessive quantities, these highly reactive radicals can set off a chain reaction that damages all cell components (López-Pedrouso et al., 2020).

Antioxidants may be described as compounds that can inhibit or delay the oxidation of oxidizable compounds *via* the scavenging of free radicals and the reduction of oxidative stress (Niki, 2010; Miguel, 2011; Budiene et al., 2021). Anthocyanins are phenolic compounds with strong antioxidant properties compared to synthetic phenolic antioxidants, such as butylated hydroxytoluene (BHT) and butylated hydroxyanisole (BHA). Their antioxidant activity is particularly crucial in facilitating oxidation of phenolic hydroxyl groups, where para- and ortho-phenolic positions play a key role in semiquinone formation and the stabilization of one-electron oxidation products. Although 1,4- and 1,2-conjugation systems contribute to radical flavonoid stabilization, other electronic frameworks also support this process. In anthocyanidins, substituents at positions 3, 5, 7, and 3′, 4′ on the C-, A-, and B-rings influence the formation of delocalized electronic and oxidized structures (Mattioli et al., 2020). In addition to providing color to foods, anthocyanins can reduce oxidation through free radical scavenging and prevent lipid peroxidation, thanks to their strong antioxidant properties (López-Pedrouso et al., 2020).

Antioxidants exhibit mechanisms of action at various levels. These mechanisms include reducing localized oxygen concentrations; scavenging ROS and RNS such as hydrogen peroxide and peroxynitrite to prevent chain initiation; binding metal ions, like iron ions ($Fe^{(3+)}$, Fe^{3+}), in a manner that prevents the formation of species such as hydroxyl radicals; decomposing lipid peroxides into peroxyl and alkoxyl radicals; breaking down peroxides into non-radical products, such as alcohols; and scavenging intermediate radicals, like peroxyl and alkoxyl, to terminate the chain reaction and inhibit further hydrogen abstraction (Miguel, 2011; López-Pedrouso et al., 2020).

9.2.1 Free Radical Scavenging

Acylated anthocyanins are widely recognized for their potent antioxidant properties, particularly their ability to scavenge free radicals (Lu et al., 2024). Their antioxidant mechanisms are primarily dictated by structural modifications resulting from acylation, which influences their stability, reactivity, and effectiveness in different environmental conditions. These compounds neutralize reactive species through several interrelated pathways, including hydrogen atom transfer (9.1), single electron transfer (9.2), and peroxyl radical trapping, each of which plays a significant role in mitigating oxidative stress (Lu et al., 2024; Ma et al., 2020; Matera et al., 2015). Antioxidants (AH) achieve this by donating hydrogen atoms or transferring electrons and protons, leading to the formation of a stable compound (RH) and an antioxidant-derived radical (A·) (Niki, 2010).

$$R\cdot + AH \longrightarrow RH + A \tag{9.1}$$

$$R\cdot + AH \longrightarrow R^{-} + AH\cdot \longrightarrow RH + A\cdot \tag{9.2}$$

One of the key mechanisms by which acylated anthocyanins exert their antioxidant effects is hydrogen atom transfer (HAT). In this process, anthocyanins donate hydrogen atoms to neutralize free radicals, forming stable radical species that do not propagate further oxidation. This mechanism is particularly effective in aqueous environments, where hydroxyl groups significantly enhance radical scavenging activity (Lu et al., 2024; Ma et al., 2020). The number and positioning of these hydroxyl groups directly influence the antioxidant potential of anthocyanins, with highly hydroxylated structures demonstrating superior reactivity toward free radicals.

Another crucial pathway is single electron transfer (SET), which involves the transfer of an electron from the anthocyanin molecule to the free radical, stabilizing it in the process (Vo et al., 2021; Ma et al., 2020). This electron donation is facilitated by the acidic moieties present in acylated anthocyanins, which enhance their ability to interact with oxidizing agents. The efficiency of the SET mechanism is particularly relevant in physiological environments, where anthocyanins function in a dynamic redox system and contribute to the maintenance of cellular homeostasis by neutralizing oxidative stressors.

In addition to these primary mechanisms, acylated anthocyanins exhibit strong peroxyl radical trapping capabilities, which are essential for protecting lipid-rich cellular structures from oxidative degradation (Rossetto et al., 2007). Peroxyl radicals, which are highly reactive intermediates in lipid peroxidation, are efficiently intercepted by anthocyanins, thereby preventing chain oxidation reactions that can lead to membrane damage. The effectiveness of this radical trapping depends on the nature of the acyl group attached to the anthocyanin structure, with sinapic acid derivatives exhibiting higher antioxidant efficacy compared to ferulic acid derivatives. The enhanced activity of sinapic acid-acylated anthocyanins is attributed to their unique electron-donating and steric properties, which increase their interaction with lipid-soluble radicals (Matera et al., 2015).

Beyond their direct radical-scavenging abilities, acylation further enhances the antioxidant potential of anthocyanins by improving their stability. Acylated anthocyanins demonstrate greater resistance to oxidative degradation and enhanced protection against environmental factors such as light and heat, which often compromise the activity of nonacylated anthocyanins (Teng et al., 2022). By increasing anthocyanins' overall structural rigidity and lipophilicity, acylation prolongs their functional integrity in both biological and food systems, sustaining their antioxidant efficacy over time.

The effectiveness of a compound's antiradical activity depends on two key factors: (1) its ability to interact with free radicals and (2) the stability of the resulting radicals, which prevents their involvement in continuous free radical reactions. In flavonoids, specific structural features known as the "Bors criteria" contribute to antioxidant efficiency. These include (1) a catechol moiety on the B-ring, which enhances radical stability through hydrogen bonding; (2) a double bond between the second and third carbon atoms along with a C4-oxo group on the C-ring, facilitating electron delocalization; and (3) hydroxyl groups at positions C3 and C5, working alongside the C4-oxo group to further stabilize electrons *via* hydrogen bonding (Bors et al., 1990; Platzer et al., 2021a). These criteria influence the response of different flavonoids in the 2,2-diphenyl-1-picrylhydrazyl (DPPH)• decolorization assay (Platzer et al., 2021b). However, their reactivity in various antioxidant tests does not always fully align with Bors' predictions (Platzer et al., 2021b, 2022). The application of Bors criteria to anthocyanins is somewhat limited due to the absence of a C4-oxo group, yet some predictions remain valid. The antioxidant capacity of anthocyanins is largely attributed to their aglycone forms. Upon interaction with oxidizing agents, anthocyanidins transform into more stable peroxyl radicals through resonance, reducing their reactivity compared to the free radicals they neutralize. This transformation effectively disrupts oxidative chain reactions.

9.2.2 Metal Ion Chelation

Acylated anthocyanins, naturally occurring pigments found in various plants, are known for their ability to bind metal ions, which can influence their color properties and stability. This chelation activity is of particular interest for applications in food coloring and nutraceuticals. Anthocyanins with *ortho*-dihydroxyl groups can scavenge hydroxyl radicals by inhibiting HO* generation *via* iron chelation (Miguel, 2011). In particular, it has been stated that anthocyanins with 3′, 4′-dihydroxy groups rapidly chelate metal ions, forming stable anthocyanin–metal complexes (Bąkowska-Barczak, 2005). Acylated anthocyanins, such as those from red cabbage and purple sweet potato, can bind metal ions like $Fe^{(3+)}$ and $Al^{(3+)}$, resulting in significant bathochromic shifts (color changes toward longer wavelengths) and hyperchromic effects (increased color intensity) (Fenger et al., 2021a; Sigurdson et al., 2016, 2017a). These shifts are influenced by the type of metal ion and the pH of the environment, with trivalent ions like $Fe^{(3+)}$ and $Al^{(3+)}$ causing more pronounced effects compared to divalent ions like $Mg(2+)$ (Sigurdson et al., 2016; Sigurdson and Giusti, 2014).

The presence and type of acyl groups significantly affect the chelation and color properties of anthocyanins. Diacylation, particularly with hydroxycinnamic acids, enhances the stability and color expression of metal–anthocyanin complexes (Sigurdson et al., 2017a, 2017b; Fenger et al., 2021b). Acylation by non-coordinating residues can also favor metal binding, leading to complexes with higher molar absorption coefficients (Fenger et al., 2021a).

Acylation improves the stability of anthocyanins by protecting the chromophore from water addition and thermal degradation. This is particularly effective with phenolic acyl groups, which enhance the stability of the pigment–metal complexes (Fenger et al., 2021b; Zhang et al., 2022). The stability is further increased with higher proportions of metal ions and specific acylation patterns (Sigurdson et al., 2017a).

Acylated anthocyanins exhibit enhanced antioxidant activity, which is beneficial for their use in functional foods. The chelation of metal ions not only stabilizes the anthocyanins but also enhances their antioxidant capacity, as seen with succinic acid acylation of mulberry anthocyanins. Zhang et al. (2022) aimed to investigate how succinic acid acylation affects the stability and antioxidant activity of anthocyanins. By modifying anthocyanins with succinic acid, the study assessed changes in Fe^{2+}-chelating ability, reducing power, and radical scavenging activity, demonstrating that acylation enhances structural stability and antioxidant properties. It was reported that succinic acid acylation introduces $C-O-C$ bonds and additional hydroxyl groups into anthocyanins. These structural changes increase the electron-donating capacity, which plays a key role in free radical scavenging. It was also emphasized that the introduction of acyl groups likely increases interaction sites for Fe^{2+} ions, making the molecules more effective in chelating metal ions. Acylation enhances hydrophobic interactions and π–π stacking, leading to improved structural stability. This stability helps preserve anthocyanins' ability to neutralize ROS over time. Acylated anthocyanins exhibit a stronger ability to donate electrons, which improves their overall reducing power and enhances antioxidant activity. Additionally, acylation significantly improved thermostability and light stability, making anthocyanins more resistant to degradation. The Fe^{2+} -chelating ability and reducing power of acylated anthocyanins were significantly higher than those of nonacylated forms.

The ability of acylated anthocyanins to produce a range of colors from red to blue, depending on pH and metal ion presence, makes them promising candidates as natural food colorants. Their enhanced stability and color intensity under various conditions are advantageous for food industry applications (Fenger et al., 2021a; Sigurdson and Giusti, 2014).

9.2.3 Inhibition of Lipid Peroxidation

Lipid peroxidation is a major contributor to oxidative deterioration in biological systems and lipid-containing food products. It is a chain reaction initiated by ROS that leads to the formation of lipid hydroperoxides, which further degrade into harmful secondary oxidation products, such as malondialdehyde (MDA). The inhibition of lipid peroxidation is of paramount importance in preventing oxidative stress-related disorders and preserving food quality (Brown and Kelly, 2007).

Lipid peroxidation can lead to damage in cell membranes, causing alterations in their characteristics, such as the surface potential and permeability. As the by-products of lipid peroxidation accumulate in tissues, they can contribute to the development of pathological conditions, potentially disrupting normal cellular functions and leading to disease (López-Pedrouso et al., 2020).

Acylated anthocyanins have emerged as promising natural antioxidants capable of mitigating lipid oxidation due to their enhanced lipid solubility and effective radical scavenging properties. The inhibition of peroxidation is enhanced by the acylation process, which increases the lipophilicity and stability of anthocyanins. The acylation process involves attaching fatty acids like lauric or cinnamic acids to anthocyanins, which increases their effectiveness in lipid-rich environments (Yang et al., 2019; Cai et al., 2020).

Acylated anthocyanins exhibit strong antioxidant activities, which contribute to their ability to inhibit lipid peroxidation. This is due to their enhanced capacity to scavenge free radicals and reduce oxidative stress (Tsuda et al., 1996; Narayan et al., 1999; Seeram and Nair, 2002). Acylated anthocyanins inhibit lipid oxidation through multiple mechanisms, primarily by improving their partitioning into lipid-rich environments, scavenging free radicals, and enhancing endogenous antioxidant defense systems.

Acylated anthocyanins have been found to be more effective than their nonacylated counterparts in preventing lipid peroxidation. This is attributed to their improved solubility in lipids and enhanced antioxidant properties (Matera et al., 2015; Yang et al., 2019; Cai et al., 2020). Native anthocyanins are highly hydrophilic, which limits their functionality in lipid-rich matrices. The lipid solubility of anthocyanins is significantly increased through acylation with fatty acids such as valeric acid, decanoic acid, myristic acid, and lauric acid. This modification alters the octanol–water partition coefficient (Log P), leading to better affinity for lipid phases where oxidative processes occur. Enhanced lipid solubility allows acylated anthocyanins to be more effective in preventing lipid peroxidation in both food systems and biological membranes.

Acylated anthocyanins maintain their ability to neutralize lipid radicals, although some loss in DPPH radical scavenging activity has been observed due to the reduction of free hydroxyl groups available for hydrogen donation. Despite this, acylated anthocyanins exhibit superior antioxidant efficacy in lipid systems compared to their native counterparts. This is evidenced by their higher inhibition ratios in the β-carotene bleaching assay, which indicates a stronger capacity to prevent lipid oxidation by interrupting free radical propagation (Wang et al., 2022).

Experimental studies using *Caenorhabditis elegans* have demonstrated that acylated anthocyanins effectively reduce MDA levels, a key marker of lipid peroxidation. The inhibition of MDA formation suggests that acylated anthocyanins suppress lipid peroxidation more efficiently than native anthocyanins. Additionally, these modified anthocyanins enhance the activity of superoxide dismutase (SOD), a crucial enzyme in antioxidant defense that converts superoxide radicals into less reactive species. By upregulating endogenous antioxidant enzymes, acylated anthocyanins contribute to the overall reduction of oxidative stress (Wang et al., 2022).

The location and nature of acylation play a significant role in determining the antioxidant efficacy of acylated anthocyanins (Luo et al., 2022). Acylation typically occurs at the hydroxyl groups of glycosyl moieties in anthocyanins, such as the 6″–OH position of glucosides or the 4‴–OH of rutinosides (Yang et al., 2019). Density functional theory (DFT) calculations suggest that shorter fatty acid chains exhibit higher reactivity for acylation, which influences the interaction of anthocyanins with lipid oxidation pathways. While longer fatty acid chains, such as myristic acid, further enhance lipophilicity, they may slightly reduce the direct radical scavenging ability due to steric hindrance effects (Wang et al., 2022).

Acylated anthocyanins serve as potent inhibitors of lipid peroxidation through enhanced lipid solubility, free radical scavenging, and the upregulation of endogenous antioxidant defenses. Their structural modifications contribute to increased affinity for lipid environments, making them highly effective in preventing oxidative damage in both food and biological systems. Given their improved stability and bioavailability, acylated anthocyanins offer a promising natural alternative for controlling lipid oxidation and mitigating oxidative stress-related disorders (Cai et al., 2020).

9.3 ACYLATED ANTHOCYANINS: STRUCTURE–ACTIVITY RELATIONSHIPS

Anthocyanin glycosyl acylation refers to the process by which the hydroxyl groups (-OH) on the sugar moieties of anthocyanins, or substituted glycosyls, are modified by organic acids through ester bonds (Kelebek et al., 2013; Zhao et al., 2017). Sugar residues may occur acylated with aromatic (phenolic) acids and/or aliphatic acids. The most prevalent phenolic acids consist of hydroxycinnamic acid derivatives, such as *p*-coumaric, caffeic, ferulic, sinapic, and 3,5-dihydroxycinnamic acids, as well as hydroxybenzoic acids like *p*-hydroxybenzoic and gallic acids, while aliphatic acids, including malonic (a frequent acyl group in acylated anthocyanins), acetic, malic, succinic, tartaric, oxalic, and erucic acids, are also commonly found (Zhao et al., 2017). These acyl substituents are usually attached to the C_3 sugar and esterified to the 6-OH group of the sugars or less frequently to the 4-OH group (Giusti and Wrolstad, 2003). Glycosyl acylation creates diverse anthocyanin molecules through variations in the types, numbers, and linkage positions of acyl groups (Sasaki et al., 2014; Zhao et al., 2017). This acylation can also result in either *cis* or *trans* isomers. According to studies, *cis*-acylated pigments have higher color durability and stability compared to their *trans*-acylated counterparts (Sigurdson et al., 2018; La and Giusti, 2022). Some anthocyanins can be acylated by both aromatic and aliphatic acids, and an anthocyanin can contain more than one kind of acyl groups simultaneously (Andersen and Jordheim, 2006). An acylated anthocyanin is usually monoacylated, meaning that it possesses only one acyl. However, the number of acyl groups in various acylated anthocyanins could be different. Acylated anthocyanins are classified as mono-, di-, tri-, and polyacylated anthocyanins based on the number of acyl groups present in the anthocyanin (Vidana Gamage et al., 2022). At times, multiple organic acids can independently acylate different hydroxyl groups on the glycosyl units, or they may sequentially attach to the same hydroxyl group on a single glycosyl, resulting in the formation of diacylated or polyacylated anthocyanins (Zhao et al., 2017).

Although anthocyanins possess remarkable antioxidant activity, they are relatively unstable pigments and can be easily affected by environmental factors such as pH, light, and temperature. Consequently, their application in the food industry is limited (Kong et al., 2023). Studies have shown that acylated anthocyanins, formed through anthocyanin glycosyl groups' acylation, exhibit enhanced stability (Sadilova et al., 2006; Fenger et al., 2019). Acylation generally enhances the antioxidant capacity of anthocyanins, with acylated anthocyanins demonstrating significantly higher antioxidant activity than their nonacylated counterparts (Hu et al., 2014; Vidana Gamage et al., 2022). A related study reported that enzymatic acylation of blueberry anthocyanins with *p*-coumaric acid and caffeic acid increased DPPH radical scavenging activity by 6.56% and 15.21%, respectively (Liu et al., 2020). In another study in which anthocyanins isolated from black rice (*Oryza sativa* L.) were acylated by enzyme catalysis using caffeic acid, it was reported that the antioxidant capacity (measured using ferric reducing antioxidant power (FRAP) and DPPH assays) of acylated anthocyanins increased significantly (Kong et al., 2023). The antioxidant activity of acylated anthocyanins is influenced by the characteristics, quantity, attachment site, and specific spatial location of intramolecular acyl units (Matsufuji et al., 2007). In a study examining the DPPH radical scavenging activities of acylated anthocyanins isolated from red radish, it was determined that all acylated anthocyanins showed significantly stronger antioxidant activity compared to the nonacylated ones. It has been stated that the activity of acylated anthocyanins mostly depends on the activity of intramolecular acyl units (caffeic acid > ferulic acid > *p*-coumaric acid), but in anthocyanins with common acyl units, the antioxidant activity is greatly affected by the binding site of intramolecular acyl units (Matsufuji et al., 2007). Research indicates that the antioxidant activity of anthocyanins primarily stems from the hydroxyl $(-OH)$ groups present in the benzene ring. However, a study investigating the antioxidant activity of blueberry anthocyanins acylated with varying levels of maleic acid reported a decrease in antioxidant activity with increasing acylation. This reduction was attributed to the interaction of maleic anhydride with the hydroxyl group $(-OH)$ in the benzene ring of anthocyanidin during the acylation process, which led to the formation of carboxylic acid $(-COOH)$. As a result, the decrease in phenolic hydroxyl groups corresponded with a decline in the antioxidant activity of the acylated anthocyanins (Fei et al., 2021).

9.3.1 Mechanisms of Anthocyanin Acylation

Acylated anthocyanins are naturally found in the roots, stems, and fruit parts of many fruits and vegetables (Luo et al., 2022). Radishes, black carrots, red cabbage, and red and purple sweet potatoes are among the food products containing acylated anthocyanins with a more stable structure (Giusti and Wrolstad, 2003). In nature, acylated anthocyanins are typically synthesized through glycosylation and acylation modifications of three common anthocyanins, pelargonidin, delphinidin, and cyanidin, with glycosyltransferase and acyltransferase enzymes playing key roles (He et al., 2010). Besides their natural occurrence, acylated anthocyanins can also be synthesized *in vitro* through chemical and enzymatic acylation. In chemical acylation, ester bonds form in anthocyanins due to the presence of multiple $-OH$ groups on both anthocyanidin and glycosyl groups. Similar to other chemical reactions, factors such as the acylation site, number and type of acyl groups, and catalyst type significantly influence the acylation process, thus affecting anthocyanin stability (Luo et al., 2022). In a previous study on this subject, lauric acid was used in the acylation of cyanidin-3-glucoside, which is abundant in blueberries, and it was reported that the stability of acylated cyanidin-3-glucoside was higher than that of nonacylated cyanidin-3-glucoside due to the higher stability of the ester group relative to the hydroxyl group (Zhao et al., 2015).

Enzymatic acylation is a process in which hydroxyl groups are esterified by organic acids, similar to chemical acylation, but it employs enzymes. Various enzymes, such as acyl transferases, proteases, and lipases, are used for enzyme acylation (Chebil et al., 2006). Typically, immobilized enzymes are used in enzymatic acylation as they improve enzyme stability and facilitate substrate access to catalytic sites, thus accelerating the reaction. However, factors such as the origin and concentration of enzymes and the structure of the substrates also play a significant role in influencing the acylation process (Luo et al., 2022). In a study conducted on this subject, enzymatic acylation of blackcurrant anthocyanins was investigated, and the thermal stability of the acylated products was compared with that of the original anthocyanins.

It was reported that the acylated anthocyanins had higher half-lives and activation energies, and also an improvement in thermal stability (Yang et al., 2019).

Finally, anthocyanins that are acylated *in vitro*, either chemically or enzymatically, demonstrate higher stability than their nonacylated counterparts. Compared to chemical acylation, enzymatic acylation provides greater regioselectivity and yield, enabling the synthesis of acylated anthocyanins with improved stability and oil solubility under controlled conditions (Luo et al., 2022). Enzymatic acylation provides the advantage of producing acylated anthocyanins with high oil solubility. While the water solubility of anthocyanins limits their application in lipophilic food matrices, those obtained through enzymatic acylation can be readily applied to such matrices (Yang et al., 2019; Luo et al., 2022).

The stability enhancement of anthocyanins *via* glycosyl acylation is largely due to the spatial configuration of acylated anthocyanins, which proves particularly effective in mildly acidic to neutral aqueous environments (Zhao et al., 2017). The primary mechanism underlying this increased stability involves stacking interactions, particularly from aromatic acyl groups, with the pyrylium ring of anthocyanidin flavylium cations. Both intra- and intermolecular co-pigmentation play essential roles in stabilizing these anthocyanins, making them less susceptible to degradation (Nakajima et al., 2003; Andersen and Jordheim, 2006). This intramolecular co-pigmentation occurs within specific anthocyanin compounds, with the anthocyanidins acting as chromophores and the acyl groups as electron-donor co-pigments (Castañeda-Ovando et al., 2009; Zhao et al., 2017). In other words, strong intramolecular π-complexes are formed between the acyl group and the anthocyanidin (Andersen and Jordheim, 2006; Fenger et al., 2021a). In general, aromatic acyl groups demonstrate a much higher ability to stabilize anthocyanins than aliphatic ones (Giusti and Wrolstad, 2003), and anthocyanins with two or more aromatic acyl groups have been known to be more stable than monoacylated ones (Malien-Aubert et al., 2001; Jokioja et al., 2021). In a related study, it was reported that diacylated anthocyanins in red radish were more stable to light at pH 3 than monoacylated anthocyanins (Matsufuji et al., 2007). Polyacylation by phenolic acids significantly enhances the color stability of anthocyanins through π-stacking interactions between the anthocyanidin chromophore and the acyl residues. This results in intramolecular co-pigmentation and/or the formation of aggregates (self-assembly), making the anthocyanidin less susceptible to degradation by various bleaching agents, such as hydrogen peroxide, water, and bisulfite (Fenger et al., 2021b).

9.3.2 Effects of Acylation on Stability

The acylation of anthocyanins involves the addition of one or more organic acids, such as sinapic, caffeic, or malic acid, to the anthocyanin molecule. Typically, the acylation of anthocyanin glycosides occurs at the C3 or C5 position of the anthocyanidins, which is necessary for facilitating intramolecular co-pigmentation (Zhao et al., 2017). This process significantly impacts the color stability, solubility, and biological activity of anthocyanins. Several factors, including temperature, pH, and light exposure, influence anthocyanin stability (Li et al., 2013; Vidana Gamage et al., 2022). Acylated anthocyanins are known to form more stable colors and exhibit greater resistance to pH fluctuations compared to their nonacylated counterparts (Keskin et al., 2021). This enhanced stability is attributed to the interaction between the acyl groups and the pyrylium ring of the flavylium cation through stacking interactions (Bąkowska-Barczak, 2005). Additionally, sugar chains can facilitate the stacking of acyl groups on the surface of the 2-phenyl benzopyran skeleton, resembling a strip formation. This arrangement can help prevent color loss in anthocyanins due to changes in hydration balance, thus preserving their color stability (Liu et al., 2020). A study on this subject reported that the highest acylated anthocyanin (malvidin-3-glucoside acetate and malvidin-3-glucoside-*p*-coumarate) contents were observed on the sixth day in different types of red wine subjected to various maceration periods (3, 6, and 10 days) and that the color intensity of the wines increased in parallel from day 3 to day 6 (Kelebek et al., 2006). Furthermore, the stability of acylated anthocyanins tends to increase with the number of acyl groups present; however, it remains influenced by the characteristics, quantity, and binding sites of these acyl units (Vidana Gamage et al., 2022). A number of investigations have found that anthocyanin glycosyl acylation influences the chemical stability of acylated anthocyanins both *in vitro* and *in vivo*, resulting in improved stability during processing and storage compared to other natural pigments (Bąkowska-Barczak, 2005; Sadilova et al., 2006; Fenger et al., 2019).

Acylated anthocyanins exhibit improved stability against color fading with increasing pH levels compared to their nonacylated counterparts. Nonacylated anthocyanins remain stable only within the pH range where the flavylium cation is predominant (Bąkowska-Barczak, 2005). In aqueous solutions, anthocyanins exist in a dynamic equilibrium among four structural forms: the flavylium cation, carbinol pseudobase, quinoidal base, and chalcone. In the case of acylated anthocyanins, the red-colored flavylium cation predominates in very strong acidic conditions (pH 1), representing the most stable form (Oliveira et al., 2020). Between pH values of 2 and 4, the uncharged blue quinonoid species are more prevalent; however, as the pH increases, the ionization of hydroxyl groups leads to the formation of anionic blue quinonoid species, which are also unstable. Nonacylated anthocyanins demonstrate instability at pH 5 and 6, where they rapidly undergo decolorization due to hydration at the 2-position of the anthocyanidin skeleton, resulting in the formation of carbinol pseudobase and chalcone structures (*cis* and *trans*) (Bąkowska-Barczak, 2005; Jokioja et al., 2021). As highlighted, anthocyanins exhibit a color spectrum ranging from red to blue-purple across different pH levels. This color characteristic makes anthocyanins valuable as natural food colorants, providing a preferable alternative to synthetic dyes in food applications (Fenger et al., 2021b).

The temperatures applied in thermal processes such as pasteurization, sterilization, and blanching significantly affect the stability of anthocyanins, and even temperatures exceeding 50°C can lead to the degradation of all or part of the anthocyanins (Li et al., 2013; Vidana Gamage et al., 2022). As with pH, acylated anthocyanins are more stable to heat treatment than their nonacylated counterparts. In a previous study, it was reported that the thermal stability of red cabbage extract, which contains high amounts of diacylated anthocyanins acylated with *p*-coumaric, sinapic, and ferulic acids, was higher at 80°C and pH 3 compared to nonacylated anthocyanin sources such as grape skin, black currants, and elderberries (Dyrby et al., 2001). The thermal stability of acylated anthocyanins can be attributed to intramolecular co-pigmentation. The hydrophobic interactions between the anthocyanin backbone and acyl groups, along with the overlapping of delocalized π-electron clouds, help shield anthocyanin molecules from nucleophilic water molecules, thereby preventing the formation of the colorless carbinol (Sari et al., 2015). In a study conducted on this subject, nonacylated anthocyanins obtained from purple corn husk, trueno fruit, and bottlebrush flowers were acylated by enzymatic esterification to form five new acylated compounds, and it was reported that these new acylated anthocyanins exhibited better thermal stability at 85°C (Fernandez-Aulis et al., 2020). Another study stated that acylated anthocyanins from red cabbage were more stable to heating at 80°C than the nonacylated anthocyanins from black currants, red grapes, and elderberries (Dyrby et al., 2001). The thermal stability of acylated anthocyanins is influenced by other components in the environment, such as ascorbic acid. In a study on this topic, it was reported that the stability of anthocyanins in heated sweet potatoes decreased with the addition of ascorbic acid. This reduction in stability is attributed to ascorbic acid's ability to decrease anthocyanin stability through the formation of hydrogen peroxide during the condensation reaction and the oxidation of ascorbic acid (Xu et al., 2019). However, although acylated anthocyanins generally exhibit higher thermal stability, this stability is influenced by the pH of the medium and may vary depending on the anthocyanin source. Some anthocyanins demonstrate higher thermal stability at low pH, while others show decreased stability under similar conditions. A previous study found that anthocyanins extracted from purple sweet potato were more resistant to thermal stress during heating at 80–100°C when added to apple juice (pH 3.5) and pear juice (pH 4.2) but exhibited instability in lemon juice (pH 2.2) (Li et al., 2013).

Light has a two-fold impact on the stability of anthocyanins. It is crucial for their synthesis in plants but can also trigger degradation processes when these compounds are present *in vitro*, within biological tissues, or in different food products. This duality emphasizes the necessity of managing light exposure carefully during the storage and handling of foods and biological samples to maintain the integrity and functionality of bioactive compounds (Oliveira et al., 2020).

Anthocyanin glycosyl acylation exerts physical effects on the properties of acylated anthocyanins. Generally, organic acids that acylate the hydroxyl groups (–OH) of anthocyanin glycosides possess relatively low polarity, resulting in a decrease in the overall polarity of the anthocyanins. This modification not only alters their structure but also increases their molecular size. Additionally, various acyl groups have carbon chains of differing lengths and volumes; as the length of the acyl group's carbon chain increases, the water solubility of the anthocyanins tends to decrease (Zhao et al., 2017).

Acylation of anthocyanins is a widely researched topic in the food industry in terms of preserving and enhancing the colors of anthocyanins. The specific type and extent of acylation can significantly influence the physicochemical properties of anthocyanins, increasing their potential as functional food ingredients. Acylated anthocyanins are becoming more popular in the food sector due to their greater stability compared to their nonacylated counterparts, which contributes to a longer shelf life and improved color retention during processing and storage. Their enhanced fat solubility, especially in those produced through enzymatic acylation, makes them particularly suitable for incorporation into lipophilic food matrices, where conventional water-soluble anthocyanins may be less effective. This allows food manufacturers to use acylated anthocyanins not only as natural colorants but also to enhance the nutritional value of a wide range of products.

9.4 CONCLUSION

Anthocyanins are water-soluble pigments classified under flavonoids, a subgroup within the phenolic family. They are responsible for the red, purple, and blue colors observed in various fruits, vegetables, and flowers, and their appealing hues have made them valuable in the food industry. These pigments possess notable antioxidant properties, primarily attributed to the flavylium cation in their structure. They act as effective scavengers of free radicals, preventing oxidative stress and offering potential health benefits. Some anthocyanins are bonded to aromatic or aliphatic organic acids through ester bonds, a process known as acylation of anthocyanins. In the literature, studies have determined that acylated anthocyanins are more stable against various environmental factors such as light, temperature, and pH. Additionally, the anthocyanins' stability, polarity, spatial structure, and bioactive qualities are also influenced by variables like the acylation site and the number and type of acyl groups. Acylation of anthocyanin glycosyl moieties modifies their chemical properties and enhances stability. Acylated anthocyanins have a more stable color due to efficient π-stacking interactions between acyl residues and the anthocyanidin nucleus. Their ability to impart vibrant colors can enhance the visual appeal of foods, thereby attracting consumers who prefer natural ingredients over synthetic additives. As the demand for clean-label products continues to rise, acylated anthocyanins present an innovative solution for food formulation, offering both aesthetic and health benefits while supporting the growing trend toward natural food ingredients.

REFERENCES

Andersen, Ø. M., & Jordheim, M. (2006). The anthocyanins. In Andersen, Ø. M., & Markham, K. R. (Eds.), *Flavonoids: Chemistry, Biochemistry and Applications*, pp. 471–553. Boca Raton: CRC Press.

Bąkowska-Barczak, A. (2005). Acylated anthocyanins as stable, natural food colorants – a review. *Polish Journal of Food and Nutrition Sciences*, 14–55(2), 107–116.

Bors, W., Heller, W., Michel, C., & Saran, M. (1990). Flavonoids as antioxidants: Determination of radical-scavenging efficiencies. In Packer, L., & Glazer, A. N. (Eds.), *Methods in Enzymology*, pp. 343–355. New York, NY, USA: Academic Press.

Brown, J. E., & Kelly, M. F. (2007). Inhibition of lipid peroxidation by anthocyanins, anthocyanidins and their phenolic degradation products. *European Journal of Lipid Science and Technology*, 109(1), 66–71.

Budiene, J., Guclu, G., Oussou, K. F., Kelebek, H., & Selli, S. (2021). Elucidation of volatiles, anthocyanins, antioxidant and sensory properties of cv. Caner pomegranate (*Punica granatum* L.) juices produced from three juice extraction methods. *Foods*, 10(7), 1497.

Cai, J., Zeng, F., Zheng, S., Huang, X., Zhang, J., Zhang, P., & Fei, P. (2020). Preparation of lipid-soluble bilberry anthocyanins through acylation with cinnamic acids and their antioxidation activities. *Journal of Agricultural and Food Chemistry*, 68(28), 7467–7473.

Castañeda-Ovando, A., Pacheco-Hernández, M., de Lourdes, P. H., Elena, M., Rodríguez, J. A., & Galán-Vidal, C. A. (2009). Chemical studies of anthocyanins: A review. *Food Chemistry*, 113(4), 859–871.

Chebil, L., Humeau, C., Falcimaigne, A., Engasser, J. M., & Ghoul, M. (2006). Enzymatic acylation of flavonoids. *Process Biochemistry*, 41(11), 2237–2251.

Dyrby, M., Westergaard, N., & Stapelfeldt, H. (2001). Light and heat sensitivity of red cabbage extract in soft drink model systems. *Food Chemistry*, 72(4), 431–437.

Fei, P., Zeng, F., Zheng, S., Chen, Q., Hu, Y., & Cai, J. (2021). Acylation of blueberry anthocyanins with maleic acid: Improvement of the stability and its application potential in intelligent color indicator packing materials. *Dyes and Pigments*, 184, 108852.

Fenger, J. A., Moloney, M. P., Robbins, R., Collins, T., & Dangles, O. (2019). The influence of acylation, metal binding and natural antioxidants on the thermal stability of red cabbage anthocyanins in neutral solution. *Food & Function*, 10, 6740.

Fenger, J. A., Roux, H., Robbins, R. J., Collins, T. M., & Dangles, O. (2021b). The influence of phenolic acyl groups on the color of purple sweet potato anthocyanins and their metal complexes. *Dyes and Pigments*, 185, 108792.

Fenger, J. A., Sigurdson, G. T., Robbins, R. J., Collins, T. M., Giusti, M. M., & Dangles, O. (2021a). Acylated anthocyanins from red cabbage and purple sweet potato can bind metal ions and produce stable blue colors. *International Journal of Molecular Sciences*, 22(9), 4551.

Fernandez-Aulis, F., Torres, A., Sanchez-Mendoza, E., Cruz, L., & Navarro-Ocana, A. (2020). New acylated cyanidin glycosides extracted from underutilized potential sources: Enzymatic synthesis, antioxidant activity and thermostability. *Food Chemistry*, 309, 125796.

Giusti, M. M., & Wrolstad, R. E. (2003). Acylated anthocyanins from edible sources and their applications in food systems. *Biochemical Engineering Journal*, 14(3), 217–225.

He, F., Mu, L., Yan, G.-L., Liang, N.-N., Pan, Q.-H., Wang, J., Reeves, M. J., & Duan, C.-Q. (2010). Biosynthesis of anthocyanins and their regulation in colored grapes. *Molecules*, 15(12), 9057–9091.

Hu, N., Zheng, J., Li, W. C., & Suo, Y. R. (2014). Isolation, stability, and antioxidant activity of anthocyanins from *Lycium ruthenicum* Murray and *Nitraria tangutorum* Bobr of Qinghai-Tibetan plateau. *Separation Science and Technology*, 49, 2897–2906.

Hu, Y., Deng, L., Chen, J., Zhou, S., Liu, S., Fu, Y., Yang, C., Liao, Z., & Chen, M. (2016). An analytical pipeline to compare and characterise the anthocyanin antioxidant activities of purple sweet potato cultivars. *Food Chemistry*, 194, 46–54.

Jokioja, J., Yang, B., & Linderborg, K. M. (2021). Acylated anthocyanins: A review on their bioavailability and effects on postprandial carbohydrate metabolism and inflammation. *Comprehensive Reviews in Food Science and Food Safety*, 20, 5570–5615.

Kelebek, H., Canbas, A., Selli, S., Saucier, C., Jourdes, M., & Glories, Y. (2006). Influence of different maceration times on the anthocyanin composition of wines made from *Vitis vinifera* L. cvs. Boğazkere and Öküzgözü. *Journal of Food Engineering*, 77(4), 1012–1017.

Kelebek, H., Jourdes, M., Selli, S., & Teissedre, L. (2013). Comparative evaluation of the phenolic content and antioxidant capacity of sun-dried raisins. *Journal of the Science of Food and Agriculture*, 93(12), 2963–2972.

Keskin, M., Guclu, G., Sekerli, Y. E., Soysal, Y., Selli, S., & Kelebek, H. (2021). Comparative assessment of volatile and phenolic profiles of fresh black carrot (*Daucus carota* L.) and powders prepared by three drying methods. *Scientia Horticulturae*, 287, 110256.

Kong, Y., Wang, X., Wu, Z., Li, Y., Xu, F., & Xie, F. (2023). Enzymatic acylation of black rice anthocyanins and evaluation of antioxidant capacity and stability of their derivatives. *Foods*, 12, 4505.

La, E. H., & Giusti, M. M. (2022). Ultraviolet-visible excitation of *cis*- and *trans*-*p*-coumaric acylated delphinidins and their resulting photochromic characteristics. *ACS Food Science and Technology*, 2(5), 878–887.

Li, J., Li, X. D., Zhang, Y., Zheng, Z. D., Qu, Z. Y., Liu, M., & Qu, L. (2013). Identification and thermal stability of purple-fleshed sweet potato anthocyanins in aqueous solutions with various pH values and fruit juices. *Food Chemistry*, 136(3–4), 1429–1434.

Liu, J. N., Zhuang, Y. H., Hu, Y. H., Xue, S., Li, H., Chen, L., & Fei, P. (2020). Improving the color stability and antioxidation activity of blueberry anthocyanins by enzymatic acylation with *p*-coumaric acid and caffeic acid. *LWT Food Science and Technology*, 130, 109673.

López-Pedrouso, M., Kovačević, D. B., Oliveira, D., Putnik, P., Moure, A., Lorenzo, J. M., Domínguez, H., & Franco, D. (2020). Chapter 7: In vitro and in vivo antioxidant activity of anthocyanins. In Lorenzo, J. M., Barba, F. J., & Munekata, P. E. S. (Eds.), *Anthocyanins*. Nova Science Publishers, Inc. ISBN: 978-1-53617-816-6.

Lu, X. Q., Li, J., Wang, B., & Qin, S. (2024). Computational insights into the radical scavenging activity and xanthine oxidase inhibition of the five anthocyanins derived from grape skin. *Antioxidants*, 13(9), 1117.

Luo, X., Wang, R., Wang, J., Li, Y., Luo, H., Chen, S., Zeng, X., & Han, Z. (2022). Acylation of anthocyanins and their applications in the food industry: Mechanisms and recent research advances. *Foods*, 11, 2166.

Ma, Y., Feng, Y., Diao, T., Zeng, W., & Zuo, Y. (2020). Experimental and theoretical study on antioxidant activity of the four anthocyanins. *Journal of Molecular Structure*, 1204, 127509.

Malien-Aubert, C., Dangles, O., & Amiot, M. J. (2001). Color stability of commercial anthocyanin-based extracts in relation to the phenolic composition. Protective effects by intra- and intermolecular copigmentation. *Journal of Food and Agricultural Chemistry*, 49(1), 170–176.

Martín, J., Kuskoski, E. M., Navas, M. J., & Asuero, A. G. (2017). Chapter 11: Antioxidant capacity of anthocyanin pigments. In Justino, G. (Ed.), *Flavonoids-from Biosynthesis to Human Health*, pp. 205–256. Croatia: National and University Library in Zagreb.

Matera, R., Gabbanini, S., Berretti, S., Amorati, R., De Nicola, G. R., Iori, R., & Valgimigli, L. (2015). Acylated anthocyanins from sprouts of *Raphanus sativus* cv. Sango: Isolation, structure elucidation and antioxidant activity. *Food Chemistry*, 166, 397–406.

Matsufuji, H., Kido, H., Misawa, H., Yaguchi, J., Otsuki, T., Chino, M., Takeda, M., & Yamagata, K. (2007). Stability to light, heat, and hydrogen peroxide at different pH values and DPPH radical scavenging activity of acylated anthocyanins from red radish extract. *Journal of Agricultural and Food Chemistry*, 55(9), 3692–3701.

Mattioli, R., Francioso, A., Mosca, L., & Silva, P. (2020). Anthocyanins: A comprehensive review of their chemical properties and health effects on cardiovascular and neurodegenerative diseases. *Molecules*, 25(17), 3809.

Miguel, M. G. (2011). Anthocyanins: Antioxidant and/or anti-inflammatory activities. *Journal of Applied Pharmaceutical Science*, 1(6), 7–15.

Nakajima, N., Sugimoto, M., Yokoi, H., Tsuji, H., & Ishihara, K. (2003). Comparison of acylated plant pigments: Light-resistance and radical-scavenging ability. *Bioscience Biotechnology and Biochemistry*, 67(8), 1828–1831.

Narayan, M. S., Naidu, K. A., Ravishankar, G. A., Srinivas, L., & Venkataraman, L. V. (1999). Antioxidant effect of anthocyanin on enzymatic and non-enzymatic lipid peroxidation. *Prostaglandins, Leukotrienes and Essential Fatty Acids (PLEFA)*, 60(1), 1–4.

Nieto, G. (2020). Chapter 1: Anthocyanins: Health benefits and potential use as functional ingredient: A review. In Lorenzo, J. M., Barba, F. J., & Munekata, P. E. S. (Eds.), *Anthocyanins*. Nova Science Publishers, Inc. ISBN: 978-1-53617-816-6.

Niki, E. (2010). Assessment of antioxidant capacity in vitro and in vivo. *Free Radical Biology and Medicine*, 49(4), 503–515.

Oliveira, H., Correia, P., Pereira, A. R., Araújo, P., Mateus, N., de Freitas, V., Oliveira, J., & Fernandes, I. (2020). Exploring the applications of the photoprotective properties of anthocyanins in biological systems. *International Journal of Molecular Sciences*, 21, 7464.

Platzer, M., Kiese, S., Herfellner, T., Schweiggert-Weisz, U., & Eisner, P. (2021b). How does the phenol structure influence the results of the Folin-Ciocalteu assay? *Antioxidants*, 10, 811.

Platzer, M., Kiese, S., Herfellner, T., Schweiggert-Weisz, U., Miesbauer, O., & Eisner, P. (2021a). Common trends and differences in antioxidant activity analysis of phenolic substances using single electron transfer based assays. *Molecules*, 26, 1244.

Platzer, M., Kiese, S., Tybussek, T., Herfellner, T., Schneider, F., Schweiggert-Weisz, U., & Eisner, P. (2022). Radical scavenging mechanisms of phenolic compounds: A quantitative structure–property relationship (QSPR) study. *Frontiers in Nutrition*, 9, 882458.

Rossetto, M., Vanzani, P., Lunelli, M., Scarpa, M., Mattivi, F., & Rigo, A. (2007). Peroxyl radical trapping activity of anthocyanins and generation of free radical intermediates. *Free Radical Research*, 41(7), 854–859.

Sadilova, E., Stintzing, F. C., & Carle, R. (2006). Thermal degradation of acylated and nonacylated anthocyanins. *Journal of Food Science*, 71(8), C504–C512.

Sari, P., Setiawan, A., & Siswoyo, T. A. (2015). Stability and antioxidant activity of acylated jambolan (*Syzygium cumini*) anthocyanins synthesized by lipase-catalyzed transesterification. *International Food Research Journal*, 22(2), 671–676.

Sasaki, N., Nishizaki, Y., Ozeki, Y., & Miyahara, T. (2014). The role of acyl-glucose in anthocyanin modifications. *Molecules*, 19(11), 18747–18766.

Sasmaz, H. K., Kilic-Buyukkurt, O., Selli, S., Bouaziz, M., & Kelebek, H. (2024). Antioxidant capacity, sugar content, and tandem HPLC–DAD–ESI/MS profiling of phenolic compounds from *Aronia melanocarpa* fruits and leaves (Nero and Viking cultivars). *ACS Omega*, 9(13), 14963–14976.

Seeram, N. P., & Nair, M. G. (2002). Inhibition of lipid peroxidation and structure–activity-related studies of the dietary constituents anthocyanins, anthocyanidins, and catechins. *Journal of Agricultural and Food Chemistry*, 50(19), 5308–5312.

Selli, S., Perestrelo, R., Kelebek, H., Sevindik, O., Travaglia, F., Coïsson, J. D., Câmara, J. S., & Bordiga, M. (2023). Impact of Japanese beetles (*Popillia japonica* Newman) on the chemical composition of two grape varieties (Nebbiolo and Erbaluce) grown in Italy. *Food Research International*, 165, 112575.

Sigurdson, G. T., & Giusti, M. M. (2014). Bathochromic and hyperchromic effects of aluminum salt complexation by anthocyanins from edible sources for blue color development. *Journal of Agricultural and Food Chemistry*, 62(29), 6955–6965.

Sigurdson, G. T., Robbins, R. J., Collins, T. M., & Giusti, M. M. (2016). Evaluating the role of metal ions in the bathochromic and hyperchromic responses of cyanidin derivatives in acidic and alkaline pH. *Food Chemistry*, 208, 26–34.

Sigurdson, G. T., Robbins, R. J., Collins, T. M., & Giusti, M. M. (2017a). Spectral and colorimetric characteristics of metal chelates of acylated cyanidin derivatives. *Food Chemistry*, 221, 1088–1095.

Sigurdson, G. T., Robbins, R. J., Collins, T. M., & Giusti, M. M. (2017b). Effects of hydroxycinnamic acids on blue color expression of cyanidin derivatives and their metal chelates. *Food Chemistry*, 234, 131–138.

Sigurdson, G. T., Tang, P., & Giusti, M. M. (2018). Cis–trans configuration of coumaric acid acylation affects the spectral and colorimetric properties of anthocyanins. *Molecules*, 23(3), 598.

Tanguler, H., Bozdogan, A., Aksay, S., Kelebek, H., & Erten, H. (2021). Comparison of anthocyanin profiles in Şalgams (Shalgams) produced with different production procedures. *Journal of Food Processing and Preservation*, 45(8), e14770.

Teng, H., Mi, Y., Cao, H., & Chen, L. (2022). Enzymatic acylation of raspberry anthocyanin: Evaluations on its stability and oxidative stress prevention. *Food Chemistry*, 372, 130766.

Tsuda, T., Shiga, K., Ohshima, K., Kawakishi, S., & Osawa, T. (1996). Inhibition of lipid peroxidation and the active oxygen radical scavenging effect of anthocyanin pigments isolated from *Phaseolus vulgaris* L. *Biochemical Pharmacology*, 52(7), 1033–1039.

Vidana Gamage, G. C., Lim, Y. Y., & Choo, W. S. (2022). Sources and relative stabilities of acylated and nonacylated anthocyanins in beverage systems. *Journal of Food Science and Technology*, 59(3), 831–845.

Vo, Q. V., Hoa, N. T., Thong, N. M., & Mechler, A. (2021). The hydroperoxyl and superoxide anion radical scavenging activity of anthocyanidins in physiological environments: Theoretical insights into mechanisms and kinetics. *Phytochemistry*, 192, 112968.

Wang, P., Liu, J., Zhuang, Y., & Fei, P. (2022). Acylating blueberry anthocyanins with fatty acids: Improvement of their lipid solubility and antioxidant activities. *Food Chemistry: X*, 15, 100420. https://doi.org/10.1016/j.fochx.2022.100420

Wu, H. Y., Yang, K. M., & Chiang, P. Y. (2018). Roselle anthocyanins: Antioxidant properties and stability to heat and pH. *Molecules*, 23(6), 1357.

Xu, X. J., Fang, S., Li, Y. H., Zhang, F., Shao, Z. P., Zeng, Y. T., & Meng, Y. C. (2019). Effects of low acyl and high acyl gellan gum on the thermal stability of purple sweet potato anthocyanins in the presence of ascorbic acid. *Food Hydrocolloids*, 86, 116–123.

Yang, W., Kortesniemi, M., Ma, X., Zheng, J., & Yang, B. (2019). Enzymatic acylation of blackcurrant (*Ribes nigrum*) anthocyanins and evaluation of lipophilic properties and antioxidant capacity of derivatives. *Food Chemistry*, 281, 189–196.

Zafra-Stone, S., Yasmin, T., Bagchi, M., Chatterjee, A., Vinson, J. A., & Bagchi, D. (2007). Berry anthocyanins as novel antioxidants in human health and disease prevention. *Molecular Nutrition & Food Research*, 51, 675–683.

Zhang, B., Jiang, X., Huang, G., Xin, X., Attaribo, T., Zhang, Y., . . . Gui, Z. (2022). Enhancement of stability and antioxidant activity of mulberry anthocyanins through succinic acid acylation. *Food Technology and Biotechnology*, 60(3), 321–329.

Zhao, C. L., Yu, Y. Q., Chen, Z. J., Wen, G. S., Wei, F. G., Zheng, Q., Wang, C. D., & Xiao, X. L. (2017). Stability-increasing effects of anthocyanin glycosyl acylation. *Food Chemistry*, 214, 119–128.

Zhao, L. Y., Chen, J., Wang, Z. Qiang, Shen, R. Meng, Cui, N., & Sun, A. Dong. (2015). Direct acylation of cyanidin-3-glucoside with lauric acid in blueberry and its stability analysis. *International Journal of Food Properties*, 19(1), 1–12.

Cancer and Neurodegeneration 10

Raffaella Colombo and Adele Papetti

10.1 INTRODUCTION

Anthocyanins (ACNs) belong to the flavonoid group and are plant pigments that provide a characteristic color (red, purple, and blue) in fruits (berries), vegetables (cabbages), and grains (blue/purple corn and black rice). Dietary consumption of ACNs is generally very high.

To date, over 600 ACNs have been identified. They have an aryl-substituted chroman ring with different sugars linked to the aglycon in the C3 and C5 positions. The most common natural compounds are cyanidin-3-*O*-glucoside and various anthocyanidin-3,5-*O*-diglucosides, particularly those of pelargonidin (Pg), cyanidin (Cy), delphinidin (Dp), peonidin (Pn), petunidin (Pt), and malvidin (Mv). Their structures provide ACNs the ability to donate electrons, thus justifying their main known bioactivity, i.e., the antioxidant activity. Their antiproliferative and neuroprotective activities are certainly strictly correlated to this main property and also to their anti-inflammatory capacity; in fact, they act against reactive oxygen species (ROS; free radicals, singlet oxygen, and hydrogen peroxides), which are responsible for cell damage and inflammation – factors that potentially contribute and lead to the development of cancer and neurodegenerative disorders (He & Giusti, 2010; Saini et al., 2024; de Aguiar Cipriano et al., 2022; Kowalczyk et al., 2024; Panchal & Brown, 2023).

Over the years, many *in vitro* and *in vivo* studies have confirmed ACNs' cell growth inhibition and antiproliferative effects on different cancer cell lines, mainly intestinal, prostate, and breast cancers. It is also well known that ACNs' anticarcinogenic activity could conceivably be the result of a combined effect with other molecules present in food, e.g., other phenolic compounds (He & Giusti, 2010; Urias-Lugo et al., 2015; Majhi, 2022).

ACNs have known central nervous system (CNS) potential due to their antioxidant and anti-inflammatory properties, with direct improvement in neuronal signaling, mainly at the level of signal transcription and transduction, as highlighted in many studies. ACNs are also potentially involved in the prevention of oxidative processes and neuroinflammation, which are responsible, e.g., for endothelial dysfunctions; cerebral ischemia; different neurological disorders, from depression and anxiety to cognitive disorders, such as Alzheimer's disease (AD) and Parkinson's disease (PD); prion diseases such as Creutzfeldt–Jakob disease (CJD); and amyloidosis (Banji et al., 2022; Panchal & Brown, 2023; Christoudia et al., 2024). Other potential neuroprotective mechanisms of ACNs involve different pathways such as regulation of mitochondrial functions, calcium homeostasis, autophagy, and protein aggregation (Henriques et al., 2020; Panchal & Brown, 2023). In addition, according to the gut–brain axis theory, the beneficial effects of ACNs on intestinal barrier inflammation and on microbiota modulation also have a positive impact on neuroinflammation, highlighting an effective connection between the gastrointestinal

DOI: 10.1201/9781003519591-13

tract and the CNS (Oteiza et al., 2023; Chen et al., 2023; Henriques et al., 2020; Panchal & Brown, 2023). All these mechanisms underline the wide range of health-promoting effects of ACNs related to numerous systemic and chronic diseases (He & Giusti, 2010; Panchal & Brown, 2023).

Whether acylation can positively affect ACNs' bioactivities still remains controversial, because it depends on the type of bioactivity considered, anthocyanin source, and acylated form. In fact, in the literature, a large number of studies refer to ACNs' bioactivities without discussing the characterization and composition of extracts containing ACNs. This makes it difficult to obtain specific information on acylated ACNs (AACNs) and to compare their bioactivities with the aglycones and/or glycosylated/acylated derivatives.

After a brief general discussion on the acylation process and its general effects on the hydrophobicity, stability features, and bioactivities of ACNs, in the main section two topics will be discussed: cancer and neurodegeneration. Anticarcinogenic activity will be discussed dividing the chapter into sections, with each section specifically dedicated to different carcinomas (colorectal, oral/esophageal, prostate, and breast) on which a large number of studies are present in the literature, and a last paragraph summarizing the few studies on other cancer types (hepatic, lung, and skin cancers). The section on neuroprotective activity will discuss ACNs' specific effect related to different neurological diseases: cognitive disorders such as AD and PD; depression and anxiety disorders; cerebral ischemia; and other neurological disorders (autism, schizophrenia, CJD, and prion diseases). For both topics, specific *in vitro* and *in vivo* studies will be critically discussed, elucidating the putative molecular mechanisms and pathways of ACNs involved in each activity, their potential healthy role, and their applications in preventing the risk of developing or mitigating cancer and neurodegenerative diseases.

10.2 THE ACYLATION MECHANISM AND ITS IMPACT ON THE PROPERTIES AND BIOACTIVITIES OF ACNS

ACNs are mainly mono- or diglycosylated at the C3 position (3-*O*-monoglycoside and 3-*O*-diglycosides) or differently glycosylated at C3 and C5 positions (3,5 di-*O*-glycosides and 3-*O*-diglycoside-5-*O*-monoglycoside). Glycosylation at C7, C3′, and C5′ positions is less frequent. The glycosidic moiety can be glucose, galactose, arabinose, rhamnose, xylose, rutinose, sambubiose, or other diglycosides. The type of aglycone or glycosylation and acylation sequence can influence ACN physical properties, such as stability and polarity; glycosylation increases hydrophilicity, while acylation enhances hydrophobicity. Bioactivities also are strictly correlated to ACNs' structure: e.g., the antioxidant activity seems to be positively linked to methoxylation, hydroxylation, and glycosylation of the B-ring structure (Giusti & Wrolstad, 2003; Jokioja et al., 2021; Asakura et al., 2024; Banji et al., 2022).

The different sugar hydroxyl groups in ACNs can be acylated with aliphatic (acetic, oxalic, succinic, and malonic) or aromatic acids (caffeic, *p*-coumaric, *p*-hydroxybenzoic, gallic, ferulic, and/or sinapic acids) through ester bonds, mainly in the 6″ position. Figure 10.1 shows ACNs' chemical structures and the common organic acids that can acylate the sugar moieties (the main acyl residues with antiproliferative and neuroprotective effects are highlighted). Acylation generally improves solubility, but only the presence of an aromatic acid residue is able to turn ACNs' color to blue and probably increases both thermostability and light resistivity (Majhi, 2022; Yan et al., 2016).

Many food matrices are rich in AACNs, mainly vegetables (red radishes, black carrots, red cabbages, red corn) and root tubers (red potatoes and purple sweet potatoes); in contrast, less amount is present in fruits (berries and grapes) (Giusti & Wrolstad, 2003; Jokioja et al., 2021; Delgado-Povedano et al., 2021). Acylation confers unique chemical and biochemical properties to ACNs, and although AACNs exist naturally in many plant species, it can be also obtained *in vivo* by genetic modifications (gene editing and regulation) or *in vitro* by chemical or enzymatic acylation (Saini et al., 2024; Leonarski et al., 2023). Recently, enzymatic processes have proved to be efficient and selective methods for the acylation

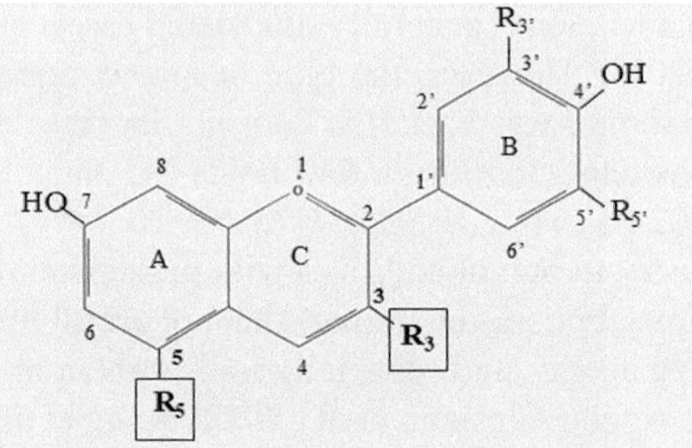

Anthocyanidin	$R_{3'}$	$R_{5'}$	R_3	R_5
Pelargonidin (Pg)	H	H	OH	OH
Cyanidin (Cy)	OH	H	OH	OH
Delphinidin (Dp)	OH	OH	OH	OH
Peonidin (Pn)	OCH_3	H	OH	OH
Petunidin (Pt)	OH	OCH_3	OH	OH
Malvidin (Mv)	OCH_3	OCH_3	OH	OH

Glycosilation	R_3	R_5
3-*O*-monoglycosides	monosaccharide	OH
3-*O*-diglycosides	disaccharide	OH
3,5-di-*O*-glycosides	monosaccharide	monosaccharide
3-*O*-diglycoside-5-*O*-monoglycoside	disaccharide	monosaccharide

Potential acylation sites

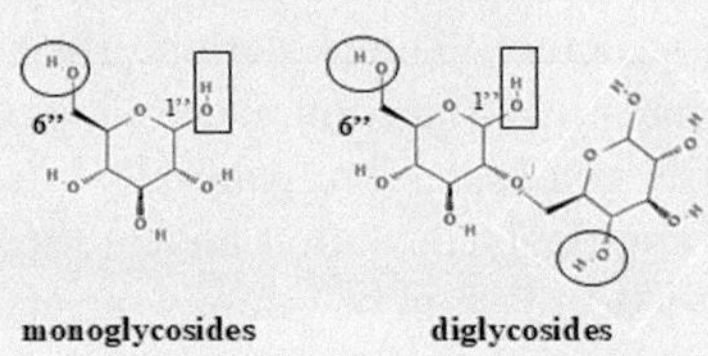

Main acyl residues

Acetic acid
Caffeic acid
Cinnamic acid
Coumaric acid
Ferulic acid
Gallic acid
Oxalic acid
Malonic acid
p-hydroxybenzoic acid
Sinapic acid
Succinic acid

FIGURE 10.1 Chemical structures of the most common ACNs indicating glycosylation (square) and acylation (circle) sites. The most common acyl residues possessing antiproliferative and neuroprotective properties are highlighted in italics.

of ACNs; in particular, the selectivity regards the glucose residue and not the rutinose residue because of the absence of primary alcohol groups (Luo et al., 2022; Wang et al., 2022; Yang et al., 2019; Leonarski et al., 2023).

Acylation with lipase and lauric acid produces derivatives with high stability to light, heat, and oxidation and can increase the extract's antioxidant capacity, an activity linked to the antiproliferative and neuroprotective properties (Teng et al., 2022; Yang et al., 2019). This is evident, e.g., for the acylated forms with lauric acid of Dp-3-*O*-glucoside, Dp-3-*O*-rutinoside, Cy-3-*O*-glucoside, and Cy-3-*O*-rutinoside in black currant (*Ribes nigrum* L.) extract (Yang et al., 2019). Similarly, the acylation of blueberry (*Vaccinium* sect. *Cyanococcus*) ACNs with fatty acids (valeric, n-decanoic, and myristic acids) causes the reduction of antioxidant activity when tested using the 2,2-diphenyl-1-picrylhydrazyl (DPPH) assay but causes an increase when evaluated using the β-carotene bleaching assay, suggesting a different mechanism of action of ACNs. In fact, in the DPPH assay, ACNs act on radicals releasing electrons; in contrast, in the β-carotene blenching assay, ACNs have an inhibitory effect on lipid oxidation (Wang et al., 2022), and the improvement in the antioxidant activity seems to be strictly related with the increase in lipophilicity and by consequence with the inhibition of lipid peroxidation (Yang et al., 2019; Wang et al., 2022).

Generally, AACNs exhibit greater stability with respect to temperature, light, pH, oxygen, and hydrolysis, with higher solubility in organic solvents (lipophilicity) than the nonacylated and glycosylated forms, as evident for black carrots (*Daucus carota* ssp. *sativus* var. *atrorubens*) which are rich in acylated cyanidins in contrast to berries, in which nonacylated ACNs are prevalent (Zozio et al., 2011; Kong et al., 2023; Berentzen et al., 2024). Regarding pH stability, AACNs are well-known for their higher resistance to pH changes in the range of acidic to slightly alkaline values, and this could be useful to selectively optimize their extraction yields. For example, the use of extremely acidic conditions (pH 1.0 by adding hydrochloric acid) in ethanolic extraction increases the concentration of Cy-3-*O*-(6″-*O*-disuccinyl)-glucoside and different Cy-3-*O*-(6″-*O*-malonyl)-glucoside isoforms, but not of Cy-3-*O*-(6″-*O*-succinyl)-glucoside, for which these acidic conditions reduce the ACN content (Kong et al., 2023; Urias-Lugo et al., 2015). Therefore, it is evident that ACNs' stability strictly depends on the acylation site, acyl moiety (aliphatic or aromatic acid and its structure and size), and number of acylations (generally diacylated forms are more stable than the monoacylated derivatives) (Zhao et al., 2017). However, the type of matrix represents a fundamental point dealing with AACNs' stability, as in some cases acylation does not increase stability because of the presence of other compounds, such as prooxidant vitamins or flavonoids and phenolic acids (Gérard et al., 2019; Gras et al., 2018; Berentzen et al., 2024).

Acylation process is widely applied in the food industry to obtain additives with higher stability and specific bioactivities (antioxidant, anticarcinogenic, and metabolism-regulating) that are useful for different applications, such as food natural dyes, functionalizing agents, and especially indicators in intelligent packaging and packaging materials (Luo et al., 2022; de Aguiar Cipriano et al., 2022; Wang et al., 2022; Leonarski et al., 2023).

AACNs' metabolization, absorption efficiency, gastrointestinal transit, and excretion are strictly dependent on the ACN structure. The acylation process protects ACNs from degradation/modification occurring during gastrointestinal digestion, increasing their bioavailability (Zhang & Jing, 2022). Clinical studies indicate that ACN glycosides are generally adsorbed both at the gastric level by specific transporters without undergoing deglycosylation and in the small intestine as aglycones (Saini et al., 2024). Acylation contributes to a decrease in or inhibits the affinity of AACNs for the gastric membrane carrier (bilitranslocase) responsible for their transport into the gastric mucosa, while they can be adsorbed at the intestinal level as occurring for the aglycones or with a lower transport efficiency (Jokioja et al., 2021). Therefore, as AACNs can be adsorbed only by a short gastrointestinal tract, their absorption and excretion processes are faster with a great impact on their bioactivities (Novotny et al., 2012; Jokioja et al., 2021; Saini et al., 2024).

Generally, the increased stability due to the acylation process leads to increased lipophilicity and simultaneously higher bioactivities, as demonstrated for the antioxidant effect of black rice (*Oryza sativa* L.) ACNs (Kong et al., 2023). In addition, anti-inflammatory properties can also increase in the acylated derivatives in comparison to nonacylated ACNs, as evident for AACNs from different sources (black rice, purple sweet potatoes, purple carrots, and berries) (Chen et al., 2023). This could have important consequences on both antiproliferative and neuroprotective properties directly linked to oxidation processes, radical species formation, and inflammation development. Conversely, AACNs can also exert similar or lower antioxidant, anti-inflammatory, and antiproliferative properties than ACNs, depending on the source, extract concentration, and system models used for the tests (de Aguiar Cipriano et al., 2022; Wang et al., 2022; Leonarski et al., 2023; Yañez-Apam et al., 2023).

10.3 ANTICARCINOGENIC ACTIVITY

In the literature, many studies on the anticarcinogenic activities of ACNs (mainly deriving from blueberry, bilberry, cranberry (*Vaccinium* sp.) (Neto, 2007; Martău et al., 2023; Zhao et al., 2023), or pomegranate (*Punica granatum* L.) fruits (Sharma et al., 2017) are reported.

Regarding the ACNs' antiproliferative effect evaluated on different cell lines, it is strictly correlated to acylation as well as aglycone structure and glycosylation pattern, which can positively or negatively affect the activity, often depending on the food matrix. For example, Pg-3-*O*-sophoroside-5-*O*-glucoside acylated with one or two cinnamic acid residues detected in red radish (*Raphanus sativus* L.) had a lower effect on the proliferation of colon HT-29 cells than the AACNs present in grapes (*Vitis vinifera* L.) and purple carrots (*Daucus carota* L.) (Jing et al., 2008).

Multiple mechanisms have been hypothesized for ACNs' antiproliferative activity, such as the inhibition of oxidative DNA damage and of carcinogen activation through the interaction with DNA, antimutagenic and antiangiogenesis effects, apoptosis, direct action on different enzymes (phase II and COX-2 enzymes involved in detoxification and inflammatory processes, respectively), and also antioxidant/antiradical effects (He & Giusti, 2010).

It is also known that a connection among cell membrane fluidity, cancer development, and the effect of a compound's potential bioactives occurs. Acylated cyanidins exhibit a stronger membrane affinity with a consequent greater membrane rigidity in model systems mimicking membranes in comparison to nonacylated compounds, thus improving antioxidant and anticarcinogenic activities. In particular, glycosylation at the C3 and C5 positions on C (heterocyclic) and A (benzyl) rings, respectively, and acetylation at C6″ of the glycosyl moiety could be fundamental for the interaction with membranes and, therefore, for the bioactivity; the action mechanism begins with the increase in membrane rigidity that inhibits lipid peroxidation, which is usually responsible for proliferative progression (Strugała et al., 2016; Dudek et al., 2023). The carcinogenesis process seems to be mainly correlated to a significant increase in the membrane fluidity and permeability of cancer cells in comparison to nonmalignant cells, involving lipid biosynthetic pathways (Sok et al., 1999; Preta, 2020). In fact, cell adhesion is promoted by high levels of cholesterol and unsaturated fatty acids, and especially AACNs have a higher affinity for hydrophobic regions of the lipid membrane, reducing fluidity, permeability, and therefore cell adhesion with an important effect on cancer progression and metastasis; this action is also supported by an antioxidant action against ROS and reactive nitrogen species (RNS) and the inhibition of lipid oxidation processes (Dudek et al., 2023; Matsuzaki et al., 2024). Acetylated Pg-3-*O*-glucoside is specifically involved in the reduction of lipid deposition in the liver cells, with a consequent reduction of oxidative stress through the activation of AMP-activated protein kinase (AMPK) phosphorylation. This involves the lysosome–autophagy pathway that interferes with lipid synthesis, accelerating lipid oxidation. The higher lipophilicity and stability of the acetylated form in comparison to the corresponding glycosylated derivative are the key factors that positively affect the antioxidant properties and the lipid reduction. This pathway could prevent pathological conditions such as nonalcoholic fatty liver disease, which is one of the main risk factors for the development of different types of cancers, mainly liver and gastrointestinal cancers (Xie et al., 2024).

Some studies on the effect of ACNs on colorectal cancer (CRC) indicate that ACNs stimulate the expression of tumor suppressor genes, whereas ACN metabolites, such as gallic acid, methylgallic acid, and trihydroxybenzaldehyde, decrease some transcription factors (NF-κB, AP-1, STAT-1, and OCT-1) that are normally involved in the inflammation pathway and cancer genesis (Sehitoglu et al., 2014).

Recently, Teng et al. hypothesized a direct inhibitory effect of AACNs on the formation of heterocyclic aromatic compounds mainly derived from high-temperature food processing methods, which are well-known mutagenic and carcinogenic agents (Teng et al., 2022, 2023).

It is important to highlight that ACNs and AACNs exhibit high specificity toward the antiproliferative activity against adenocarcinoma (gastrointestinal, prostate, and breast) cells. In fact, AACNs abundant in blue/purple corn (*Zea mays* L.) and sweet potato (*Ipomoea batatas*) have high apoptotic activity against cell lines of specific cancer types: colon (Caco-2, HT-29, HCT-116, HCT-15, SW480, and SW620), prostate (PC3, RWPE-1, RWPE-2, 22Rv1, LNCaP, and DU145), and breast (MCF7, MC7–10A, N202/1A, N202/1E, and MDA-MB-453) (He & Giusti, 2010; Majhi, 2022). *In vitro* and *in vivo* studies on AACNs' mechanism of action against different cancer cell types are discussed in the next section, and the specific pathways involved in their antiproliferative activity are summarized in Figure 10.2.

10.3.1 Colorectal Cancer

CRC is one of the most common cancers worldwide, and over 75% of cases occur in people with no known risk factors. In contrast, it is well known that a correct diet and lifestyle and prevention have an important role. In fact, obesity, sedentary lifestyle, and cigarette smoke can be important risk factors. In addition, environmental pollution (mainly polycyclic aromatic hydrocarbons and their metabolites) and consequent dietary contaminants can increase the risk of CRC characterized by mortality mainly due to metastatic lesions (Pan et al., 2018b; Mudd et al., 2020; Bars-Cortina et al., 2022). In particular, some environmental pollutants can have a negative impact on the composition of gut microbiota, and some bacteria seem to be associated with the development of inflammation and carcinogenic metabolites with an increased risk of CRC (Mudd et al., 2020).

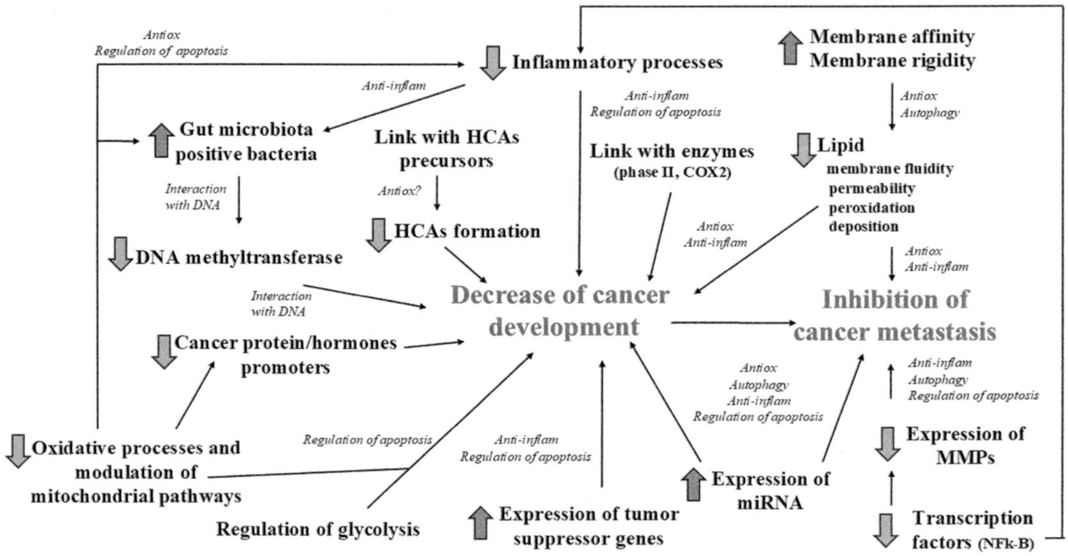

FIGURE 10.2 Specific pathways involved in the antiproliferative activity of AACNs.

Nowadays, the effective role of ACNs in the prevention of CRC is well-known, as they are able to interfere with cancer development through multiple pathways. Many clinical studies indicate that high dietary consumption of ACNs is associated with a low risk of developing CRC, thanks to their capacity to interfere with colon cancer, from formation to angiogenesis and the metastasis process (Wang et al., 2018). In fact, ACNs extracted from different sources exert antiproliferative activity against CRC when in direct contact with the colon epithelial layer, with important implications on neurotensin and epidermal growth factor (EGF), which are strictly implicated in the pathophysiology of colon cancer. ACNs also inhibit the increase in intracellular calcium concentration induced in cancer cells by neurotensin (Briviba et al., 2001; Lila et al., 2016; Henriques et al., 2020). In addition, ACNs can modulate the colon microbiota, which has a significant role in cancer pathogenesis, particularly at the luminal level (Bars-Cortina et al., 2022) where ACN metabolites (conjugates and phenolic acid degradation products) stay for a longer time than ACNs (well-known for their instability in aqueous solution and during digestion), thus exhibiting their anticarcinogenic properties (Lila et al., 2016; Prado et al., 2022).

An increase in ROS and COX-2 levels is associated with epithelial barrier dysfunction and cancer development, and the main mechanisms involved are local inflammation and an electrolyte imbalance. Therefore, compounds such as ACNs that are active on radical species and COX-2 can be promising preventive agents, as demonstrated using a rat model in which a colon carcinogen azoxymethane was used to induce rat colon cancer; the administration of ACNs resulted in a direct effect on the colon lumen by exhibiting a protective effect against radical species and inflammation. The anticarcinogenic activity of ACNs against CRC is widely confirmed by *in vitro*, *in vivo*, and clinical studies (He & Giusti, 2010).

ACNs extracted from different matrices have a proven antiproliferative effect on different colon cancer cellular models; e.g., ACNs from bilberry (*Vaccinium myrtillus* L.), chokeberry (*Aronia melanocarpa* E.), and grape extracts exhibited antiproliferative effects on a colorectal adenocarcinoma model (HT-29 cell line); ACNs from muscadine grapes (*Vitis rotundifolia*), blueberries, and blue corn exhibited antiproliferative effects on both Caco-2 and HT-29 cell lines; and ACNs from strawberry (Alba cultivar), bilberry, and cranberry (*Vaccinium* subg. *Oxycoccus*) exhibited antiproliferative effects on HT-29 and HCT-116 colon cancer cells. Generally, the dose required for 50% cell proliferation inhibition (IC50) ranges from 0.5 to 100 μg/mL (He & Giusti, 2010; Urias-Lugo et al., 2015). ACNs derived from other sources, such as black lentil and sorghum, are active on HT-29 and HCT-116 cell lines but with higher IC50 values (0.9–2.0 mg/mL) (Mazewski et al., 2018).

Comparing the antiproliferative effect of bilberry ACNs on colon cancer cells (HT-29 and HCT-116) and on normal colon cells (CCD-18Co), it is evident that ACNs in the aglycones exhibit a specific cytotoxicity against cancer cells. In addition, these compounds in free and formulated forms (exosomes) were administered to mice (8.6 mg/kg/day, 3 days a week for 4 weeks) previously treated with enterotoxigenic bacteria, giving a positive effect on liver metabolization, modulating phase I and II enzymes. This mechanism highlights how ACNs can prevent the development of CRC, acting directly on intestinal bacteria responsible for microbiota dysbiosis (Mudd et al., 2020). The involvement of intestinal microflora as a target pathway for ACNs has been confirmed by using black raspberry (*Rubus occidentalis*) extract powder on HCT-116 cells and on a mice model of colitis-associated cancer (Chen et al., 2018b). Black raspberry is particularly rich in Cy-3-*O*-xylosyl-rutinoside (49%–58%) and also contains AACNs that have been previously considered responsible for the extract's antioxidant properties and that could also contribute to the maintenance of a healthy gut microbiota (Tulio et al., 2008). ACNs promote an increase in Gram-positive bacterial growth (butyrate-producing bacteria and *Neisseria*) and a decrease in toxic bacteria (*Helicobacter pylori* and *Prevotella*), and a correlation between gut microbiota and DNA methyltransferase has been demonstrated. In fact, this enzyme is downregulated, thus reducing the risk of cancer development. In this ACN mechanism, secreted frizzled-related protein 2 gene is involved. This protein gene plays an important role in the pathogenesis of tumors, and ACNs upregulate the expression of this gene by reducing its methylation, resulting in the overexpression of beneficial proteins that help suppress tumor development (Chen et al., 2018b).

However, some authors indicated that glycosylation, deoxylation, and acylation of ACNs could negatively influence their anticarcinogenic activity (Sousa et al., 2016; Jing et al., 2008). In fact, the gastrointestinal microbiota metabolizes the glycosylated and acetylated ACNs, although the presence of β-glucosidase, which has an important role in ACN metabolization at the oral level, in the human intestinal epithelium is controversial (Mallery et al., 2011). Considering the ACNs from purple/blue corn, chokeberry, bilberry, purple carrot, grape, and elderberry (*Sambucus nigra* L.), the acylated derivatives (mainly Pg-3-*O*-sophoroside-5-*O*-glucoside acylated with one or two cinnamic acid residues, and Cy-3-*O*-(xylosyl)-glucosyl-galactoside acylated with one cinnamic acid) have lower activity than the nonacylated monoglycosylated forms against HT-29 cell proliferation. In general, red radish ACNs, containing triglucoside compounds acylated with more than one cinnamic acid have a HT-29 cell growth inhibitory value (GI_{50}) higher than 50% (Jing et al., 2008). Conversely, blue corn AACNs (5 mg/ml extract) maintain their action on HT-29 (cell viability almost 50%) and Caco-2 (cell viability 60%–70%) cells. In fact, this extract contains nine ACNs, of which seven are acylated, and Cy-3-*O*-(6″-*O*-malonyl)-glucoside is the compound present in blue corn in the highest concentration and is also the most active compound in Caco-2 cells (Urias-Lugo et al., 2015). Therefore, the antiproliferative activity is confirmed to be strictly dependent both on the type of acylated form and the source.

Other studies about the activity of black raspberry extracts rich in AACNs on human HCT-116 cells, Caco-2 cells, and also *in vivo* on mice demonstrate an overexpression of microRNA (miR)-24-1-5p, which is able to regulate the autophagy and apoptosis of malignant melanoma cells. This overexpression could be related to a consequent decrease in β-catenin expression, involved in cell–cell adhesion and CRC development and metastasis. The effect of miR-24-1-5p on β-catenin is directly through the overexpression of E-cadherin, phosphoglycogen synthase kinase 3 β (*p*-GSK3-β), and secreted frizzled-related protein 5 and the underexpression of B-cell lymphoma-2 (BCL-2) with the consequent inhibition of inflammation and apoptosis (Zhang et al., 2019a). Also in this case, the prevalence of acylated derivatives in black raspberry extracts could have an important impact against ROS and the resulting oxidative stress, which play a pivotal role in apoptosis (Tulio et al., 2008).

Cancer stem cells (CSCs) represent another interesting cellular line used to study ACNs' antiproliferative effect, as this cell model is involved in many cancer progressions, including CRC. Baked purple-fleshed potato extract (5 μg/ml) rich in AACNs, mainly 3,5-di-*O*-glycosides, is active in colon cell proliferation with and without the activation of p53, as demonstrated on CSCs and an azoxymethane (a DNA alkylating agent)-induced colon cancer mice model. The extract increases proapoptotic protein (Bax) involved in the release of cytochrome c, which promotes mitochondria-mediated apoptosis and suppresses β-catenin and other cancer protein promoters (c-Myc and cyclinD1) (Charepalli et al., 2015).

Anticarcinogenic activity on HCT-116 cells has been demonstrated for purple root tuber and sweet potato leaf AACNs (rich in acylated Pn and Cy derivatives). Sweet potato ACNs' (100 μg/mL) action mechanism involves the G0/G1 phase cell-cycle arrest with an induction of apoptosis *via* caspase-3 activation and a modulation of mitochondrial pathways and membrane permeability (Vishnu et al., 2019). This action is also reported for red grape extracts, containing phenolic compounds, tannins, ACNs, and one acylated form; in fact, the extracts directly arrest G0/G1 phase, downregulating the apoptotic proteins (cIAP-2, HSP70, survivin, XIAP, and IGFBPs) and activating the caspase cascade that drives apoptosis (Mazewski et al., 2018).

ACNs' activity on colon cancer cells could also not be related to the antioxidant property, as evident for black mulberry (*Morus nigra* L.) ACNs after heat treatment (100°C, 45 min), which has a negative impact on the extract's antioxidant activity but increases its antiproliferative effect on HCT-116 colon cancer cells. In fact, heat treatment promotes the reduction of ACN content but simultaneously improves the total amount of phenols and flavonoids; therefore, the antiproliferative effect could be due to these molecules or the formation of new ACNs that are more stable to heat, e.g., the acylated forms. The mechanism involves an increase in calcium intracellular concentration, which leads to the activation of mitochondria-mediated apoptosis (Cui et al., 2020). ACNs isolated from *Vitis coignetiae* Pulliat and tested on HCT-116 cells induce apoptosis and inhibit NF-kB, a transcription factor involved in many processes from cell proliferation and migration to immune responses and inflammation, through the suppression of metalloproteinase (MMP-2 and MMP-9) expression (which has a key role in cancer and neurodegenerative progression) and with the involvement of mitogen-activated protein kinase (MAPK) isoform p38 (which regulates the expression of pro-inflammatory mediators and leukocyte adhesion) (Sehitoglu et al., 2014). The authors of this research do not characterize the chemical structure of ACNs, but other studies in the literature indicate that this species is rich in glycosylated derivatives and that the inhibition of NF-κB-regulated proteins (MMP-2 and MMP-9) could be attributable to mono- and diglycosides (Yun et al., 2010). Specific studies on acylated forms and their effect on MMP expression are not present in the literature for CRC, but sources of acylated derivatives have a direct impact on MMPs in oral cancer (Par. 10.3.2), so this needs further investigation.

Experiments performed on HT-29 cells and *in vivo* studies involving healthy subjects treated with anthocyanin-rich grape/bilberry juice (0.33 L daily consumption) for almost 1 month confirmed an important effect of ACNs on cancer cell migration and metastasis through the same mechanism. In fact, in addition to an antioxidant effect that has a positive impact on the increased oxidative stress in CRC cells, ACNs and their metabolites reduce tumor cell migration, suppressing the NF-kB pathway and MMP expression (Behrendt et al., 2022). This effect could be attributed to the abundance of Dp derivatives in the extract, which specifically degrade and phosphorylate IkBa (an inhibitory protein) with consequent NF-kB translocation to the nucleus (Iqbal et al., 2024). In the work of Behrendt et al., the ACN composition of the juice constituting 80% red grape juice (grape variety Accent) and 20% bilberry juice is not reported, but red grape is known to contain glycosylated and acylated derivatives (Cahyana et al., 2019).

ACNs' apoptotic and anti-inflammatory activities are also strictly correlated to a real reduction of colorectal tumor as shown in mice models treated with food supplemented with 10% ACN extract for 10 weeks. In fact, mice treated with this extract (composition not declared) showed less inflammation and reduction of colon length than the control mice (Lippert et al., 2017).

As reported previously, AACNs, such as Cy-3-*O*-(6″-*O*-cinnamoyl)-glycoside, have a significant inhibitory effect against heterocyclic amines (HCAs), such as 2-amino-3-methylimidazo [4,5-f]quinoline (IQ), 2-amino-3,8-dimethylimidazo[4,5-f]quinoxaline (MeIQx), 2-amino-3,4,8-trimethylimidazo[4,5-f]quinoxaline (4,8-DiMeIQx), 9H-pyrido[3,4-b]indole (Norharman), 1-methyl-9H-pyrido[3,4-b]indole (Harman), 2-amino-1-methyl-6-phenylimidazo[4,5-b]pyridine (PhIP), and (2-amino-9H-pyrido[2,3-b] indole) (AαC), which are agents potentially involved in the development of CRC and pancreatic cancer. The putative action mechanism can be related to a direct effect on HCA formation, acting on precursors (glucose and creatinine) or on the active carbonyl group, thus reducing its reaction with creatinine (Teng et al., 2023).

10.3.2 Oral/Esophageal Cancer

Oral and esophageal cancers are less common than CRC, but they are very dangerous as poor diagnosis contributes to treating cancer only in advanced stages and with a relatively low survival rate. Therefore, prevention and early diagnosis are fundamental. Multiple oral carcinoma types have been characterized to date (basal cell carcinoma, nasopharyngeal carcinoma, ameloblastoma, and mucoepidermoid carcinoma), among which oral squamous cell carcinoma (OSCC) is the most common. Esophageal cancers can be classified into adenocarcinoma (AC) or squamous cell carcinoma (SCC), which is the predominant form (Abati et al., 2020; Rai et al., 2023).

In the literature, there are a few research about the activity of ACNs on these tumor types, and some studies highlight the potential activity of aglycone metabolites, since β-glucosidase and lactase phlorizin hydrolase present in saliva and oral microflora are responsible for ACN deglycosylation. This metabolization could question the potential role and activity of acylated forms (Mallery et al., 2011). In contrast, other studies point out an activity also for extracts rich in glycosylated and acylated derivatives, but without specifying ACNs' chemical composition. For example, berry (strawberry, blueberry, blackberry, cranberry, and raspberry) and black rice extracts, which contain also glycosylated/acylated derivatives, are active on CAL-27 (a common cell line used in the study of OSCC), exhibiting mainly antiproliferative and pro-apoptotic effects. In fact, ACNs act by inhibiting the NF-kB expression and pathway, reducing MMP-2 and MMP-9 (as in CRC), and interfering with cancer proliferation (invasion and migration processes) (He & Giusti, 2010; Fan et al., 2015).

OSCC is a cancer particularly malignant for its metastasis, and ACNs tested on specific cell lines derived from SCC of the human tongue (Tca8113 and SCC15 cells) had an interesting activity on oral cancer growth and metastasis. The mechanism involved in this anticarcinogenic activity is specific for the tumor cells and is probably due to an anti-inflammatory effect, involving caspase-1. In fact, ACNs (250 μg/ml) suppress OSCC cell viability, migration, and invasion, activating pyroptosis, which is a form of inflammatory caspase-1-dependent programmed cell death. The activation of pyroptosis is a consequence of the increase in pyroptosis-associated factors, such as NLRP3, caspase-1, and interleukin (IL)-1β (Yue et al., 2019). This mechanism is also confirmed in clinical studies using a black raspberry (particularly rich in Cy-3-*O*-xylosyl-rutinoside)-supplemented diet, which causes a reduction in pro-inflammatory (NFKB1 and PTGS2) and antiapoptotic (AURKA, BIRC5, and EGFR) gene expressions (Knobloch et al., 2016). Another mechanism that involves glucose metabolism and AMPK has been proposed by studying the effect of a diet supplemented with black raspberries (5%–10 %) on oral cancer development in rats. In fact, in rats treated with a carcinogenic compound, a direct effect of ACNs on glycolytic enzymes and AMPK pathway genes with the metabolic regulation of glycolysis, normally upregulated in cancer cells, is evident (Knobloch et al., 2019).

In other studies, bilberry powder (containing 24 ACNs, all aglycones or glycosylated forms, with the exception of Pg-*p*-coumaroyl 3-*O*-glucose) was tested on HSC-3 human OSCC and in an animal model (zebrafish). These studies confirm ACNs' effect against cancer growth and migration at 10 and 25 mg/mL in the cellular model and at 10 and 25 μg/ml in zebrafish (Mauramo et al., 2021).

Regarding esophageal cancer forms (AC and SCC), most studies concern ACNs from black raspberries and strawberries, which contain acylated forms (Tulio et al., 2008; da Silva et al., 2007).

Studies on animal models highlight ACNs' role in the cancer inflammation process, decreasing the expression of pro-inflammatory cytokines (IL-1β) and increasing the expression of the anti-inflammatory cytokines (IL-10), with effects also higher than those obtained using a combination of anti-inflammatory drugs (COX-2 and iNOS inhibitors). These results suggest that berry extracts can also be combined with anti-inflammatory drugs, inhibiting esophageal tumor development more strongly (Peiffer et al., 2016; Shi et al., 2016; Pan et al., 2018a; Shi and Chen, 2022). Clinical studies with such fruits (30–60 g/day, 6 months) confirmed that both antioxidant and anti-inflammatory (reduction of pro-inflammatory mediators) effects are the key processes to reduce the progression of esophageal precancerous lesions in potential precancer conditions, such as esophageal dysplasia and Barrett's esophagus (Chen et al., 2012; Kresty et al., 2018).

10.3.3 Prostate Cancer

Prostate cancer is the most common cancer among men, and early diagnosis remains the most important approach. Aging, race, and a family history of prostate cancer are well-known risk factors, while its etiology remains partially unknown. Nowadays the involvement of androgens, sexual factors, and diets rich in sugars and/or lipids are factors promoting the carcinoma development (Kaiser et al., 2019). ACNs, especially Cy derivatives and their glycosylated derivatives, have well-known documented activities against prostate cancer, mainly involving androgen receptor (AR), inflammatory, and retinoblastoma pathways with NF-kB factor, and with consequent cell proliferation as the final target (Mottaghipisheh et al., 2022).

Berry (strawberry and cranberry) extracts and their ACNs have been largely investigated and exhibit anticarcinogenic activity on different prostate cellular (LNCaP, DU145, RWPE-1, RWPE-2, and 22Rv1) and animal models and also in clinical trials (He & Giusti, 2010; Mottaghipisheh et al., 2022). ACNs present in blue corn and tortilla extracts (quite rich in acylated forms) have very high antiproliferative activity against PC-3 cells in the range of 125–500 μg/ml (Herrera-Sotero et al., 2017). In particular, an isomeric form of Cy-malonyl-glucoside, present in blue corn extracts (5 mg/ml), is active on PC-3 cells by reducing the viability to 50% (Urias-Lugo et al., 2015).

ACNs from blue corn and tortilla extracts also decrease the viability of LNCaP cells by almost 40% and 50%, respectively, when tested in the range of 500–1,000 μg/ml; in addition, tortilla ACNs (400 μg/ml) reduce the expression of the AR, which nowadays is considered the most important target for prostate (and also breast) cancer (Herrera-Sotero et al., 2020; Rampurwala et al., 2016).

Studies on prostatic hyperplasia-induced rat models have indicated that extracts rich in AACNs (from mainly cranberry and grape seeds) have a direct effect on benign prostatic hyperplasia (BPH), whose association with prostate cancer is controversial. The extract acts by decreasing prostate gland volume and weight as a consequence of its antioxidant and antiproliferative activities, preventing the potential development of prostate cancer (Jang et al., 2010; An et al., 2020; Iannuzzo et al., 2025). The mechanism could involve the suppression of steroid 5-alpha reductase (5AR) that converts testosterone into dihydrotestosterone, a BPH promotor (An et al., 2020).

10.3.4 Breast Cancer

Nowadays, notwithstanding the research progress, the incidence of breast cancer is continuously increasing, and it remains the leading cause of cancer-related disease for women. Genetic and environmental factors are involved in the prognosis, and the search for predictive biomarkers remains an ongoing and essential mandate (Sarhangi et al., 2022).

The first results reported in the literature regard the antiproliferative effect of ACN aglycones. In fact, Cy, Dp, Pn, Pt, and Mv specifically inhibit MCF-7 cell proliferation in the range 47%–74.7% when tested at 200 μg/ml (Zhang et al., 2005). The same cellular model was used by Suganyadevi P et al. to test the effect of 3-deoxyanthocyanin (250 μg/ml) extracted from red sorghum bran to demonstrate that this molecule induces apoptosis through the stimulation of p53 gene, decreasing the expression of the antiapoptotic protein Bcl-2 (Suganyadevi et al., 2013).

Regarding AACNs, native blue corn extract (5 mg/ml) acts as a cancer cell growth inhibitor on MCF-7 cells (cell viability 50%–80%), and an important role could be played by Cy-succinyl-glucoside (Urias-Lugo et al., 2015). Similarly, AACNs from sweet potato leaves (100 μg/ml) exhibit apoptosis during the growth phase and before mitosis of MCF-7 cells (Vishnu et al., 2019).

Other studies highlight a direct role of ACNs on human epidermal growth factor receptor 2 (HER-2, also known as ErbB2)-positive breast cancer, and this is widely evident on the MDA-MB-453 model (HER-2-positive). For example, some authors suggest a potential pathway for the antiproliferative activity of black rice (a rich source of different ACNs) associated with HER-2 involved in cell growth and tumorigenesis. In fact, *HER-2* gene amplification and protein overexpression represent markers of metastasis. Black rice ACNs act specifically on HER-2-positive breast cancer cells (MDA-MB-453), decreasing the migration distance, the invasive ability, and the adhesion to fibronectin in the extracellular matrix (ECM), essential for the metastasis process. HER-2 seems to be involved in the focal adhesion kinase (FAK) pathway across different human cancer cells (breast, but also colon, pancreas, prostate and liver), and black rice ACNs reduce the epithelial–mesenchymal transition (EMT) that regulates FAK and β-catenin phosphorylation, which are involved in cancer metastasis. ACNs from black rice decrease the expression of proteolytic enzymes such as MMPs and urokinase-type plasminogen activator, responsible for the degradation of collagen and ECM invasion (Sehitoglu et al., 2014; Hui et al., 2010; Zhou et al., 2017).

The metastasis inhibitory effects of black rice ACNs on the HER-2 pathway have also been demonstrated on the same MDA-MB-453 cells by Li-Ping Luo et al. In addition, the authors performed *in vivo* studies on BALB/c nude mice bearing HER-2-positive breast cancer cells, highlighting ACNs' role in decreasing tumor volumes, without preventing carcinogenesis (Luo et al., 2014). The activity on the MDA-MB-453 model is evident also for blue corn and tortilla extracts rich in AACNs, as they cause a 70% inhibition of cell viability at doses in the range of 500–1,000 μg/ml. ACNs act by arresting the cell cycle and inducing apoptosis, with important cellular morphological changes. The mechanism involves a decrease in AR expression when tortilla ACNs are used at 300 μg/ml, as for prostate cancer (Herrera-Sotero et al., 2020).

Two murine cancer cellular models (N202/1A and N202/1E), named HER-2 models, clarify the role of ACNs related to HER-2, as they exhibit high and low levels of the *HER2/neu* oncogene, respectively. In fact, a strawberry ACN extract containing different types of ACNs (aglycones, glycosylated, and acylated forms) (da Silva et al., 2007) exhibits a pro-apoptotic effect on cancer cells, specifically acting on the AMPK signaling pathway that controls the cell growth/metabolism and autophagy mechanisms, with the downregulation of the expression of both AMPK and *p*-AMPK proteins. In addition, downregulation of NRF2 expression could be involved, especially for the N202/1A cell line, with an important role in the regulation of the expression of antioxidant proteins that are protective agents against oxidative damage (Mazzoni et al., 2019).

10.3.5 Other Cancer Types

10.3.5.1 Hepatic Cancer

Hepatocellular carcinoma (HCC) is the most common type of cancer with poor diagnosis and median survival times. Risk factors are obesity, alcohol consumption, smoking, and chronic liver diseases or infections (nonalcoholic fatty liver disease and hepatitis B) (Petrick et al., 2016). In the literature, only few research on the effect of ACNs on HCC are present, but many *in vitro* and *in vivo* studies report that ACN extracts reduce hepatocyte oxidative stress by promoting antioxidant response and modulate lipid homeostasis in the liver by decreasing lipogenesis and lipid accumulation. In addition, they increase lipolysis with important implications in many chronic liver disorders such as steatosis and hepatitis, which can easily evolve into HCC. Therefore, ACNs could be useful for HCC chemoprevention (Mohammed & Khan, 2022). Recent studies on Cy and Cy-3-*O*-glucoside indicate the involvement of mTOR protein kinase and metastasis-associated lung adenocarcinoma transcript 1 (*MALAT1*) RNA gene, which controls the cellular growth and metabolism, as potential mechanisms of action. They are overexpressed in HCC, stimulating cell proliferation, migration, and metastasis. In addition, Cy has also an effect against the autophagy protein ATG16L1, which can control cancer progression by an apoptotic mechanism that probably involves a pro-apoptotic caspase (caspase-3), an inhibitor of autophagy (Safdar et al., 2023; Matboli et al., 2021; Zabady et al., 2022; Wu et al., 2019).

Regarding acylated forms, blue corn and tortilla extracts have also a strong antiproliferative effect on HepG2 cells, exhibiting their highest activity at 125–250 μg/ml (Herrera-Sotero et al., 2017). Urias-Lugo et al. highlighted an interesting activity of 5 mg/mL blue corn extract, too. In particular, the acylated form Cy-3-*O*-(6″-*O*-malonyl)-glucoside (isomer I) promotes cell viability of about 50%–70% on the same cellular model (Urias-Lugo et al., 2015).

Pg-3-*O*-(6″-*O*-acetyl)-glucoside could also have important effects on hepatic health. In fact, this AACN, isolated from strawberry, has been tested on normal human hepatocyte cell line L02 and is able to regulate lipid metabolism, interfering with lipid deposition and free fatty acid-induced oxidative stress through the AMPK pathway. AMPK is a regulator of metabolism and metabolic disorders, strictly associated with the development of cancer (Xie et al., 2024).

10.3.5.2 Lung Cancer

The most prevalent type (85% of all cases) of lung cancer is non-small cell lung cancer (NSCLC). It is an adenocarcinoma characterized by high morbidity and mortality because of poor prognosis and low survival rate (5 years). It is strictly correlated to cigarette smoking. ACN dietary intakes, mainly in the aglycone form (less than 20 mg/day), could be related to a reduced cancer risk, with a protective effect possibly through the mitigation of oxidative stress (Zhang et al., 2022; Papavassiliou et al., 2024).

Many *in vitro* studies indicate an apoptotic activity of ACNs, mainly Pg, Dp, and Pn, and their glycosylated forms on lung cancer cells (A549, NCI-H441, SK-MES-1, H1299, and BEAS-2B cells) (Alsharairi, 2022; Pal et al., 2013; Ho et al., 2010; Lu et al., 2017; Yue et al., 2024; Amararathna et al., 2020), as well as *in vivo* experiments (Alsharairi, 2022; Pal et al., 2013; Ho et al., 2010). In these studies, Dp and Pn promote oxidative stress-regulated apoptosis, interfering with inflammation by modulating MMPs and inhibiting the MAPK pathway with implications in cell migration and invasion processes (Pal et al., 2013; Ho et al., 2010; Lu et al., 2017). Dp also acts by inhibiting the epidermal growth factor receptor (EGFR) and vascular endothelial growth factor receptor 2 (VEGFR2), which are normally overexpressed in many cancers, such as NSCLC (Pal et al., 2013; Kim et al., 2017). Studies using Haskap (*Lonicera caerulea* L.) berry ACNs, mainly glycosides, highlight a potential preventive role against the tobacco smoke carcinogen 4-[(methyl)nitrosamino]-1–(3-pyridyl)-1-butanone (NNK), acting directly on NKK carcinogenic metabolites with the inhibition of cytochrome P450 (P450) enzymes. Therefore, ACN-rich haskap berry supplementation could be used to reduce the risk of lung cancer in smokers (Amararathna et al., 2020). ACNs and especially the glycosylated forms are active and involve AMPK signaling and a reduction of lipid deposition, as demonstrated by *in vitro* and *in vivo* studies on mice with urethane-induced lung cancer fed with a 0.5% ACN (Cy-3-glucoside chloride) diet for 20 weeks (Luo et al., 2024; Chen et al., 2021). The activity of this molecule against proliferation and migration is related to the inhibition of the activation of phosphatidylinositol 3-kinase (PI3K)/protein kinase B (AKT)/mammalian target of rapamycin (mTOR) pathway (Chen et al., 2021). This could also have important implications in the lipid metabolism and inhibition of membrane fluidity and by consequence in the reduction of cell adhesion and migration (Sok et al., 1999; Strugała et al., 2016).

No specific articles on acylated derivatives and lung cancer are present in the literature – only aglycones and glycosides are commented. Therefore, the role of acylated forms is not clear and deserves a closer inspection. In fact, e.g., AMPK pathway could be a common target also for acylated forms as in the case of black raspberry extracts useful in oral cancer (Par. 10.3.2) (Knobloch et al., 2019). A research on grain potentially rich in acylated forms (acyl residues: succinic, malonic, dimalonic, and sinapic acids) indicates that the administration of ACN-rich grain diet in mice has a positive effect on lung tumor size probably due to a prolonged autophagic response and partially or completely inhibits the number of metastasis, mainly through an activity against tumor necrosis factor alpha (TNF-α) and toll-like receptor 2, essential factors in the metastatic process (Zykin et al., 2018; Tikhonova et al., 2024).

10.3.5.3 Skin Cancer

Basal cell carcinoma (BCC) and SCC are the most common skin cancers and, together with melanoma, are correlated with UV exposure. Polyphenols (flavonoids, resveratrol, and ACNs) are widely used in cosmetic preparations for the prevention of skin tumors because they support the inhibition of tumor cell proliferation and the activation of tumor cell apoptosis (Sajadimajd et al., 2020).

misfolding remain the main etiologic factors, leading to the degeneration of dopaminergic neurons in the substantia nigra and the formation of alpha-synuclein (α-syn) fibrillar aggregates, named Lewy bodies. A neurotrophic factor such as insulin-like growth factor-1 (IGF-1) plays a critical role in PD: in fact, lower IGF-1 levels in the early stages of PD are associated with worse disease progression and prognosis and correlated not only with cognitive impairment but also with depression and anxiety (Trist et al., 2019; Fan et al., 2018). In addition, gut microbiota dysbiosis is associated with an increase in the aforementioned factors with a potential acceleration of PD pathogenesis (Panchal & Brown, 2023).

It is known that a lower risk of PD is correlated to high consumption of berry fruits, but ACN mechanisms need to be clarified. Recently, studies on black currant ACNs administered in PD patients highlight a potential specific target such as IGF-1. After the administration of the supplement for 28 days, an IGF-1 metabolite, i.e., cyclic glycine-proline (cGP), which has important neuroprotective and neurotrophic functions, increases in the plasma and cerebrospinal fluid. This could be important to treat patients in the early stage of PD (Fan et al., 2018). The composition of black currant capsules administered in this study is not declared by the authors, but *Ribes nigrum* L. is known to contain also AACNs such as Cy- and Dp-3-*O*-(6″-*O*-*p*-coumaroyl)-glucosides (Slimestad & Solheim, 2002).

A specific study on acylated forms regards black carrot extract tested on neuronal model cells treated with 1-methyl-4-phenylpyridinium (MPP+) to induce PD. Black carrots are particularly rich (60.25%) in acylated Cy derivatives (Cy-3-*O*-xylosyl-glucosyl-galactosides acylated with ferulic, sinapic, and coumaric acids), which exhibit an antioxidant effect on intracellular ROS generation induced by MPP+, as previously demonstrated for the only cyanidin aglycone (Zaim et al., 2021; Chen et al., 2018a).

Another important pathway related to PD is represented by AACNs' COMT inhibitory activity, as already mentioned for purple yam extract ACNs (mainly mono/diacylated derivatives), as nowadays COMT inhibitors (tolcapone, entacapone, and opicapone) represent the first PD treatment (Miyata et al., 2022).

10.4.3 Depression and Anxiety Disorders

Depression and anxiety, which are mood disorders that are often correlated, increased in the past decade, with a global (year 2020/2021) prevalence of 27.6% and 25.6%, respectively (Panchal & Brown, 2023; Álvarez et al., 2022). Among the first targets of ACNs are monoamine oxidases (MAO) A and B. Their inhibition represents a key factor in many neurodegenerative disorders' treatments, including depression and anxiety disturbances. In fact, e.g., berry ACNs act against MAO. It is important to highlight that berry extracts are rich in acylated derivatives, which contribute by interfering with the processes involved in depression/anxiety (Delgado-Povedano et al., 2021). Among ACNs, Cy and Cy-3-*O*-glucoside inhibit mainly MAO A, which preferentially metabolizes monoammines (serotonin and noradrenaline) and, to a lesser extent, MAO B, as evident in *in vitro* studies (Dreiseitel et al., 2009). Recent clinical studies report that ACN intake (without specifying the ACN composition of different supplementations) is strictly related to a significant reduction of depression symptoms, which is useful both in prevention and in early treatment in children, young people, and adults (Fisk et al., 2020; Khalid et al., 2017; Mestrom et al., 2024). Cy seems to have major effects on the reduction of the risk of depression also in clinical studies (Chen & Zhao, 2023).

In vivo studies in mice with LPS-induced depressive-like behavior confirmed that blueberry (*Vaccinium virgatum* var.) ACNs (100 or 200 mg/kg, orally) (not characterized) act on MAO-A in specific regions such as the hippocampus but not in the cerebral cortex. In the hippocampus, blueberry extract also decreases acetylcholinesterase (AChE) activity (Spohr et al., 2023).

Another important pathway correlated with these disorders is neuroinflammation. Berry extracts inhibit the expression of TNF-α, IL-1ß, and IL-10 in the cerebral cortex, attenuating inflammatory responses in the microglia (Spohr et al., 2023). Among ACNs, Mv-3-*O*-glucoside together with its metabolites is particularly promising to treat depression-related conditions and anxiety, acting mainly on inflammasomes, which are essential promotors of inflammation and pro-inflammatory cytokines, as demonstrated by *in vivo* studies in mice treated with this ACN (12.5 mg/kg bw/day for 7 days). It reduces the secretion of caspase-1 and IL-1β upon activation of inflammasomes (Iban-Arias et al., 2022).

A recent hypothesis points out that depression and anxiety can be associated with obesity and its consequent gut inflammation and changes in the gut microbiota. In fact, gut microbiota alteration is associated with different neurological and psychiatric diseases. The positive effect of polyphenols, ACNs included, on these disorders seems to be mainly due to their direct effect on the gut microbiota through the modulation of oxidative stress, systemic inflammation, and the immune system (Panchal & Brown, 2023; Álvarez et al., 2022). Recently, *in vivo* studies on different sources of ACNs confirm their effect on MAO/AChE activity and inflammatory mediators (e.g., IL-6), while their potential modulation on microbiota composition (*Lactobacillus* and *Bifidobacterium* spp.) remains controversial and needs further investigation (Natthee et al., 2024; Peng et al., 2023).

A potential specific effect of acylated forms on depression and anxiety disorders could be correlated to their activity on COMT, as some COMT genetic variations are associated with the development of depression (Miyata et al., 2022; Ni et al., 2021).

10.4.4 Cerebral Ischemia

All different ischemic events (hepatic, myocardial, and cerebral) involve oxygen species and inflammatory mediators. Cerebral ischemia, causing 5.2% of global stroke deaths, results from the occlusion of a cerebral artery with a consequent reduction of cerebral blood flow, and it is induced by ischemia–reperfusion injury with hemorrhagic transformation and brain edema (Zhao et al., 2022). ROS formation is the key factor of traumatic brain, ischemic, and hypoxic injuries and the main target of ACN activity (Banji et al., 2022). In fact, *in vivo* (animal and clinical) studies indicate that ACNs (300 mg/kg orally administered) have a preventive role in cerebral ischemia by blocking the oxidative stress signaling pathways in which stress-activated protein kinase phospho-c-Jun N-terminal kinase (*p*-JNK), protein p53, and glutathione peroxidase are involved. In addition, microglial reactions, implicated in BBB formation and neurotransmission and stimulated during ischemic events, are inactivated. All these mechanisms act by reducing brain edema and infarct volume (Shin et al., 2006; Hwang et al., 2018).

Among the specific ACNs, Cy (20 mg/kg) and Cy-3-*O*-glucoside (10 mg/kg) intraperitoneal injection in rats reduce ROS, leading to positive consequences on arteriolar dilation and cerebral microcirculation with a potential role both in cerebral ischemia and during reperfusion following the ischemic event (Mastantuono et al., 2018; Di Giacomo et al., 2012; Cui et al., 2018).

Regarding acylated forms, Pt-3-*O*-rutinoside-(6″-*O*-*p*-coumaroyl)-5-*O*-glucoside (50–200 μg/ml) increases neuronal SH-SY5Y cell viability, reducing oxidative stress, ROS formation, and apoptosis, thus decreasing the levels of neuroinflammatory promoters. The promotion of autophagy, which possesses a protective effect on cerebral ischemia, can also be involved in ACN mechanism (Cai et al., 2020). The same molecule (200 mg/Kg) tested in animals (rats) reduces brain infarct volume and cerebral edema, decreasing MMP-9, which is the promoter of BBB damage. The antiapoptotic activity of this ACN is due to the regulation of Bcl-2 family proteins, which are mytochondrial apoptosis regulators; conversely, the anti-inflammatory effect occurs through the inhibition of the expression of different inflammatory factors, such as TNF-α, IL-1β, IL-6, and also *p*-NF-kB p50, *p*-NF-kB p65, and *p*-IkB (Pan et al., 2018c). Recently, studies with purple highland barley bran (PHBB) extract (<1600 μg/mL), which is mainly rich in Cy-succinyl glucoside, Cy-malonyl glucoside, Cy-acetyl galactoside, Pg-succinyl glucoside, and Pn-acetyl galactoside, confirm that these compounds act by upregulating autophagy, protecting the brain from hypoxic damage with important implications in hypoxic–ischemic brain injury, an unresolved consequence that can occur after a cardiac arrest. In fact, PHBB ACNs (mainly 400 g/ml), in addition to a specific effect on some intracellular antioxidant enzymes (catalase and superoxide dismutase), increase the level of an autophagy marker protein (LC3-II), inducing autophagy with protective mechanisms against hypoxic and oxidant injuries (Zhang et al., 2021b).

Additional studies on these acylated derivatives and their involvement in autophagy may have implications for potential therapeutic applications in the treatment of other neurodegenerative diseases (e.g., AD and PD), in which upregulation of autophagy is involved (Menzies et al., 2015).

10.4.5 Other Neurological Disorders

10.4.5.1 Autism

Autism spectrum disorders (ASDs) are mainly associated with oxidative and inflammation pathways, in addition to changes in gastrointestinal functions (diarrhea, constipation abdominal pain, and high gut permeability) and gut microbiota; in fact, the use of prebiotics and probiotics has advantages not only for gastrointestinal but also for behavioral symptoms in subjects with ASD (Saurman et al., 2020; Serra et al., 2022). Recently, a work by Serra et al. highlights the positive effect of an extract obtained from Portuguese blueberries (containing 16 ACNs, of which eight are acylated derivatives, mainly acetyl and coumaroyl residues) in an ASD mouse model when orally administered at 30 mg/kg/day for 3 weeks. In these conditions, the main target of ACNs is the microbiota–gut–brain axis with a modulation of gut microbiota composition (increase in Lactobacillales and decrease in Clostridiales) and a direct effect on gut inflammation and neuroinflammation with a reduction in the pro-inflammatory mediators (Serra et al., 2022). An increase in the serotonin levels occurs both in the gut (ileum and colon) and in the prefrontal cortex, with potential positive implications in autism-like behaviors (Serra et al., 2022; Walsh et al., 2021). Additional studies on the effect of ACNs on gut microbiota in ASD patients could provide evidence of their potential activity in all gut–brain axis disorders, from AD and PD to depression, as mentioned previously in other subsections.

In addition, ACNs reduce the synaptic dysfunctions involved in autism-like behaviors, indicating a possible application of ACN extracts to improve different ASD symptoms (Serra et al., 2022).

10.4.5.2 Schizophrenia

Schizophrenia is a severe psychiatric condition. Different events in the uterus, e.g., infections, can represent important risk factors for this disease that is characterized by reality distortion and cognitive disturbance. Oxidative pathways and neuroinflammation are key processes behind the development of this chronic mental health condition. In particular, neuroinflammation has a critical role in the development of schizophrenia and mood disorders (Jauhar et al., 2022; Banji et al., 2022).

A recent work of Asakura et al. highlights that ACNs could directly act by strongly inhibiting pentosidine (PEN) synthesis; PEN is an advanced glycation end product (AGE), which increases in schizophrenia (Asakura et al., 2024; Arai et al., 2010). AGEs are the final heterogeneous products derived from nonenzymatic reactions among reducing sugars and proteins able to accelerate the aging process through the promotion of oxidative and inflammatory processes (Twarda-Clapa et al., 2022). PEN inhibition is strictly correlated with the antioxidant capacity of ACNs. To date, among 93 different tested compounds, five ACNs were found to exhibit the most potent inhibitory activity. Although the authors hypothesize that the highest activity could be precisely related to the basic flavonoid structure, no acylated forms have been tested (Asakura et al., 2024). Obviously for future clarifications, testing acylated forms will also be necessary, but this study opens new pathways to be investigated. In fact, the glycation process and AGE formation have a very important role in the pathogenesis of all age-related disorders (e.g., cardiovascular diseases, diabetes, and neurodegenerative disorders) (Reddy et al., 2022).

Regarding the role of acylation, the COMT gene is one of the most investigated candidate genes and by consequence the target for schizophrenia, and as for PD and depression/anxiety disorders, the work by Miyata et al. highlights that AACNs, especially the diacylated forms, can have a direct inhibitory effect on COMT due to the catechol structure of B-ring (Miyata et al., 2022).

10.4.5.3 Creutzfeldt–Jakob and Prion Diseases

Creutzfeldt–Jakob Disease (CJD) and prion disease are a group of rare viral diseases characterized by spongiform neuronal degeneration caused by mutations of the human prion protein (*PrP*) gene with consequent Prp misfolding, aggregation, and accumulation, correlated to amyloid processes. Neuronal loss and cognitive dysfunctions are the main clinical consequences of this infection (Zerr et al., 2024).

Recently, only glycosylated compounds (Mv-3-*O*-glucoside and Dn-3-*O*-glucoside) have been tested in *in vitro* studies for their anti-prion activity. One of the involved mechanisms is the ACN antioxidant effect, upregulating the expression of genes involved in the antioxidant response, and reducing ROS. The second hypothesized process is a potential direct effect on Prp aggregates and fibrils, which are associated with this infectious disease, with important future implications in proteinopathies and amyloidogenic diseases (Christoudia et al., 2024). Considering this second hypothesized positive interference with the amyloid aggregation process, it could be interesting to test Mv-3-*O*-glucoside and Dp-3-*O*-glucoside in other amyloidosis, such as AD and PD, and investigate how acylation could impact the ACN anti-aggregation effect.

Table 10.2 summarizes the most active AACNs with neuroprotective effect with their main specific *in vitro* and *in vivo* studies and mechanisms.

TABLE 10.2 Specific AACNs from Different Food Matrices and Their Neuroprotective Properties

AACN	*SOURCE*	*MODEL*	*EFFECT/MECHANISM*	*REFERENCES*
Cy-3-*O*-diglycoside-5-*O*-glucoside acylated with one or two hydroxycinnamic acids	Red cabbage	Normal aging mice	↓ Inflammation, ↑ Gram-positive gut microbiota	Zhang & Jing (2022)
AACN-rich source (n.s.) (acylated forms with aliphatic acids?)	Strawberry	Neuropathological patients	↓ Oxidative stress, ↓ neurofibrillary tangles	Agarwal et al. (2020)
AACN-rich source (n.s.) (acylated forms of Cy and Dp-3-*O*-(6″-*O*-*p*-coumaroyl)-glucosides?)	Black currant	PD patients	↑ IGF-1, ↑ Gram-positive gut microbiota	Fan et al. (2018)
Cy-3-*O*-xylosyl(glucosyl) galactosides acylated with ferulic, sinapic, and coumaric acids	Black carrot	SH-SY5Y cells treated with MPP+	↓ Oxidative stress	Zaim et al. (2021)
Mono/diacetylated ACNs	Purple yam extract	Chemical COMT inhibitory assay	↓ COMT activity	Miyata et al. (2022)
Pt-3-*O*-rutinoside-(6″-*O*-*p*-coumaroyl)-5-*O*-glucoside	*Lycium ruthenicum* Murr.	SH-SY5Y cell rat models	↑ Cell viability, ↓ oxidative stress, ↓ inflammation, ↑ autophagy, ↓ NF-kB pathway, ↓ MMP expression, ↓ mitochondria-mediated apoptosis, ↓ inflammation	Cai et al. (2020);Pan et al. (2018c)

TABLE 10.2 (Continued) Specific AACNs from Different Food Matrices and Their Neuroprotective Properties

AACN	*SOURCE*	*MODEL*	*EFFECT/MECHANISM*	*REFERENCES*
Cy-3-*O*-succinyl-glucoside, Cy-3-*O*-malonyl-glucoside, Cy-3-*O*-acetyl-galactoside, Pg-3-*O*-succinyl-glucoside, Pn-3-*O*-acetyl-galactoside	Purple highland barley bran	PC12 and U2OS cells	↓ Oxidative stress, ↑ autophagy,	Zhang et al. (2021b)
Cy, Dp, Mv, Pt-3-*O*-acetyl-galactoside; Mv, Pn, Pt-3-*O*-acetyl-glucoside; Dp-3-*O*-coumaroyl-hexose	Blueberry	ASD mouse model	↑ Gram-positive gut microbiota, ↓ inflammation, ↑ serotonin	Serra et al. (2022)

Note: n.s., not specified.

REFERENCES

Abati, S., Bramati, C., Bondi, S., Lissoni, A., & Trimarchi, M. (2020). Oral cancer and precancer: A narrative review on the relevance of early diagnosis, *Int J Environ Res Public Health*, 17, 9160. https://doi.org/10.3390/ijerph17249160.

Afzal, M., Redha, A., & AlHasan, R. (2019). Anthocyanins potentially contribute to defense against Alzheimer's disease, *Molecules*, 24, 4255. https://doi.org/10.3390/molecules24234255.

Agarwal, P., Holland, T., Wang, Y., Schneider, J., & Martha, M. (2020). Pelargonidin in strawberries may reduce Alzheimer's disease neuropathology: A community-based study, *Curr Dev Nutr,* 4 (Supplement_2), 1186. https://doi.org/10.1093/cdn/nzaa057_002.

Alsharairi, N.A. (2022). Insights into the mechanisms of action of proanthocyanidins and anthocyanins in the treatment of nicotine-induced non-small cell lung cancer, *Int J Mol Sci*, 23, 7905. https://doi.org/10.3390/ijms23147905.

Álvarez, S.A., Rocha-Guzmán, N.E., González-Laredo, R.F., Gallegos-Infante, J.A., Moreno-Jiménez, M.R., & Bravo-Muñoz, M. (2022). Ancestral food sources rich in polyphenols, their metabolism, and the potential influence of gut microbiota in the management of depression and anxiety, *J Agric Food Chem*, 70, 944–956. https://doi.org/10.1021/acs.jafc.1c06151.

Amararathna, M., Hoskin, D.W., & Rupasinghe, H.P.V. (2020). Anthocyanin-rich haskap (*Lonicera caerulea* L.) berry extracts reduce nitrosamine-induced DNA damage in human normal lung epithelial cells in vitro, *Food Chem Toxicol*, 141, 111404. https://doi.org/10.1016/j.fct.2020.111404.

An, Y.J., Lee, J.Y., Kim, Y., Jun, W., & Lee, Y.-H. (2020). Cranberry powder attenuates benign prostatic hyperplasia in rats, *J Med Food*, 23, 1296–1302. https://doi.org/10.1089/jmf.2020.4779.

Andres-Lacueva, C., Shukitt-Hale, B., Galli, R.L., Jauregui, O., Lamuela-Raventos, R.M., & Joseph, J.A. (2005). Anthocyanins in aged blueberry-fed rats are found centrally and may enhance memory, *Nutr Neurosci*, 8, 111–120. https://doi.org/10.1080/10284150500078117.

Arai, M., Yuzawa, H., Nohara, I., Ohnishi, T., Obata, N., Iwayama, Y., Haga, S., Toyota, T., Ujike, H., Arai, M., Ichikawa, T., Nishida, A., Tanaka, Y., Furukawa, A., Aikawa, Y., Kuroda, O., Niizato, K., Izawa, R., Nakamura, K., Mori, N., Matsuzawa, D., Hashimoto, K., Iyo, M., Sora, I., Matsushita, M., Okazaki, Y., Yoshikawa, T., Miyata, T., & Itokawa, M. (2010). Enhanced carbonyl stress in a subpopulation of schizophrenia, *Arch Gen Psychiatr*, 67, 589–597. https://doi.org/10.1001/archgenpsychiatry.2010.62.

Asakura, M., Toriumi, K., Nozaki, A., Yang, J., Tatsuzaki, J., Suzuki, K., Miyashita, M., Itokawa, M., & Arai, M. (2024). Anthocyanins as potent inhibitors of pentosidine synthesis: Antioxidant-mediated effects, *Biochem Biophys Res Commun,* 740, Art.n. 151007. https://doi.org/10.1016/j.bbrc.2024.151007.

Banji, O.J.F., Banji, D., Makeen, H.A., Alqahtani, S.S., & Alshahrani, S. (2022). Neuroinflammation: The role of anthocyanins as neuroprotectants, *Curr Neuropharmacol*, 20, 2156–2174. https://doi.org/10.2174/1570159X20666220119140835.

Bars-Cortina, D., Sakhawat, A., Piñol-Felis, C., & Motilva, M.-J. (2022). Chemo preventive effects of anthocyanins on colorectal and breast cancer: A review, *Semin Cancer Biol*, 81, 241–258. https://doi.org/110.1016/j.semcancer.2020.12.013.

Behrendt, I., Röder, I., Will, F., Mostafa, H., Gonzalez-Dominguez, R., Meroño, T., Andres-Lacueva, C., Fasshauer, M., Rudloff, S., & Kuntz, S. (2022). Influence of plasma-isolated anthocyanins and their metabolites on cancer cell migration (HT-29 and Caco-2) in vitro: Results of the ATTACH study, *Antioxidants*, 11, Art.n. 1341. https://doi.org/10.3390/antiox11071341.

Berentzen, E.I., Hauer Møller, A., Danielsen, M., Jensen, M., Joernsgaard, B., & Kastrup Dalsgaard, T. (2024). Stability of individual anthocyanins from black carrots stored in light and darkness – impact of acylation, *Food Res Int*, 186, Art.n. 114382. https://doi.org/10.1016/j.foodres.2024.114382.

Boespflug, E.L., Eliassen, J.C., Dudley, J.A., Shidler, M.D., Kalt, W., Summer, S.S., Stein, A.L., Stover, A.N., & Krikorian, R. (2018). Enhanced neural activation with blueberry supplementation in mild cognitive impairment, *Nutr Neurosci*, 21, 297–305. http://dx.doi.org/10.1080/1028415X.2017.1287833.

Briviba, K., Abrahamse, S.L., Pool-Zobel, B.L., & Rechkemmer, G. (2001). Neurotensin-and EGF-induced metabolic activation of colon carcinoma cells is diminished by dietary flavonoid cyanidin but not by its glycosides, *Nutr Cancer*, 41, 172–179. https://doi.org/10.1080/01635581.2001.9680629.

Bunea, A., Ruginã, D., Sconţa, Z., Po, R.M., Pintea, A., Socaciu, C., Tăbăran, F., Grootaert, C., Karin Struijs, K., & VanCamp, J. (2013). Anthocyanin determination in blueberry extracts from various cultivars and their antiproliferative and apoptotic properties in B16-F10 metastatic murine melanoma cells, *Phytochem*, 95, 436–444. https://doi.org/10.1016/j.phytochem.2013.06.018.

Cahyana, Y., Gordon, M.H., & Gibson, T.M. (2019). Urinary excretion of anthocyanins following consumption of strawberry and red grape juice, *Int J Vitam Nutr Res*, 89, 29–36. https://doi.org/10.1024/0300-9831/a000546.

Cai, Y., Li, X., Pan, Z., Zhu, Y., Tuo, J., Meng, Q., Dai, G., Yang, G., & Pan, Y. (2020). Anthocyanin ameliorates hypoxia and ischemia induced inflammation and apoptosis by increasing autophagic flux in SH-SY5Y cells, *Eur J Pharmacol*, 883, 173360. https://doi.org/10.1016/j.ejphar.2020.173360.

Charepalli, V., Reddivarib, L., Radhakrishnana, S., Vaddea, R., Agarwald, R., & Vanamala, J.K. (2015). Anthocyanin-containing purple-fleshed potatoes suppress colon tumorigenesis via elimination of colon cancer stem cells, *J Nutr Biochem*, 26, 1641–1649. https://doi.org/10.1016/j.jnutbio.2015.08.005.

Chen, J., Sun, J., Jiang, J., & Zhou, J. (2018a). Cyanidin protects SHSY5Y human neuroblastoma cells from 1-methyl-4-phenylpyridinium-induced neurotoxicity, *Pharmacol*, 102, 126–132. https://doi.org/10.1159/000489853.

Chen, K., Kortesniemi, M.K., Linderborg, K.M., & Yang, B. (2023). Anthocyanins as promising molecules affecting energy homeostasis, inflammation, and gut microbiota in type 2 diabetes with special reference to impact of acylation. *J Agric Food Chem*, 71, 1002–1017. https://doi.org/10.1021/acs.jafc.2c05879.

Chen, L., Jiang, B., Zhong, C., Guo, J., Zhang, L., Mu, T. Zhang, Q., & Bi, X. (2018b). Chemoprevention of colorectal cancer by black raspberry anthocyanins involved the modulation of gut microbiota and SFRP2 demethylation, *Carcinogenesis*, 39, 471–481. https://doi.org/10.1093/carcin/bgy009.

Chen, T., Yan, F., Qian, J., Guo, M., Zhang, H., Tang, X., Chen, F., Stoner, G.D., & Wang, X. (2012). Randomized phase II trial of lyophilized strawberries in patients with dysplastic precancerous lesions of the esophagus, *Cancer Prev Res*, 5, 41–50. https://doi.org/10.1158/1940-6207.CAPR-11-0469.

Chen, W.-L., & Zhao, J. (2023). Association between dietary anthocyanidins intake and depression among US adults: A cross-sectional study (NHANES, 2007–2010 and 2017–2018), *BMC Psychiatry*, 23, 525. https://doi.org/10.1186/s12888-023-05029-8.

Chen, X., Zhang, W., & Xu, X. (2021). Cyanidin-3-glucoside suppresses the progression of lung adenocarcinoma by downregulating TP53I3 and inhibiting PI3K/AKT/mTOR pathway, *World J Surg Onc*, 19, 232. https://doi.org/10.1186/s12957-021-02339-7.

Christoudia, N., Bekas, N., Kanata, E., Chatziefsthathiou, A., Petta, S., Karagianni, K., Da Silva Correia, S.M., Schmitz, M., Zerr, I., Tsamesidis, I., Xanthopoulos, K., & Dafou, D. (2024). Anti-prion effects of anthocyanins, *Redox Biology,* 72, Art.n. 103133. https://doi.org/10.1016/j.redox.2024.103133.

Cui, H.-X., Chen, J.-H., Li, J.-W., Cheng, F.-R., & Yuan, K. (2018). Protection of anthocyanin from *Myrica rubra* against cerebral ischemia-reperfusion injury via modulation of the TLR4/NF-B and NLRP3 pathways, *Molecules*, 23, 1788. https://doi.org/10.3390/molecules23071788.

Cui, W.-S., Zhang, Q., & Zhao, X.-H. (2020). Impact of heat treatment on anti-oxidative and anti-colon cancer activities of the soluble extracts from black mulberry (*Morus nigra* L.) using water and ethanol–water solvents, *RSC Adv*, 10, 30415. https://doi.org/10.1039/D0RA05598K.

da Silva, F.L., Escribano-Bailón, M.T., Pérez Alonso, J.J., Rivas-Gonzalo, J.C., & Santos-Buelga, C. (2007). Anthocyanin pigments in strawberry, *LWT-Food Sci Technol*, 40, 374–382. https://doi.org/10.1016/j.lwt.2005.09.018.

de Aguiar Cipriano, P., Kim, H., Fang, C., Venancio, V.P., Mertens-Talcott, S.U., & Talcott, S.T. (2022). In vitro digestion, absorption and biological activities of acylated anthocyanins from purple sweet potatoes (*Ipomoea batatas*), *Food Chem*, 374, Art.n. 131076. https://https://doi.org/10.1016/j.foodchem.2021.131076.

Delgado-Povedano, M.M., de Villiers, A., Hann, S., & Causon, T. (2021). Identity confirmation of anthocyanins in berries by LC-DAD-IM-QTOFMS, *Electrophoresis*, 42, 473–481. https://doi.org/10.1002/elps.202000274.

Di Giacomo, C., Acquaviva, R., Santangelo, R., Sorrenti, V., Vanella, L., Li Volti, G., D'Orazio, N., Vanella, A., & Galvano, F. (2012). Effect of treatment with cyanidin-3-O-β-D-glucoside on rat ischemic/reperfusion brain damage, *Evid Based Complement Alternat Med*, 2012, 285750. https://doi.org/10.1155/2012/285750.

Diaconeasa, Z., Ayvaz, H., Ruginã, D., Leopold, L., Stănilă, A., Socaciu, C., Tăbăran, F., Luput, L., Mada, D.C., Pintea, A., & Jefferson, A. (2017). Melanoma inhibition by anthocyanins is associated with the reduction of oxidative stress biomarkers and changes in mitochondrial membrane potential, *Plant Foods Hum Nutr*, 72, 404–410. https://doi.org/10.1007/s11130-017-0638-x.

do Rosario, V.A., Fitzgerald, Z., Broyd, S., Paterson, A., Roodenrys, S., Thomas, S., Bliokas, V., Potter, J., Walton, K., Weston-Green, K., Yousefi, M., Williams, D., Wright, I.M.R., & Charlton, K. (2021). Food anthocyanins decrease concentrations of TNF-α in older adults with mild cognitive impairment: A randomized, controlled, double blind clinical trial, *Nutr Metab Cardiovasc Dis*, 31, 950–960. https://doi.org/10.1016/j.numecd.2020.11.024.

Dreiseitel, A., Korte, G., Schreier, P., Oehme, A., Locher, S., Domani, M., Hajaka, G., & Sand, P.G. (2009). Berry anthocyanins and their aglycons inhibit monoamine oxidases A and B, *Pharmacological Research*, 59, 306–311. https://doi.org/10.1016/j.phrs.2009.01.014.

Dudek, A., Strugała-Danak, P., Kral, T., Hof, M., & Pruchnik, H. (2023). An analysis of interactions between three structurally diverse anthocyanidins, as well as their glucosides, and model biological membranes, albumin, and plasmid DNA, *Sci Rep*, 13, Art.n. 12883. https://doi.org/10.1038/s41598-023-39470-2

Fan, D., Alamri, Y, Liu, K., MacAskill, M., Harris, P., Brimble, M., Dalrymple-Alford, J., Prickett, T., Menzies, O., Laurenson, A., Anderson, T., & Guan, J. (2018). Supplementation of blackcurrant anthocyanins increased cyclic glycine-proline in the cerebrospinal fluid of Parkinson patients: Potential treatment to improve insulin-like growth factor-1 function, *Nutrients*, 10, 714. https://doi.org/10.3390/nu10060714.

Fan, M.J., Wang, I.C., Hsiao, Y.-T., Lin, H.-Y., Tang, N.-Y., Hung, T.-C., Quan, C., Lien, J.-C., & Chung, J.-G. (2015). Anthocyanins from black rice (*Oryza sativa* L.) demonstrate antimetastatic properties by reducing MMPs and NF-κB expressions in human oral cancer CAL 27 cells, *Nutr Cancer*, 67, 327–338. https://doi.org/10.1080/01635581.2015.990576.

Fisk, J., Khalid, S., Reynolds, S.A., &. Williams, C.M. (2020). Effect of 4 weeks daily wild blueberry supplementation on symptoms of depression in adolescents, *Br J Nutr*, 124, 181–188. https://doi.org/10.1017/S0007114520000926.

Forni, C., Braglia, R., Mulinacci, N., Urbani, A., Ronci, M., Gismondi, A., Tabolacci, C., Provenzano, B., Lentini, A., & Beninati, S. (2014). Antineoplastic activity of strawberry (*Fragaria ananassa* Duch.) crude extracts on b16-f10 melanoma cells, *Mol Biosyst*, 10, 1255–1263. https://doi.org/10.1039/c3mb70316a.

Galli, R.L., Bielinski, D.F., Szprengiel, A., Shukitt-Hale, B., & Joseph, J.A. (2006). Blueberry supplemented diet reverses age-related decline in hippocampal HSP70 neuroprotection, *Neurobiol Aging*, 27, 344–350. https://doi.org/10.1016/j.neurobiolaging.2005.01.017.

Gérard, V., Ay, E., Morlet-Savary, F., Graff, B., Galopin, C., Ogren, T., Mutilangi, W., & Lalevée, J. (2019). Thermal and photochemical stability of anthocyanins from black carrot, grape juice, and purple sweet potato in model beverages in the presence of ascorbic acid, *J Agric Food Chem*, 67, 5647–5660. https://doi.org/10.1021/acs.jafc.9b01672.

Giusti, M.M., & Wrolstad, R.E. (2003). Acylated anthocyanins from edible sources and their applications in food systems, *Biochem Eng J*, 14, 217–225. https://doi.org/10.1016/S1369-703X(02)00221-8.

Gras, C.C., Bause, K., Leptihn, S., Carle, R., & Schweiggert, R.M. (2018). Effect of chlorogenic acid on spectral properties and stability of acylated and non-acylated cyanidin-3-O-glycosides, *Food Chem*, 240, 940–950. https://doi.org/10.1016/j.foodchem.2017.07.137.

Grasso, M., Caruso, G., Godos, J., Bonaccorso, A., Carbone, C., Castellano, S., Currenti, W., Grosso, G., Musumeci, T., & Caraci, F. (2021). Improving cognition with nutraceuticals targeting TGF-1 signaling, *Antioxidants*, 10, 1075. https://doi.org/10.3390/antiox10071075.

He, J., & Giusti, M.M. (2010). Anthocyanins: Natural colorants with health-promoting properties, *Annu Rev Food Sci Technol*, 1, 163–187. https://https://doi.org/10.1146/annurev.food.080708.100754.

Henriques, J.F., Serra, D., Dinis, T.C.P., & Almeida, L.M. (2020). The anti-neuroinflammatory role of anthocyanins and their metabolites for the prevention and treatment of brain disorders, *Int J Mol Sci*, 21, 1–31. https://doi.org/10.3390/ijms21228653.

Herrera-Sotero, M.Y., Cruz-Hernández, C.D., Oliart-Ros, R.M., Chávez-Servia, J.L., Guzmán-Gerónimo, R.I., González-Covarrubias, V., Cruz-Burgos, M., & Rodríguez-Dorantes, M. (2020). Anthocyanins of blue corn and tortilla arrest cell cycle and induce apoptosis on breast and prostate cancer cells, *Nutr Cancer*, 2020, 72, 768–777. https://doi.org/10.1080/01635581.2019.1654529.

Herrera-Sotero, M.Y., Cruz-Hernández, C.D., Trujillo-Carretero, C., Rodríguez-Dorantes, M., García-Galindo, H.S., Chávez-Servia, J.L., Oliart-Ros, R.M., & Guzmán-Gerónimo, R.I. (2017). Antioxidant and antiproliferative activity of blue corn and tortilla from native maize, *Chem Cent J*, 11, Art.n. 110. https://doi.org/10.1186/s13065-017-0341-x.

Ho, M.-L., Chen, P.-N., Chu, S.-C., Kuo, D.-Y., Kuo, W.-H., Chen, J.-Y., & Hsieh, Y.-S. (2010). Peonidin 3-glucoside inhibits lung cancer metastasis by downregulation of proteinases activities and MAPK pathway, *Nutr Cancer*, 62, 505–516. https://doi.org/10.1080/01635580903441261.

Hui, C., Bin, Y., Xiaoping, Y., Long, Y., Chunye, C., Mantian, M., & Wenhua, L. (2010). Anticancer activities of an anthocyanin-rich extract from black rice against breast cancer cells in vitro and in vivo, *Nutr Cancer*, 62, 1128–1136. https://doi.org/10.1080/01635581.2010.494821.

Hwang, S.-N., Kim, J.-C., Bhuiyan, M.I.H., Kim, J.Y., Yang, J.S., Yoon, S.H., Yoon, K.D., & Kim, S.Y. (2018). Black rice (*Oryza sativa* L., Poaceae) extract reduces hippocampal neuronal cell death induced by transient global cerebral ischemia in mice, *Exp Neurobiol*, 27, 129–138. https://doi.org/10.5607/en.2018.27.2.129.

Iannuzzo, F., Schiano, E., Maisto, M., Schettino, A., Marigliano, N., Saviano, A., Abo Mansour, A., Iqbal, A.J., Maione, F., Tenore, G.C., & Novellino, E. (2025). Effect of proanthocyanidins from grape seed extract on benign prostatic hyperplasia, *Nutrients*, 17, Art. n. 73. https://doi.org/10.3390/nu17010073.

Iban-Arias, R., Sebastian-Valverde, M., Wu, H., Lyu, W., Wu, Q., Simon, J., & Pasinetti, G.M. (2022). Role of polyphenol-derived phenolic acid in mitigation of inflammasome-mediated anxiety and depression, *Biomedicines*, 10, 1264. https://doi.org/10.3390/biomedicines10061264.

Iqbal, S., Omara, T., Kahwa, I., & Khan, U.M. (2024). Anticancer potential of delphinidin and its derivatives: Therapeutic and mechanistic insights, *Med Chem Res*, 33, 1769–1786. https://doi.org/10.1007/s00044-024-03296-y.

Jang, H., Ha, U.-S., Kim, S., Yoon, B.-I., Han, D.-S., Yuk, S.-M., & Kim, S.-W. (2010). Anthocyanin extracted from black soybean reduces prostate weight and promotes apoptosis in the prostatic hyperplasia-induced rat model, *J Agric Food Chem*, 58, 12686–12691. https://doi.org/10.1021/jf102688g.

Jauhar, S., Johnstone, M., & McKenna, P.J. (2022). Schizophrenia, *Lancet*, 339, 473–486. https://doi.org/10.1016/S0140-6736(21)01730-X.

Jing, P., Bomser, J.A., Schwartz, S.J., He, J., Magnuson, B.A., & Giusti, M.M. (2008). Structure-function relationships of anthocyanins from various anthocyanin-rich extracts on the inhibition of colon cancer cell growth, *J Agric Food Chem*, 56, 9391–9398. https://doi.org/10.1021/jf8005917.

Jokioja, J., Yang, B., & Linderborg, K.M. (2021). Acylated anthocyanins: A review on their bioavailability and effects on postprandial carbohydrate metabolism and inflammation, *Compr Rev Food Sci Food Saf*, 20, 5570–5615. https://doi.org/10.1111/1541–4337.12836.

Kaiser, A., Haskins, C., Siddiqui, M.M., Hussain, A., & D'Adamo, C. (2019). The evolving role of diet in prostate cancer risk and progression, *Curr Opin Oncol*, 31, 222–229. https://doi.org/10.1097/CCO.0000000000000519.

Kaur, R., Sood, A., Lang, D.K., Bhatia, S., Al-Harrasi, A., Aleya, L., & Behl, T. (2022). Potential of flavonoids as anti-Alzheimer's agents: Bench to bedside, *Environ Sci Pollut Res*, 29, 26063–26077. https://doi.org/10.1007/s11356-021-18165-z.

Kent, K., Charlton, K., Roodenrys, S., Batterham, M., Potter, J., Traynor, V., Gilbert, H., Morgan, O., & Richards, R. (2017). Consumption of anthocyanin-rich cherry juice for 12 weeks improves memory and cognition in older adults with mild-to-moderate dementia, *Eur J Nutr*, 56, 333–341. http://dx.doi.org/10.1007/s00394-015-1083-y.

Khalid, S., Barfoot, K.L., May, G., Lamport, D.J., Reynolds, S.A., & Williams, C.M. (2017). Effects of acute blueberry flavonoids on mood in children and young adults, *Nutrients*, 9, 158. https://doi.org/10.3390/nu9020158.

Khan, M.S., Ali, T., Kim, M.W., Jo, M.H., Jo, M.G., Badshah, H., & Kim, M.O. (2016). Anthocyanins protect against LPS-induced oxidative stress-mediated neuroinflammation and neurodegeneration in the adult mouse cortex, *Neurochem Int*, 100, 1–10. https://doi.org/10.1016/j.neuint.2016.08.005.

Kim, M.-H., Jeong, Y.-J., Cho, H.-J, Hoe, H.-S., Park, K.-K., Park, Y.-Y., Choi, Y.H., Kim, C.-H., Chang, H.-W., Park, Y.-J., Chung, I.-K., & Chang, Y.-C. (2017). Delphinidin inhibits angiogenesis through the suppression of HIF-1α and VEGF expression in A549 lung cancer cells, *Oncolol Rep*, 37, 777–784. https://doi.org/10.3892/or.2016.5296.

Knobloch, T.J., Ryan, N.M., Bruschweiler-Li, L., Wang, C., Bernier, M.C., Somogyi, A., Yan, P.S., Cooperstone, J.L., Mo, X., Brüschweiler, R.P., Weghorst, C.M., & Oghumu, S. (2019). Metabolic regulation of glycolysis and AMP activated protein kinase pathways during black raspberry-mediated oral cancer chemoprevention, *Metabolites*, 9, 140. https://doi.org/10.3390/metabo9070140.

Knobloch, T.J., Uhrig, L.K., Pearl, D.K., Casto, B.C., Warner, B.M., Clinton, S.K., Sardo-Molmenti, C.L., Ferguson, J.M., Daly, B.T., Riedl, K., Schwartz, S.J., Vodovotz, Y., Buchta, A.J., Schuller, D.E., Ozer, E., Agrawal, A., & Weghorst, C.M. (2016). Suppression of proinflammatory and prosurvival biomarkers in oral cancer patients consuming a black raspberry phytochemical-rich troche, *Cancer Prev Res (Phila)*, 9, 159–171. https://doi.org/10.1158/1940-6207.CAPR-15-0187.

Kong, Y., Wang, X., Wu, Z., Li, Y., Xu, F., & Xie, F. (2023). Enzymatic acylation of black rice anthocyanins and evaluation of antioxidant capacity and stability of their derivatives, *Foods*, 12, Art.n. 4505. https://doi.org/10.3390/foods12244505.

Kowalczyk, T., Muskała, M., Merecz-Sadowska, A., Sikora, J., Picot, L., & Sitarek, P. (2024). Anti-inflammatory and anticancer effects of anthocyanins in in vitro and in vivo studies, *Antioxidants*, 13, Art.n. 1143. https://https://doi.org/10.3390/antiox13091143.

Kresty, L.A., Fromkes, J.J., Frankel, W.L., Hammond, C.D., Seeram, N.P., Baird, M., & Stoner, G.D. (2018). A phase I pilot study evaluating the beneficial effects of black raspberries in patients with Barrett's esophagus, *Oncotarget*, 9, 35356–35372. https://doi.org/10.18632/oncotarget.10457.

Lane, C.A., Hardy, J., & Schott, J.M. (2018). Alzheimer's disease, *Eur J Neurol*, 25, 59–70. https://doi.org/10.1111/ene.13439.

Leonarski, E., Cesca, K., de Oliveira, D., & Zielinski, A.A.F. (2023). A review on enzymatic acylation as a promising opportunity to stabilizing anthocyanins, *Crit Rev Food Sci Nutr*, 63(24), 6777–6796. https://doi.org/10.1080/10408398.2022.2041541.

Lila, M.A., Burton-Freeman, B., Grace, M., & Kalt, W. (2016). Unraveling anthocyanin bioavailability for human health, *Annu Rev Food Sci Technol*, 7, 375–393. https://doi.org/10.1146/annurev-food-041715-033346.

Lippert, E., Ruemmele, P., Obermeier, F., Goelder, S., Kunst, C., Rogler, G., Dunger, N., Messmann, H., Hartmann, A., & Endlicher, E. (2017). Anthocyanins prevent colorectal cancer development in a mouse model, *Digestion*, 95, 275–280. https://doi.org/10.1159/000475524.

Lu, J.N., Panchanathan, R., Lee, W.S., Kim, H.J., Kim, D.H., Choi, Y.H., Kim, G., Shin, S.C., & Hong, S.C. (2017). Anthocyanins from the fruit of *Vitis coignetiae* Pulliat inhibit TNF-augmented cancer proliferation, migration, and invasion in A549 cells, *Asian Pac J Cancer Prev*, 18, 2919–2923. https://doi.org/10.22034/APJCP.2017.18.11.2919.

Luo, H., Gao, M., Lu, H., Chen, Q., & Lian, X. (2024). Anthocyanins prevent the development and progression of urethane-induced lung cancer by regulating energy metabolism in mice, *Food Nutr Res*, 68, 10434. https://doi.org/10.29219/fnr.v68.10434.

Luo, L.-P., Han, B., Yu, X.-P, Chen, X.-Y., Zhou, J., Chen, W., Zhu, Y.-F., Peng, X.-L., Zou, Q., & Li, S.-Y. (2014). Anti-metastasis activity of black rice anthocyanins against breast cancer: Analyses using an ErbB2 positive breast cancer cell line and tumoral xenograft model, *Asian Pac J Cancer Prev*, 15, 6219–6225. https://doi.org/10.7314/APJCP.2014.15.15.6219.

Luo, X., Wang, R., Wang, J., Li, Y. Luo, H., Chen, S., Zeng, X., & Han, Z. (2022). Acylation of anthocyanins and their applications in the food industry: Mechanisms and recent research advances, *Foods*, 11, Art.n. 2166. https://doi.org/10.3390/foods11142166.

Majhi, S. (2022). Discovery, development and design of anthocyanins-inspired anticancer agents: A comprehensive review, *Anti-Cancer Agents Med Chem*, 22, 3219–3238. https://doi.org/10.2174/187152062166621101514 2310.

Mallery, S.R., Budendorf, D.E., Larsen, M.P., Pei, P., Tong, M., Holpuch, A.S., Larsen, P.E., Stoner, G.D., Fields, H.W., Chan, K.K., Ling, Y., & Liu, Z. (2011). Effects of human oral mucosal tissue, saliva, and oral microflora on intraoral metabolism and bioactivation of black raspberry anthocyanins, *Cancer Prev Res*, 4, 1209–1221. https://doi.org/10.1158/1940–6207.CAPR-11–0040.

Martău, G.A., Bernadette-Emőke, T., Odocheanu, R., Soporan, D.A., Bochis, M., Simon, E., & Vodnar, D.C. (2023). *Vaccinium* species (Ericaceae): Phytochemistry and biological properties of medicinal plants, *Molecules,* 28, Art.n. 1533. https://doi.org/10.3390/molecules28041533.

Mastantuono, T., Di Maro, M., Chiurazzi, M., Battiloro, L., Muscariello, E., Nasti, G., Starita, N., Colantuoni, A., & Lapi, D. (2018). Rat pial microvascular changes during cerebral blood flow decrease and recovery: Effects of cyanidin administration, *Front Physiol*, 9, 540. https://doi.org/10.3389/fphys.2018.00540.

Matboli, M., Hasanin, A.H., Hussein, R., El-Nakeep, S., Habib, E.K., Ellackany, R., & Saleh, L.A. (2021). Cyanidin 3-glucoside modulated cell cycle progression in liver precancerous lesion, in vivo study, *World J Gastroenterol*, 27, 1435–1450. https://doi.org/10.3748/wjg.v27.i14.1435.

Matsuzaki, T., Fujii, M., Noro, H., Togo, S., Watanabe, M., Suganuma, M., Sharma, S., Kobayashi, N., Kawamura, R., Nakabayashi, S., & Yoshikawa, H.Y. (2024). Simultaneous visualization of membrane fluidity and morphology defines adhesion signatures of cancer cells, *Proc Natl Acad Sci USA*, 121, Art.n. e2412914121. https://doi.org/10.1073/pnas.2412914121.

Mattioli, R., Francioso, A., Mosca, L., & Silva, P. (2020). Anthocyanins: A comprehensive review of their chemical properties and health effects on cardiovascular and neurodegenerative diseases, *Molecules*, 25, 3809, https://doi.org/10.3390/molecules25173809.

Mauramo, M., Onali, T., Wahbi, W., Vasara, J., Lampinen, A., Mauramo, E., Kivimäki, A., Martens, S., Häggman, H., Sutinen, M., & Salo, T. (2021). Bilberry (*Vaccinium myrtillus* L.) powder has anticarcinogenic effects on oral carcinoma in vitro and in vivo, *Antioxidants*, 10, 1319. https://doi.org/10.3390/antiox10081319.

Mazewski, C., Liang, K., & Gonzalez de Mejiaa, E. (2018). Comparison of the effect of chemical composition of anthocyanin-rich plant extracts on colon cancer cell proliferation and their potential mechanism of action using in vitro, in silico, and biochemical assays, *Food Chem*, 242, 378–388. https://doi.org/10.1016/j.foodchem.2017.09.086.

Mazzoni, L., Giampieri, F., Alvarez Suarez, J.M., Gasparrini, M., Mezzetti, B., Forbes Hernandez, T.Y., & Battino, M.A. (2019). Isolation of strawberry anthocyanin-rich fractions and their mechanisms of action against murine breast cancer cell lines, *Food Funct*, 10, 7103–7120. https://doi.org/10.1039/c9fo01721f.

Menzies, F.M., Fleming, A., & Rubinsztein, D.C. (2015). Compromise autophagy and neurodegenerative diseases, *Nat Rev Neurosci*, 16, 345–357. https://doi.org/10.1038/nrn3961.

Mestrom, A., Charlton, K.E., Thomas, S.J., Larkin, T.A., Walton, K.L., Elgellaie, A., & Kent, K. (2024). Higher anthocyanin intake is associated with lower depressive symptoms in adults with and without major depressive disorder, *Food Sci Nutr*, 12, 2202–2209. https://doi.org/10.1002/fsn3.3850.

Miyata, R., Sano, H., Hoshino, S., & Kumazawa, S. (2022). Thermostability and catechol-O-methyltransferase inhibitory activity of acylated anthocyanins from purple yam, *Biosci Biotechnol Biochem*, 86, 916–921. https://doi.org/10.1093/bbb/zbac060.

Mohammed, H.A., & Khan, R.A. (2022). Anthocyanins: Traditional uses, structural and functional variations, approaches to increase yields and products' quality, hepatoprotection, liver longevity, and commercial products, *Int J Mol Sci*, 23(2), 2149. https://doi.org/10.3390/ijms23042149.

Mottaghipisheh, J., Doustimotlagh, A.H., Irajie, C., Tanide, N., Barzegar, A., & Iraji, A. (2022). The promising therapeutic and preventive properties of anthocyanidins/anthocyanins on prostate cancer, *Cells*, 11, 1070. https://doi.org/10.3390/cells11071070.

Mudd, A.M., Gu, T., Munagala, R., Jeyabalan, J., Egilmez, N.K., & Gupta, R.C. (2020). Chemoprevention of colorectal cancer by anthocyanidins and mitigation of metabolic shifts induced by dysbiosis of the gut microbiome, *Cancer Prev Res*, 13, 41–52. https://doi.org/10.1158/1940-6207.CAPR-19-0362.

Murakami, A., Nesumi, A., Maeda-Yamamoto, M., Yamaguchi, H., Yashima, K., Miura, M., Nakano, T., & Nekoshima, K. (2015). Anthocyanin-rich tea Sunrouge upregulates expressions of heat shock proteins in the gastrointestinal tract of ICR mice: A comparison with the conventional tea cultivar Yabukita, *J Food Drug Anal*, 23, 407–416. https://doi.org/10.1016/j.jfda.2014.11.004.

Natthee, P., Wattanathorn, J., Thukham-mee, W., Paholpak, P., Rangseekajee, P., Piyavhatkul, N., Wattanathorn, S., Muchimapura, S., & Tong-Un, T. (2024). A black sticky rice-derived functional ingredient improves anxiety, depression, and stress perception in adult volunteers, *Foods*, 13, 3884. https://doi.org/10.3390/foods13233884.

Neto, C.C. (2007). Cranberry and its phytochemicals: A review of in vitro anticancer studies, *J Nutr*, 137, 186S–193S. https://doi.org/10.1093/jn/137.1.186s.

Ni, P., Liu, M., Wang, D., Tian, Y., Zhao, L., Wei, J., Yu, X., Qi, X., Li, X., Yu, H., Ni, R., Ma, X., Deng, W., Guo, W., Wang, Q., & Li, T. (2021). Association analysis between catechol-*O*-methyltransferase expression and cognitive function in patients with schizophrenia, bipolar disorder, or major depression, *Neuropsychiatr Dis Treat*, 17, 567–574. https://doi.org/10.2147/NDT.S286102.

Novotny, J.A., Clevidence, B.A., & Kurilich, A.C. (2012). Anthocyanin kinetics are dependent on anthocyanin structure, *Br J Nutr*, 107, 504–509. https://doi.org/10.1017/S000711451100314X.

Oteiza, P.I., Cremonini, E., & Fraga, C.G. (2023). Anthocyanin actions at the gastrointestinal tract: Relevance to their health benefits, *Mol Asp Med*, 89, 101156. https://doi.org/10.1016/j.mam.2022.101156.

Pal, H.C., Sharma, S., Strickland, L.R., Agarwal, J., Athar, M., Elmets, C.A., & Afaq, F. (2013). Delphinidin reduces cell proliferation and induces apoptosis of non-small-cell lung cancer cells by targeting EGFR/VEGFR2 signaling pathways, *PLoS One*, 8, e77270. https://doi.org/10.1371/journal.pone.0077270.

Pan, P., Peiffer, D.S., Huang, Y.-W., Oshima, K., Stoner, G.D., & Wang, L.-S. (2018a). Inhibition of the development of N-nitrosomethylbenzylamine-induced esophageal tumors in rats by strawberries and aspirin, alone and in combination, *J Berry Res*, 8, 137–146. https://doi.org/10.3233/JBR–170291.

Pan, P., Yu, J., & Wang, L.-S. (2018b). Colon cancer: What we eat, *Surg Oncol Clin N Am*, 27, 243–267. https://doi.org/10.1016/j.soc.2017.11.002.

Pan, Z., Cui, M., Dai, G., Yuan, T., Li, Y., Ji, T., & Pan, Y. (2018c). Protective effect of anthocyanin on neurovascular unit in cerebral ischemia/reperfusion injury in rats, *Front Neurosci*, 12, 947. https://doi.org/10.3389/fnins.2018.00947.

Panchal, S.K., & Brown, L. (2023). Potential benefits of anthocyanins in chronic disorders of the central nervous system, *Molecules*, 28, Art.n. 80. https://doi.org/10.3390/molecules28010080.

Papavassiliou, K.A., Sofianidi, A.A., & Papavassiliou, A.G. (2024). Anthocyanins in non-small cell lung cancer (NSCLC) treatment and prevention, *Nutrients*, 16, 1458. https://doi.org/10.3390/nu16101458.

Peiffer, D.S., Wang, L.-S., Zimmerman, N.P., Ransom, B.W.S., Carmella, S.G., Kuo, C.-T., Chen, J.-H., Oshima, K., Huang, Y.-W., Hecht, S.S., & Stoner, G.D. (2016). Dietary consumption of black raspberries or their anthocyanin constituents alters innate immune cell trafficking in esophageal cancer, *Cancer Immunol Res*, 4, 72–82. https://doi.org/10.1158/2326-6066.CIR-15-0091.

Peng, Y., Dong, W., Chen, G., Mi, J., Lu, L., Xie, Z., Xu, W., Zhou, W., Sun, Y., Zeng, X., Cao, Y., & Yan, Y. (2023). Anthocyanins from *Lycium ruthenicum* Murray ameliorated high-fructose diet-induced neuroinflammation through the promotion of the integrity of the intestinal barrier and the proliferation of lactobacillus, *J Agric Food Chem*, 71, 2864–2882. https://doi.org/10.1021/acs.jafc.2c06713.

Pereira, A.R., Fernandes, V.C., Delerue-Matos, C., de Freitas, V., Mateus, N., & Oliveira, J. (2024). Exploring acylated anthocyanin-based extracts as a natural alternative to synthetic food dyes: Stability and application insights, *Food Chem*, 461, 140945. https://doi.org/10.1016/j.foodchem.2024.140945.

Petrick, J.L., Kelly, S.P., Altekruse, S.F., McGlynn, K.A., & Rosenberg, P.S. (2016). Future of hepatocellular carcinoma incidence in the United States forecast through 2030, *J Clin Oncol*, 34, 1787–1794. https://doi.org/10.1200/JCO.2015.64.7412.

Prado, G., Pierattini, I, Villarroel, G., Fuentes, F., Silva, A., Echeverria, F., Valenzuela, R., & Bustamante, A. (2022). Bioaccessibility of anthocyanins on in vitro digestion models: Factors implicated and role in functional foods development, *Curr Med Chem*, 29, 1124–1141. https://doi.org/10.2174/0929867328666211123102536.

Preta, G. (2020). New insights into targeting membrane lipids for cancer therapy, *Front Cell Dev Biol*, 8, Art.n. 571237. https://doi.org/10.3389/fcell.2020.571237.

Rai, V., Abdo, J., & Agrawal, D.K. (2023). Biomarkers for early detection, prognosis, and therapeutics of esophageal cancers, *Int J Mol Sci*, 24, Art.n. 3316. https://doi.org/10.3390/ijms240433.

Rampurwala, M., Wisinski, K.B., & O'Regan, R. (2016). Role of the androgen receptor in triple-negative breast cancer, *Clin Adv Hematol Oncol*, 14, 186–193.

Reddy, V.P., Aryal, P., & Darkwah, E.K. (2022). Advanced glycation end products in health and disease, *Microorganisms*, 10, 1848. https://doi.org/10.3390/microorganisms10091848.

Rugină, D., Hanganu, D., Diaconeasa, Z., Tăbăran, F., Coman, C., Leopold, L., Bunea, A., & Pintea, A. (2017). Antiproliferative and apoptotic potential of cyanidin-based anthocyanins on melanoma cells, *Int J Mol Sci*, 18, 949. https://doi.org/10.3390/ijms18050949.

Safdar, M.A., Aslam, R.M.N., Shakeel, A., Shiza, Waqar, M., Jmail, A., Mehmood, M.H., & Gul, H. (2023). Cyanidin as potential anticancer agent targeting various proliferative pathways, *Chem Biol Drug Des*, 101, 438–452. https://doi.org/10.1111/cbdd.14173.

Saini, R.K., Khan, M.I., Shang, X., Kumar, V., Kumari, V., Kesarwani, A., & Ko, E.-Y. (2024). Dietary sources, stabilization, health benefits, and industrial application of anthocyanins-a review, *Foods*, 13, Art.n. 1227. https://doi.org/10.3390/foods13081227.

Sajadimajd, S., Bahramsoltani, R., Iranpanah, A., Kumar Patra, J., Das, G., Gouda, S., Rahimi, R., Rezaeiamiri, E., Cao, H., Giampieri, F., Battino, M., Tundis, R., Campos, M.G., Farzaei, M.H., & Xiao, J. (2020). Advances on natural polyphenols as anticancer agents for skin cancer, *Pharmacol Res*, 151, Art.n. 104584. https://doi.org/10.1016/j.phrs.2019.104584.

Sanjay, Shin, J.H., Park, M., & Lee, H.-J. (2022). Cyanidin-3-*O*-glucoside regulates the M1/M2. Polarization of microglia via PPARγ and Aβ42 phagocytosis through TREM2 in an Alzheimer's disease model, *Mol Neurobiol*, 59, 5135–5148. https://doi.org/10.1007/s12035-022-02873-9.

Sarhangi, N., Hajjari, S., Heydari, S.F., Ganjizadeh, M., Rouhollah, F., & Hasanzad, M. (2022). Breast cancer in the era of precision medicine, *Mol Biol Rep*, 49, 10023–10037. https://doi.org/10.1007/s11033-022-07571-2.

Saurman, V., Margolis, K.G., & Luna, R.A. (2020). Autism spectrum disorder as a brain–gut–microbiome axis disorder, *Dig Dis Sci*, 65, 818–828. https://doi.org/10.1007/s10620-020-06133-5.

Sehitoglu, M.H., Farooqi, A.A., Qureshi, M.Z., Butt, G., & Aras, A. (2014). Anthocyanins: Targeting of signaling networks in cancer cells, *Asian Pac J Cancer Prev*, 15, 2379–2381. https://doi.org/10.7314/APJCP.2014.15.5.2379.

Serra, D., Henriques, J.F., Sousa, F.J., Laranjo, M., Resende, R., Ferreira-Marques, M., de Freitas, V., Silva, G., Peça, J., Dinis, T.C.P., & Almeida, L.M. (2022). Attenuation of autism-like behaviors by an anthocyanin-rich extract from Portuguese blueberries via microbiota–gut–brain axis modulation in a valproic acid mouse model, *Int J Mol Sci*, 23, 9259. https://doi.org/10.3390/ijms23169259.

Shah, S.A., Amin, F.U., Khan, M., Abid, M.N., Rehman, S.U., Kim, T.H., Kim, M.W., & Kim, M.O. (2016). Anthocyanins abrogate glutamate-induced AMPK activation, oxidative stress, neuroinflammation, and neurodegeneration in postnatal rat brain, *J Neuroinflam*, 13, Art.n. 286. https://doi.org/10.1186/s12974-016-0752-y.

Sharma, P., McClees, S.F., & Afaq, F. (2017). Pomegranate for prevention and treatment of cancer: An update, *Molecules*, 22, Art.n. 177. https://doi.org/10.3390/molecules22010177.

Shi, N., & Chen, T. (2022). Chemo preventive properties of black raspberries and strawberries in esophageal cancer review, *Antioxidants*, 11, 1815. https://doi.org/10.3390/antiox11091815.

Shi, N., Riedl, K.M., Schwartz, S.J., Zhang, X., Clinton, S.K., & Chen, T. (2016). Efficacy comparison of lyophilised black raspberries and combination of celecoxib and PBIT in prevention of carcinogen-induced esophageal cancer in rats, *J Funct Foods*, 27, 84–94. https://doi.org/10.1016/j.jff.2016.08.044.

Shin, W.-H., Park, S.-J., & Kim, E.-J. (2006). Protective effect of anthocyanins in middle cerebral artery occlusion and reperfusion model of cerebral ischemia in rats, *Life Sci*, 79, 130–137. https://doi.org/10.1016/j.lfs.2005.12.033.

Slimestad, R., & Solheim, H. (2002). Anthocyanins from black currants (*Ribes nigrum* L.), *J Agric Food Chem*, 50, 3228–3231. https://doi.org/10.1021/jf011581u.

Sok, M., Šentjurc, M., & Schara, M. (1999). Membrane fluidity characteristics of human lung cancer, *Cancer Lett*, 139, 215–220. https://doi.org/10.1016/S0304-3835(99)00044-0.

Sousa, A., Araújo, P., Azevedo, J., Cruz, L., Fernandes, I., Mateus, N., & de Freitas, V. (2016). Antioxidant and antiproliferative properties of 3-deoxyanthocyanidins, *Food Chem*, 192, 142–148. https://doi.org/10.1016/j.foodchem.2015.06.108.

Spohr, L., de Aguiar, M.S.S., Bona, N.P., Luduvico, K.P., Alves, A.G., Domingues, W.B., Blödorn, E.B., Bortolatto, C.F., Brüning, C.A., Campos, V.F., Stefanello, F.M., & Spanevello, R.M. (2023). Blueberry extract modulates brain enzymes activities and reduces neuroinflammation: Promising effect on lipopolysaccharide-induced depressive-like behavior, *Neurochem Res*, 48, 846–861. https://doi.org/10.1007/s11064-022-03813-8.

Strugała, P., Dudra, A., & Gabrielska, J. (2016). Interaction between mimic lipid membranes and acylated and nonacylated cyanidin and its bioactivity, *J Agric Food Chem*, 64(20), 7414–7422. https://doi.org/https://doi.org/10.1021/acs.jafc.6b03066.

Su, C.-C., Wang, C.-J., Huang, K.-H., Lee, Y.-J., Chan, W.-M., & Chang, Y.-C. (2018). Anthocyanins from *Hibiscus sabdariffa* calyx attenuate in vitro and in vivo melanoma cancer metastasis, *J Funct Foods*, 48, 614–631. https://doi.org/10.1016/j.jff.2018.07.032.

Suganyadevi, P., Saravanakumar, K.M., & Mohandas, S. (2013). The antiproliferative activity of 3-deoxyanthocyanins extracted from red sorghum (*Sorghum bicolor*) bran through P(53)-dependent and Bcl-2 gene expression in breast cancer cell line, *Life Sci*, 92, 379–382. https://doi.org/10.1016/j.lfs.2013.01.006.

Teng, H., Mi, Y., Cao, H., & Chen, L. (2022). Enzymatic acylation of raspberry anthocyanin: Evaluations on its stability and oxidative stress prevention, *Food Chem*, 372, Art.n. 130766. https://doi.org/10.1016/j.foodchem.2021.130766.

Teng, H., Mi, Y., Deng, H., He, Y., Wang, S., Ai, C., Cao, H., & Chen, L. (2023). Acylated anthocyanin inhibited the formation of heterocyclic amines in hybrid chemical model system and its underlying mechanism, *Food Chem X*, 17, Art.n. 100559. https://doi.org/10.1016/j.fochx.2023.100559.

Thummayot, S., Tocharus, C., Jumnongprakhon, P., Suksamrarn, A., & Tocharus, J. (2018). Cyanidin attenuates Aβ25_35-induced neuroinflammation by suppressing NF-κB activity downstream of TLR4/NOX4 in human neuroblastoma cells, *Acta Pharmacol Sin*, 39, 1439–1452. https://doi.org/10.1038/aps.2017.203.

Tikhonova, M.A., Shoeva, O.Y., Tenditnik, M.V., Akopyan, A.A., Litvinova, E.A., Popova, N.A., Amstislavskaya, T.G., & Khlestkina, E.K. (2024). Antitumor effects of an anthocyanin-rich grain diet in a mouse model of Lewis lung carcinoma, *Int J Mol Sci*, 25, 5727. https://doi.org/10.3390/ijms25115727.

Tran, P.H.L., & Tran, T.T.D. (2021). Blueberry supplementation in neuronal health and protective technologies for efficient delivery of blueberry anthocyanins, *Biomolecules*, 11, 102. https://doi.org/10.3390/biom11010102.

Trist, B.G., Hare, D.J., & Double, K.L. (2019). Oxidative stress in the aging substantia nigra and the etiology of Parkinson's disease, *Aging Cell*, 18, e13031. https://doi.org/10.1111/acel.13031.

Tulio Jr., A.Z., Reese, R.N., Wyzgoski, F.J., Rtnaldi, P.L., Fu, R., Scheerens, J.C., & Miller, A.R. (2008). Cyanidin 3-rutinoside and cyanidin 3-xylosylrutinoside as primary phenolic antioxidants in black raspberry, *J Agric Food Chem*, 56, 1880–1888. https://doi.org/10.1021/jf072313k.

Twarda-Clapa, A., Olczak, A., Białkowska, A.M., & Koziołkiewicz, M. (2022). Advanced glycation end-products (AGEs): Formation, chemistry, classification, receptors, and diseases related to AGEs, *Cells*, 11, 1312. https://doi.org/10.3390/cells11081312.

Urias-Lugo, D.A., Heredia, J.B., Muy-Rangel, M.D., Valdez-Torres, J.B., Serna-Saldívar, S.O., & Gutiérrez-Uribe, J.A. (2015). Anthocyanins and phenolic acids of hybrid and native blue maize (*Zea mays* L.) extracts and their antiproliferative activity in mammary (MCF7), liver (HepG2), colon (Caco2 and HT29) and prostate (PC3) cancer cells, *Plant Foods Hum Nutr*, 70, 193–199. https://doi.org/10.1007/s11130-015-0479-4.

Vishnu, V.R., Renjith, R.S., Mukherjee, A., Anil, S.R., Sreekumar, J., & Jyothi, A.N. (2019). Comparative study on the chemical structure and in vitro antiproliferative activity of anthocyanins in purple root tubers and leaves of sweet potato (*Ipomoea batatas*), *J Agric Food Chem*, 67, 2467–2475. https://doi.org/10.1021/acs.jafc.8b05473.

Walsh, J.J., Llorach, P., Cardozo Pinto, D.F., Wenderski, W., Christoffel, D.J., Salgado, J.S., Heifets, B.D., Crabtree, G.R., Malenka, R.C. (2021). Systemic enhancement of serotonin signaling reverses social deficits in multiple mouse models for ASD, *Neuropsychopharmacol*, 46, 2000–2010. https://doi.org/10.1038/s41386-021-01091-6.

Wang, E., Liu, Y., Xu, C., & Liu, J. (2017). Antiproliferative and proapoptotic activities of anthocyanin and anthocyanidin extracts from blueberry fruits on B16-F10 melanoma cells, *Food Nutr Res*, 61, 1325308. https://doi.org/10.1080/16546628.2017.1325308.

Wang, P., Liu, J., Zhuang, Y., & Fei, P. (2022). Acylating blueberry anthocyanins with fatty acids: Improvement of their lipid solubility and antioxidant activities, *Food Chem: X*, 15, Art.n. 100420. https://doi.org/10.1016/j.fochx.2022.100420.

Wang, X., Yang, D.-Y., Yang, L.-Q., Zhao, W.-Z., Cai, L.-Y., & Shi, H.-P. (2018). Anthocyanin consumption and risk of colorectal cancer: A meta-analysis of observational studies, *J Am Coll Nutr*, 38, 470–477. https://doi.org/10.1080/07315724.2018.1531084.

Wu, S., Hu, Y., Bai, W., Zhao, J., Huang, C., Wen, C., Deng, L., & Lu, D. (2019). Cyanidin-3-*O*-glucoside inhibits UVA-induced human dermal fibroblast injury by upregulating autophagy, *Photodermatol Photoimmunol Photomed*, 35, 360–368. https://doi.org/10.1111/phpp.12493.

Xie, L., Hao, X., Xie, J., Mo, J., Yuan, C., & Chen, W. (2024). Acetylated pelargonidin-3-*O*-glucoside alleviates hepatocyte lipid deposition through activating the AMPK-mediated lysosome-autophagy pathway and redox state, *Food Funct*, 15, 6929–6942. https://doi.org/10.1039/d4fo00185k.

Yan, Z., Li, C., Zhang, L., Liu, Q., Ou, S., & Zeng, X. (2016). Enzymatic acylation of anthocyanin isolated from black rice with methyl aromatic acid ester as donor: Stability of the acylated derivatives, *J Agric Food Chem*, 64, 1137–1143. https://doi.org/10.1021/acs.jafc.5b05031.

Yañez-Apam, J., Domínguez-Uscanga, A., Herrera-González, A., Contreras, J., Mojica, L., Mahady, G., & Luna-Vital, D.A. (2023). Pharmacological activities and chemical stability of natural and enzymatically acylated anthocyanins: A comparative review, *Pharmaceuticals,* 16, Art.n. 638. https://doi.org/10.3390/ph16050638.

Yang, W., Kortesniemi, M., Ma, X., Zheng, J., & Yang, B. (2019). Enzymatic acylation of blackcurrant (*Ribes nigrum*) anthocyanins and evaluation of lipophilic properties and antioxidant capacity of derivatives, *Food Chem*, 281, 189–196. https://doi.org/10.1016/j.foodchem.2018.12.111.

Yue, E., Tuguzbaeva, G., Chen, X., Qin, Y., Li, A., Sun, X., Dong, C., Liu, Y., Yu, Y., Zahra, S.M., Shan, Q., Jiang, Y., Du, Z., & Bai, Y. (2019). Anthocyanin is involved in the activation of pyroptosis in oral squamous cell carcinoma, *Phytomed*, 56, 286–294. https://doi.org/10.1016/j.phymed.2018.09.223.

Yue, L., Li, Y., Luo, Y., Alarfaj, A.A., & Shi, Y. (2024). Pelargonidin inhibits cell growth and promotes oxidative stress-mediated apoptosis in lung cancer A549 cells, *Biotechnol Appl Biochem*, 71, 1195–1203. https://doi.org/10.1002/bab.2621.

Yun, J.W., Lee, W.S., Kim, M.J., Lu, J.N., Kang, M.H., Kim, H.G., Kim, D.C., Choi, E.J., Choi, J.Y., Kim, H.G., Lee, Y.-K., & Ryu, C.H. (2010). Characterization of a profile of the anthocyanins isolated from *Vitis coignetiae* Pulliat and their anti-invasive activity on HT-29 human colon cancer cells, *Food Chem Toxicol*, 48, 903–909. https://doi.org/10.1016/j.fct.2009.12.031.

Zabady, S., Mahran, N., Soltan, M.A., Eldeen, M.A., Eid, R.A., Albogami, S., Fayad, E., Matboli, M., Habib, E.K., Hasanin, A.H., Ali, M.A., Mesbah, N.M., Abo-Elmatty, D.M., & Abdel-Hamed, A.R. (2022). Cyanidin-3-glucoside modulates hsa_circ_0001345/miRNA106b/ATG16L1 axis expression as a potential protective mechanism against hepatocellular carcinoma, *Curr Issues Mol Biol*, 44, 1677–1687. https://doi.org/10.3390/cimb44040115.

Zaim, M., Kara, I., & Muduroglu, A. (2021). Black carrot anthocyanins exhibit neuroprotective effects against MPP+ induced cell death and cytotoxicity via inhibition of oxidative stress mediated apoptosis, *Cytotechnology*, 73, 827–840. https://doi.org/10.1007/s10616-021-00500-4.

Zerr, I., Ladogana, A., Mead, S., Hermann, P., Forloni, G., & Appleby, B.S. (2024). Creutzfeldt–Jakob disease and other prion diseases, *Nat Rev Dis Primers*, 10, 14. https://doi.org/10.1038/s41572-024-00497-y.

Zhang, H., Guo, J., Mao, L., Li, Q., Guo, M., Mu, T., Zhang, Q., & Bi, X. (2019a). Up-regulation of miR-24-1-5p is involved in the chemoprevention of colorectal cancer by black raspberry anthocyanins, *Br J Nutr*, 122, 518–526. https://doi.org/10.1017/S0007114518003136.

Zhang, J., Wu, J., Liu, F., Tong, L., Chen, Z., Chen, J., He, H., Xu, R., Ma, Y., & Huang, C. (2019b). Neuroprotective effects of anthocyanins and its major component cyanidin-3-*O*-glucoside (C3G) in the central nervous system: An outlined review, *Eur J Pharmacol*, 858, 172500. http://dx.doi.org/10.1016/j.ejphar.2019.172500.

Zhang, N., & Jing, P. (2022). Anthocyanins in Brassicaceae: Composition, stability, bioavailability, and potential health benefits, *Crit Rev Food Sci Nutr*, 62, 2205–2220. https://doi.org/10.1080/10408398.2020.1852170.

Zhang, N., & Jing, P. (2023). Red cabbage anthocyanins attenuate cognitive impairment by attenuating neuroinflammation and regulating gut microbiota in aging mice, *J Agric Food Chem*, 271, 15064–15072. https://doi.org/10.1021/acs.jafc.3c03183.

Zhang, N., Zhang, Z., Xu, W., & Jing, P. (2021a). TMT-based quantitative proteomic analysis of hepatic tissue reveals the effects of dietary cyanidin-3-diglucoside-5-glucoside-rich extract on alleviating D-galactose-induced aging in mice, *J Proteomics*, 232, 104042. https://doi.org/10.1016/j.jprot.2020.104042.

Zhang, Y., Vareed, S.K., & Nai, M.G. (2005). Human tumor cell growth inhibition by nontoxic anthocyanidins, *Life Sci*, 76, 1465–1472. https://doi.org/10.1016/j.lfs.2004.08.025.

Zhang, Y., Yin, L., Huang, L., Tekliye, M., Xia, X., Li, J., & Dong, M. (2021b). Composition, antioxidant activity, and neuroprotective effects of anthocyanin-rich extract from purple highland barley bran and its promotion on autophagy, *Food Chem*, 339, 127849. https://doi.org/10.1016/j.foodchem.2020.127849.

Zhang, Y., Zhu, M., Wan, H., Chen, L., & Luo, F. (2022). Association between dietary anthocyanidins and risk of lung cancer, *Nutrients*, 14, 2643. https://doi.org/10.3390/nu14132643.

Zhao, C.-L, Yu, Y.-Q., Chen, Z.-J., Wen, G.-S., Wei, F.-G., Zheng, Q., Wang, C.-D., & Xiao, X.-L. (2017). Stability-increasing effects of anthocyanin glycosyl acylation, *Food Chem*, 214, 119–128. https://doi.org/10.1016/j.foodchem.2016.07.073.

Zhao, F., Wang, J., Wang, W., Lyu, L., Wu, W., & Li, W. (2023). The extraction and high antiproliferative effect of anthocyanin from garden blue blueberry, *Molecules*, 28, Art.n. 2850. https://doi.org/10.3390/molecules28062850.

Zhao, Y., Zhang, X., Chen, X., & Wei, Y. (2022). Neuronal injuries in cerebral infarction and ischemic stroke: From mechanisms to treatment (Review), *Int J Mol Med*, 49, 15. https://doi.org/10.3892/ijmm.2021.5070.

Zhou, J., Zhu, Y.-F., Chen, X.-Y., Han, B., Li, F., Chen, J.-Y., Peng, X.-L., Luo, L.-P., Chen, W., & Yu, X.-P. (2017). Black rice-derived anthocyanins inhibit HER-2-positive breast cancer epithelial–mesenchymal transition-mediated metastasis in vitro by suppressing FAK signaling, *Int J Mol Med*, 40, 1649–1656. https://doi.org/10.3892/ijmm.2017.3183.

Zozio, S., Pallet, D., & Dornier, M. (2011). Evaluation of anthocyanin stability during storage of a colored drink made from extracts of the Andean blackberry (*Rubus glaucus* Benth.), açai (*Euterpe oleracea* Mart.) and black carrot (*Daucus carota* L.), *Fruits*, 66, 203–215. https://doi.org/10.1051/fruits/2011030.

Zykin, P.A., Andreeva, A.A., Lykholay, A.N., Tsvetkova, N.V., & Voylokov, A.V. (2018). Anthocyanin composition and content in rye plants with different grain color, *Molecules*, 23, 948. https://doi.org/10.3390/molecules23040948.

Obesity and Cardiovascular and Gastrointestinal Diseases

11

Saroj Paudel and Joana Pico

11.1 INTRODUCTION

High blood glucose levels are a hallmark of a range of metabolic illnesses, collectively referred to as "metabolic syndrome," which increases the risk of heart disease, stroke, and type 2 diabetes (T2D) (Cho et al., 2018) and lowers the quality of life, increases mortality, and raises medical care expenses (Baena-Díez et al., 2016). Specifically, T2D is becoming a bigger public health concern around the world, with an expected impact on more than 693 million people (8.7% of the global population) by 2045 (Cho et al., 2018). The ninth edition of the *International Diabetes Federation's Diabetes Atlas* (IDF) states that, as of 2019, 463 million adults in the 20–79 age range have diabetes, and by 2030, that number is expected to increase to 578 million (see Figure 11.1). Several phenolic compounds extracted from different plant species like *Arctium*, *Morus nigra*, and *Vitex agnus-castus*; grape juice; and many others possess antioxidant activity and antitumor properties, as well as a hypoglycemic effect that helps control the glucose level in the body (Ali Asgar, 2013).

While plants have been utilized for centuries in traditional medicine to manage blood glucose levels, modern scientific research is increasingly supporting the antidiabetic properties of many plant species, which are typically attributed to a diverse array of bioactive compounds, including phenolic compounds, alkaloids, saponins, terpenoids (specially triterpenoids), and glycosides (known for delaying sugar absorption at the brush border membrane). Many of these phytochemicals are known to exert their antidiabetic effects through multiple mechanisms, often synergistically, including (1) stimulation of insulin secretion from pancreatic β-cells; (2) regeneration of pancreatic β-cells; (3) enhancement of glucose uptake; (4) glucose transport facilitation into cells; (5) inhibition of carbohydrate-digesting enzymes, which slows down glucose generation and absorption; (6) reduction of hepatic glucose production (so-called gluconeogenesis); (7) antioxidant and anti-inflammatory effects, ameliorating the damage to pancreatic β-cells and contributing to insulin resistance; and (8) improvement of lipid metabolism (i.e., lowering LDL ("bad") cholesterol and triglycerides, and increasing HDL ("good") cholesterol), which is beneficial for diabetes management and preventing associated complications.

DOI: 10.1201/9781003519591-14

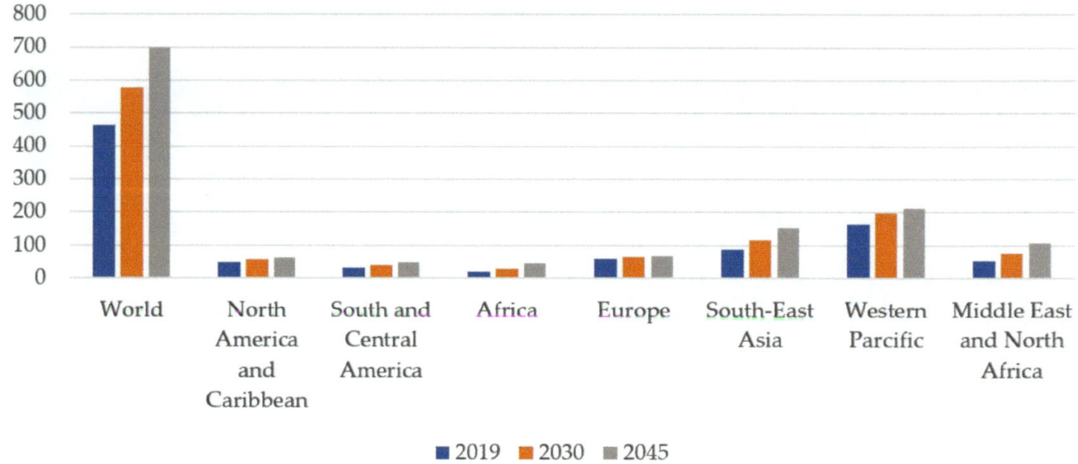

FIGURE 11.1 Number of individuals with diabetes around the world as per IDF region in 2019, 2030, and 2045.

Phenolic compounds, abundant in these medicinal plants, are known to present antioxidant and anti-inflammatory properties and regulate glucose metabolism *via* the mechanisms mentioned previously. Among the different groups of phenolic compounds, flavonoids – the largest phenolic group widely found in fruits, vegetables, and herbs – are specialized in enhancing insulin sensitivity and inhibiting carbohydrate-digesting enzymes. Anthocyanins are flavonoids often found in colorful fruits and vegetables (Koh et al., 2015) and are characterized by a hydroxyl group at position C3 of the B-ring, a positive charge in the oxygen of the C-ring, and a completely unsaturated C-ring, which makes them good antioxidants that can be used as medicinal agents (Diaconeasa et al., 2015). Anthocyanins can be categorized as either nonacylated or acylated, depending on whether the glucoside has been acylated by groups such as phenolic acyl, aromatic acid acyl, aliphatic acyl, and terpenoid-derived acyl groups. Acylation of the glycosyl moieties of anthocyanins is known to change the chemical properties of anthocyanins (Zhao et al., 2017a), making them more resistant to changes in heat, pH, and light, with a stronger potential to inhibit lipid peroxidation (Chen et al., 2022b). It has been shown that acylated anthocyanins are more stable (Xu et al., 2015) and, in some cases, have a stronger antioxidant and anti-inflammatory activity (Moriya et al., 2015) than their nonacylated counterparts. However, it is important to consider that acylated anthocyanins, having a stronger bioactivity compared to the nonacylated ones, can vary depending on the specific type of anthocyanin and the nature and number of acyl groups (Zhao et al., 2017a).

Small berries, including cranberries, blueberries, elderberries, black goji berries, and chokeberries, are the best sources of anthocyanins among the wide range of fruits and vegetables. Several studies investigated berry extracts against the most common diseases with promising outcomes. Additionally, colored vegetables and tubers such as purple sweet potatoes, purple corn, black carrots, radishes, red potatoes, and different *Brassica* vegetables, including purple cauliflower and red cabbage, are prominent sources of acylated anthocyanins (Giusti & Wrolstad, 2003). Therefore, a lot of studies have been done on the health advantages of eating colorful fruits and vegetables, especially concerning their potential to prevent diabetes (Nizamutdinova et al., 2009). Due to their superior stability and strong antioxidant and anti-inflammatory activities, acylated anthocyanins have also been hypothesized to have a more profound beneficial impact on gut microbiota than nonacylated ones, contributing probiotic features (Chen et al., 2022b). Additionally, as discussed in Chapter 10, acylated anthocyanins have been investigated for their anticancer activities and their potential positive impact on reducing the incidence of breast cancer in humans, according to some research (Adlercreutz, 1998; Fung et al., 2013). For instance, purple sweet potatoes, containing acylated anthocyanins, have been shown to possess anticancer and antimutagenic properties against acute lymphoblastic leukemia, as reported by Guo et al. (2022). Consequently, due to the promising potential health benefits of acylated anthocyanins and the abundance of colored vegetables and fruits containing them, this chapter aims to highlight their preventive and therapeutic roles in obesity, diabetes, and cardiovascular and gastrointestinal (GI) diseases.

11.2 BIOAVAILABILITY AND STABILITY ADVANTAGES OF ACYLATED ANTHOCYANINS

Although already discussed in detail in Part II of this book, a brief discussion of the bioavailability and stability of these compounds is presented here. The "bioavailability" of anthocyanins is defined as the degree and pace of absorption of anthocyanins into the stream blood *via* the intestinal system and their subsequent availability for tissue activity in the human body as intact compounds. Despite their well-established anti-inflammatory and antioxidant properties, anthocyanins – including acylated forms – generally have limited bioavailability and rarely reach a concentration in plasma that could be considered therapeutically active (see Table 11.1). There are several reasons behind anthocyanins' limited bioavailability (Braga et al., 2018), including (1) pH-dependent stability, being more stable at low gastric pH compared to the higher pH levels found in the small intestine and blood. Although anthocyanins could be rapidly absorbed in the stomach, the site of maximal absorption is the small intestine and colon, where the pH is 7–8.5 and 5–7, respectively. At pH < 3, anthocyanins are present as a stable flavylium cation, while at $3 < pH < 6$, the flavylium cation stability decreases; at pH 6–7, it becomes a quinonoid anhydro base; at pH 7–11, it becomes a slightly stable anionic quinone; and at pH > 11, it becomes an unstable dianionic form (Mattioli et al., 2020). (2) The highly polar nature and low lipid solubility of anthocyanins make it difficult for them to cross cell membranes, decreasing the effectiveness of intestinal absorption. Pelargonidin derivatives are more readily absorbed than anthocyanins with more substituents on the B-ring (Mattioli et al., 2020). (3) Anthocyanins undergo extensive metabolic transformation due to their significant biotransformation while traversing the GI tract. Biotransformation entails enzymatic hydrolysis in the small intestine, followed by extensive catabolism and bioconversion by the colonic resident microflora. It is, therefore, believed that anthocyanins may possess specific biological activities the more metabolized they are, as they give rise to many different molecular bioactive intermediates. Anthocyanins are metabolized by phase I and II enzymes (Hribar & Poklar Ulrih, 2014), which includes reactions such as hydroxylation (typically occurs on the B-ring, which can alter their color and stability), glucuronidation (which commonly occurs at the hydroxyl groups on the aglycone, although it is also possible to occur on existing sugar moieties), sulfation (following similar a process as glucuronidation but with the addition of a sulfate group), and methylation (also at the hydroxyl groups on the B-ring, influencing their color and stability). These modifications of phase I and II happen mainly in the liver, but also at the renal and enterocyte levels. As these anthocyanin modifications in the liver primarily occur at the hydroxyl groups present on their flavonoid backbone, it is expected that the more acylated the anthocyanin is, the less extensive the biotransformation will be in the small intestine during the early stages of digestion, and the larger proportion of acylated anthocyanins will reach the large intestine intact (Jokioja et al., 2021).

These authors explained that by improving interactions with the intestinal mucosa, the acylation groups may boost the effectiveness of their absorption. However, even with this enhanced absorption mechanism, a substantial amount of acylated anthocyanins is still not absorbed at the small intestine level and passes into the colon, where the gut flora further changes them. Because they can generate molecules with increased antioxidant activity, these microbial mechanisms are crucial for regulating oxidative stress in diabetes (Deng et al., 2021).

In vitro GI simulations show that acylated anthocyanins are more stable than their nonacylated counterparts (Ryu & Koh, 2022). Studies attributed this increased stability to the acyl groups' capacity to

TABLE 11.1 Reported Bioavailability and Metabolic Effects of Anthocyanins from Selected Foods

TYPE OF FOOD	*MAIN FINDINGS*	*LITERATURE SOURCE*
Purple-fleshed sweet potato	Acylated anthocyanins were more stable during digestion and degraded less in the stomach and small intestine; showed increased absorption and retention under intestinal simulations, suggesting potential metabolic benefits in managing metabolic syndrome and T2D	Oliveira et al. (2019a)
Red wine (nonacylated anthocyanins)	Less stable than acylated anthocyanins during digestion, with faster degradation in GI conditions	Oliveira et al. (2019b)
Blueberries	Gut bacteria metabolize anthocyanins into smaller phenolic acids, which are more widely absorbed and may help lower blood glucose levels	Wang & Lin (2000)
Blackberries	Acylated anthocyanins showed higher absorption efficiency than the nonacylated forms, possibly due to improved interactions with intestinal mucosa; unabsorbed fractions reach the colon where the gut flora converted them into antioxidant metabolites	McGhie & Walton (2007)
Various plant sources (general)	Acylated anthocyanins, especially those acylated with hydroxycinnamic acids (*p*-coumaric and ferulic), showed no degradation in acidic gastric environments; demonstrated improved resistance to breakdown. Overall recovery in the small intestine was ~25%, with substantial degradation due to digestive enzymes and near-neutral pH	McDougall et al. (2007)

shield the anthocyanin structure from deterioration brought on by acids. The intestinal phase, when conditions become somewhat alkaline and digestive enzymes are very active, caused a considerable decrease in the recovery of acylated anthocyanins by around 25%. This implies that while acylation enhances GI stability, total stability is hard to sustain over the whole digestion process, particularly in the colon, where further degradation takes place. This finding of a 25% drop highlights ongoing difficulties in maintaining anthocyanin bioavailability, since it contrasts with the approximately 25% total anthocyanin recovery reported by McDougall et al. (2007) for acylated anthocyanins in the small intestine.

11.3 ACYLATED ANTHOCYANINS, DIABETES, AND OBESITY

The biological effects of anthocyanins and their modes of action have been the subject of a great number of *in vitro* and *in vivo* studies in recent years. *In vivo* animal studies (on mice) have investigated the impact of anthocyanins (either pure anthocyanins as standards, or anthocyanin extracts from various plants) on the metabolic syndrome (Naseri et al., 2018), especially on obesity – and T2D – suppression, such as oxidative stress, chronic inflammation, insulin resistance, hyperlipidemia, and postprandial glucose response and hyperglycemia (Castro-Acosta et al., 2016), using both genetic and dietary models of obesity. Anthocyanins' capacity to reduce the production of free radicals, enhance insulin signaling, change the expression of genes related to the inflammatory response, or directly regulate the activities of important enzymes has also been the subject of mechanistic *in vitro* research employing cell lines and purified enzymes (Guo & Ling, 2015).

The primary line of treatment and prevention for T2D and obesity is managing weight gain. The first *in vivo* study on nonacylated anthocyanins' ability to reduce obesity was published in 2003 by Tsuda et al. For 12 weeks, C57BL/6 J mice were given a high-fat diet (HFD) with or without purple maize corn, which is rich in acylated anthocyanins (2 g/kg of diet, including acylated and nonacylated ones). Some of the acylated anthocyanins found in these purple maize extracts were cyanidin-3-(6″-malonyl)-glucoside, pelargonidin-3-(6″-malonyl)-glucoside, and peonidin-3-(6″-malonyl)-glucoside. Supplementing a HFD with anthocyanins was shown to dramatically lower body weight gain and fat formation in both brown and white adipose tissue caused by the HFD (Tsuda et al., 2003). Later, it was demonstrated that acylated anthocyanin extracts from blood oranges (*Citrus sinensis* L. Osbeck), mulberries (*Morus australis* Poir), blueberries (*Vaccinium angustifolium*), and blackberries (*Rubus* sp.) prevented obesity in rats that were given a HFD supplemented with these extracts (Prior et al., 2008, 2010; Titta et al., 2010); some of the acylated anthocyanins found in these extracts were cyanidin-3-(6″-acetyl)-galactoside, delphinidin-3-(6″-acetyl)-galactoside, malvidin-3-(6″-acetyl)-glucoside, petunidin-3-(6″-acetyl)-galactoside. After feeding mice a HFD for 8 weeks, they were given acylated anthocyanins from honeysuckle (*Lonicera caerulea* L.) at varying levels (50, 100, or 200 mg/kg diet) for an additional 8 weeks. When 100 or 200 mg/kg was added to the HFD, body weight increase was inhibited by 24% and 17%, respectively (Wu et al., 2013a). Some of the acylated anthocyanins present in honeysuckle were peonidin-3-5-dihexoside, cyanidin-3-5-dihexoside, and cyanidin-3-acetylhexoside.

Similarly, litchi flower–water extract rich in phenolic compounds like acylated anthocyanins and proanthocyanidins has shown notable anti-obesity effects in rats by decreasing adipose tissue mass, liver lipids, and inflammatory markers while simultaneously increasing fecal lipid excretion. This suggests that litchi flower–water extract may be used as a nutraceutical to manage obesity (Wu et al., 2013b). Some of the acylated anthocyanins present in litchi flower–water extract were delphinidin-3-*O*-(6″-*O*-feruloyl)-glucoside, cyanidin-3-*O*-(6″-*O*-*p*-coumaroyl)-glucoside, and peonidin-3-*O*-(6″-*O*-caffeoyl)-glucoside.

The putative mechanisms underlying acylated anthocyanins' potential to prevent diabetes may stem from their ability to inhibit body weight gain, prevent the production of free radicals and lipid peroxidation, regulate the inflammatory response, lower blood glucose and lipid levels, and improve insulin resistance, as schematically illustrated in Figure 11.2. Inhibition of the action of starch-active digestive enzymes (i.e., α-amylases and α-glucosidases) and inhibition of glucose transporters (i.e., glucose transporter type 2 (GLUT2) and sodium/glucose cotransporter 1 (SGLT1)) by acylated anthocyanins have also been reported to help in the management and prevention of T2D, as the inhibition of starch digestion results in a decrease in the postprandial glucose response and an increase in slow digestible and nondigestible starch (Pico et al., 2019; Pico & Martínez, 2019). Of note, *in vitro* studies with Caco-2 cells revealed that anthocyanin acylation and glycosylation have a greater effect on SGLT1 mRNA expression than on GLUT2, and significantly downregulated the protein expression of SGLT1 (Liu et al., 2022). The complicated interplay of many signaling pathways, transcription factors, and enzymes is responsible for the overall positive effects of acylated anthocyanins. Future research should concentrate on identifying the most promising acylated anthocyanin downstream targets *in vitro* and extrapolating these results to *in vivo* animal mechanistic investigations (Guo & Ling, 2015).

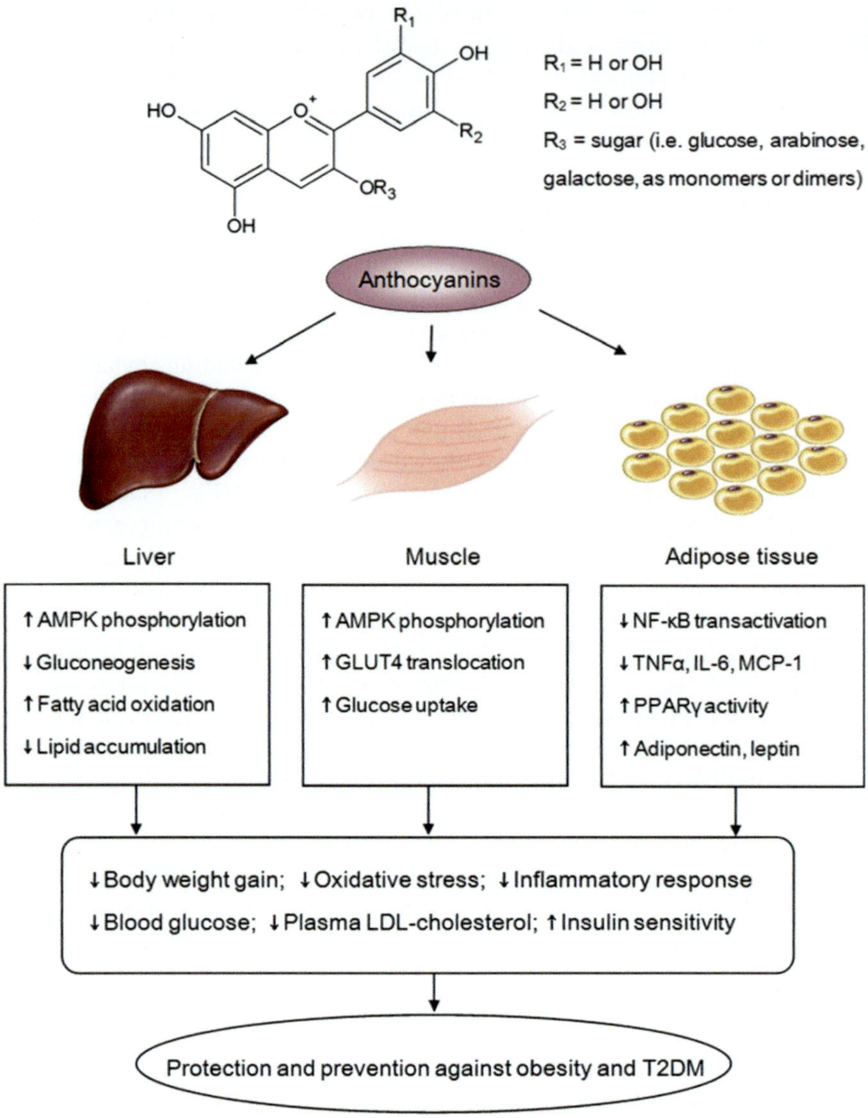

FIGURE 11.2 Some potential molecular mechanisms that underline how acylated anthocyanins affect T2D and obesity.

11.4 ACYLATED ANTHOCYANINS AND CARDIOVASCULAR DISEASE

Chronic diseases are on the rise due to factors like aging, urbanization, population expansion, and globalization. Cardiovascular disease (CVD) is the leading cause of death worldwide and accounts for almost half of noncommunicable disease deaths. Moreover, 80% of the 17.3 million deaths that occur each year – which are expected to rise to 23.6 million by 2030 – occur in low- and middle-income nations and affect younger populations with fewer means to respond (Committee et al., 2012).

There is an inverse relationship between a lower risk of CVD and fruit and vegetable consumption (Nöthlings et al., 2008), which is believed to be due to the wide range and quantity of bioactive compounds found in them. Researchers have been very interested in the use of diets rich in natural bioactive components as substitutes for pharmaceutical CVD drugs and in their role in preserving or enhancing cardiovascular health. Specifically, dietary flavonoids, like anthocyanins and acylated anthocyanins, have gained attention as possible CVD-preventive agents as epidemiological research indicates that consuming large amounts of flavonoids may lower the risk of CVD, although other research found little to no correlation (Mursu et al., 2008). A recent 16-year follow-up study of 34,489 postmenopausal women without CVD in the Iowa Women's Health Study revealed that dietary intakes of specific flavonoid classes, such as flavanones and anthocyanidins, and specific foods rich in total flavonoids were associated with a lower risk of CVD and CHD (Mink et al., 2007).

Acute myocardial infarction risk was found to be significantly inversely correlated with a prudent dietary pattern rich in fruits and vegetables, according to data from the Interheart study, which included dietary patterns from 52 nations (Iqbal et al., 2008). An analysis of the dietary intakes of 2,757 overweight US adults with type 2 diabetes, a known risk factor for CVD, revealed that fewer than half of the participants met even the minimum recommended intake of fruits and vegetables (which contain acylated anthocyanins, which contribute to the suppression of such diseases) (Vitolins et al., 2009).

Additionally, anthocyanins exhibit *in vitro* anti-thrombotic properties (Rechner & Kroner, 2005). Red grapes, cocoa, and other polyphenol-rich foods have recently been demonstrated to improve human platelet function and vascular function due to the presence of acylated anthocyanins (Murphy et al., 2003; Rein et al., 2000). Another study showed that rats fed anthocyanin-containing maize seed (20% of the diet) for 8 weeks were less prone to ischemia, which supports the anti-thrombotic effect, reperfusion damage, and infarct size decrease with elevated myocardial antioxidant enzymes (Toufektsian et al., 2008). According to Bell and Gochenaur (2006), acylated anthocyanin-rich (cyanidin, peonidin, delphinidin, petunidin, and malvidin classes) extracts of bilberries and chokeberries, but not elderberries, have the ability to relax blood vessels. In addition, the coronary reactivity to nitric oxide, a strong vasodilator, remains unchanged, which means acylated anthocyanins from bilberries and chokeberries help relax blood vessels, supporting cardiovascular health without disrupting natural vasodilation.

Researchers found that eating strawberries rich in acylated anthocyanins like pelargonidin-3-*O*-(6″-*O*-malonyl)-glucoside and pelargonidin-3-*O*-(6″-*O*-*p*-coumaroyl)-glucoside for a month improves the lipid profile and platelet function of healthy participants, which have positive implications in CVD (Alvarez-Suarez et al., 2014). However, the presence of non-anthocyanin chemicals in strawberries, like vitamin C and phenolic compounds, may be responsible for the benefits. Therefore, the positive health outcomes observed might not be solely due to acylated anthocyanins. To clearly determine the specific role of these compounds, well-designed studies including appropriate control groups are essential for accurate comparison and to isolate the effects of acylated anthocyanins from those of other bioactive compounds.

Contrary to the aforementioned studies, Curtis et al. (2009) found that consuming 500 mg of elderberry extract (with cyanidin-3-*O*-(6″-*O*-caffeoyl)-glucoside and cyanidin-3-*O*-(6″-*O*-feruloyl)-glucoside) daily for 12 weeks did not lower the incidence of CVD in postmenopausal women in good health. Additionally, when compared to acute intake, no change in metabolic processes after 12 weeks of elderberry consumption was observed (de Ferrars et al., 2014).

11.5 ACYLATED ANTHOCYANINS AND GASTROINTESTINAL DISORDER

Developing dietary plans and treatment therapies for GI illnesses require an understanding of their function in GI health (Jokioja et al., 2021). For instance, it has been reported that the uptake of black raspberry nonacylated anthocyanins in the small intestinal tissue was around 7.5% of the administered dose, which was much higher than the concentrations found in plasma and urine (He et al., 2009). The presence of higher anthocyanin concentrations in the intestine than in blood suggests that intact anthocyanins were absorbed by the GI tissues but were not efficiently transported into systemic circulation. More optimistic results were obtained *in vitro* using epithelial Caco-2 cell models by Steinert et al. (2008), who found significant translocation of intact nonacylated black currant anthocyanins across the basolateral membrane. It is noteworthy that the presence of glucose moieties, instead of galactose or arabinose, on cyanidin and peonidin seems to make them more bioavailable as a percentage of the delivered dose, as observed in the literature on cranberry juice (Wallace, 2011).

Depending on their structure, acylated anthocyanins are absorbed in the GI tract at varying rates, according to research. For example, monoacylated anthocyanins from black carrots and colored-fleshed potatoes were still present in rats in plasma in their original forms after consumption, demonstrating partial GI absorption and stability in plasma. The precise structure of the anthocyanin, including the aglycone and linked sugar moieties, affected the absorption efficiency (Ichiyanagi et al., 2022). However, it is generally recognized that acylation improves anthocyanin stability in the upper GI system (i.e., mouth, esophagus, stomach, and duodenum), facilitating their transport into the gut. In other words, and according to *in vitro* research, acylated anthocyanins are more resilient to upper GI degradation than nonacylated versions, which may result, indeed, in increased bioavailability in the gut and longer biological activity by the end of the digestion process in the colon (Jokioja et al., 2021).

This last statement is supported by the *in vitro* research by Liu et al. (2022), who stated that for the transport of anthocyanins from wine, methylation, hydroxylation, glycosylation, and acylation negatively impacted the transport of anthocyanins in Caco-2 cells. Similarly, Hélder Oliveira et al. (2019b) used MKN-28 gastric cells to compare the transport efficiency of anthocyanins with that of different structures and found that, compared with monoglycosides, the transport efficiency anthocyanins substituted with a disaccharide and subjected to acylation was hindered by 7%–8%. Human trials also indicate that acylation reduces the absorption efficiency of anthocyanins, and the absorption efficiency of anthocyanin-3-(xylose-spiroyl-glucose)-galactoside is lower than that of nonacylated anthocyanins with the same structure (Novotny et al., 2012). In conclusion, it is expected that glycosylation and acylation patterns decrease the bioavailability of anthocyanins in blood, associated with lower absorption in systemic circulation and higher transport into the gut. If acylated anthocyanins are absorbed, it will rarely happen in their intact form, as glycosidases and carboxylesterases (CES) primarily present in the GI tract (i.e., found within the brush border membrane and endoplasmic reticulum and cytosol of cells, respectively) will hydrolyze acylated anthocyanins and anthocyanins into anthocyanidins, which can be better absorbed and transported into the bloodstream. The composition and function of gut bacteria have been demonstrated to be influenced by acylated anthocyanins as well. It is recognized that several illnesses are linked to the disruption of gut microbiota, which is essential for GI health maintenance. In this context, supplementing with anthocyanins has been shown to enhance gut health biomarkers in experimental investigations. Anthocyanin consumption, for instance, decreased the Firmicutes/Bacteroidetes ratio, a measure linked to obesity-induced gut dysbiosis, and increased the synthesis of short-chain fatty acids (SCFAs), which are good for intestinal health, according to a meta-analysis of mouse studies (Kapoor et al., 2023). Although there is not much research specifically on acylated anthocyanins, anthocyanins in general have shown antibacterial qualities against a range of diseases. For example, *Lycium ruthenicum* anthocyanins have demonstrated strong inhibitory effects on *Staphylococcus aureus* by rupturing biofilm formations and compromising the integrity of its cell membrane (Dong et al., 2022). Furthermore, blueberry extracts rich in anthocyanins have demonstrated antibacterial action against strains of *Pseudomonas aeruginosa* and *Klebsiella pneumoniae*, as well as other Gram-positive and Gram-negative bacteria (Jeyaraj et al., 2023). These results imply that by inhibiting the development of pathogenic bacteria, anthocyanins support a more beneficial microbial balance. The precise consequences of acylated anthocyanin variations require more investigation.

Many other GI illnesses, such as colitis and inflammatory bowel disease (IBD), include inflammation as a common underlying cause. These conditions may be improved by the anti-inflammatory qualities of

Castro-Acosta, M. L., Lenihan-Geels, G. N., Corpe, C. P., & Hall, W. L. (2016). Berries and anthocyanins: Promising functional food ingredients with postprandial glycaemia-lowering effects. *Proceedings of the Nutrition Society*, *75*(3), 342–355. https://doi.org/10.1017/S0029665116000240

Cerf-Bensussan, N., & Gaboriau-Routhiau, V. (2010). The immune system and the gut microbiota: Friends or foes? *Nature Reviews Immunology*, *10*(10), 735–744. https://doi.org/10.1038/nri2850

Chen, K., Kortesniemi, M. K., Linderborg, K. M., & Yang, B. (2022a). Anthocyanins as promising molecules affecting energy homeostasis, inflammation, and gut microbiota in type 2 diabetes with special reference to impact of acylation. *Journal of Agricultural and Food Chemistry*, *71*(2), 1002–1017. https://doi.org/10.1021/acs.jafc.2c05879

Chen, K., Kortesniemi, M. K., Linderborg, K. M., & Yang, B. (2022b). Anthocyanins as promising molecules affecting energy homeostasis, inflammation, and gut microbiota in type 2 diabetes with special reference to impact of acylation. *Journal of Agricultural and Food Chemistry*, *71*(2), 1002–1017. https://doi.org/10.1021/acs.jafc.2c05879

Cho, N. H., Shaw, J. E., Karuranga, S., Huang, Y., da Rocha Fernandes, J. D., Ohlrogge, A., & Malanda, B. (2018). IDF diabetes atlas: Global estimates of diabetes prevalence for 2017 and projections for 2045. *Diabetes Research and Clinical Practice*, *138*, 271–281. https://doi.org/10.1016/j.diabres.2018.02.023

Choi, K. H., Lee, H. A., Park, M. H., & Han, J.-S. (2016). Mulberry (*Morus alba* L.) fruit extract containing anthocyanins improves glycemic control and insulin sensitivity via activation of AMP-activated protein kinase in diabetic C57BL/Ksj-db/db mice. *Journal of Medicinal Food*, *19*(8), 737–745. https://doi.org/10.1089/jmf.2016.3665

Committee, W., Smith, S. C., Jr., Collins, A., Ferrari, R., Holmes, D. R., Jr., Logstrup, S., McGhie, D. V., Ralston, J., Sacco, R. L., Stam, H., Taubert, K., Wood, D. A., & Zoghbi, W. A. (2012). Our time: A call to save preventable death from cardiovascular disease (heart disease and stroke). *European Heart Journal*, *33*(23), 2910–2916. https://doi.org/10.1093/eurheartj/ehs313

Curtis, P. J., Kroon, P. A., Hollands, W. J., Walls, R., Jenkins, G., Kay, C. D., & Cassidy, A. (2009). Cardiovascular disease risk biomarkers and liver and kidney function are not altered in postmenopausal women after ingesting an elderberry extract rich in anthocyanins for 12 weeks. *The Journal of Nutrition*, *139*(12), 2266–2271. https://doi.org/10.3945/jn.109.113126

Day, E. A., Ford, R. J., Steinberg, G. R. J. T. I. E., & Metabolism. (2017). AMPK as a therapeutic target for treating metabolic diseases. *Trends in Endocrinology & Metabolism*, *28*(8), 545–560. https://doi.org/10.1016/j.tem.2017.05.004

de Ferrars, R. M., Cassidy, A., Curtis, P., & Kay, C. D. (2014). Phenolic metabolites of anthocyanins following a dietary intervention study in post-menopausal women. *Molecular Nutrition & Food Research*, *58*(3), 490–502. https://doi.org/10.1002/mnfr.201300322

Deng, L., Du, C., Song, P., Chen, T., Rui, S., Armstrong, D. G., & Deng, W. (2021). The role of oxidative stress and antioxidants in diabetic wound healing. *Oxidative Medicine and Cellular Longevity, 2021*(1), 8852759. https://doi.org/10.1155/2021/8852759

Diaconeasa, Z., Leopold, L., Rugină, D., Ayvaz, H., & Socaciu, C. (2015). Antiproliferative and antioxidant properties of anthocyanin rich extracts from blueberry and blackcurrant juice. *International Journal of Molecular Sciences*, *16*(2), 2352–2365. https://doi.org/10.3390/ijms16022352

Dong, Y., Yang, C., Zhong, W., Shu, Y., Zhang, Y., & Yang, D. (2022). Antibacterial effect and mechanism of anthocyanin from *Lycium ruthenicum* Murr. *Frontiers in Microbiology*, *13*, 974602. https://doi.org/10.3389/fmicb.2022.974602

Fleschhut, J., Kratzer, F., Rechkemmer, G., & Kulling, S. E. (2006). Stability and biotransformation of various dietary anthocyanins in vitro. *European Journal of Nutrition*, *45*, 7–18. https://doi.org/10.1007/s00394-005-0557-8

Fung, T. T., Chiuve, S. E., Willett, W. C., Hankinson, S. E., Hu, F. B., & Holmes, M. D. (2013). Intake of specific fruits and vegetables in relation to risk of estrogen receptor-negative breast cancer among postmenopausal women. *Breast Cancer Research and Treatment*, *138*(3), 925–930. https://doi.org/10.1007/s10549-013-2484-3.

Giusti, M. M., & Wrolstad, R. E. (2003). Acylated anthocyanins from edible sources and their applications in food systems. *Biochemical Engineering Journal*, *14*(3), 217–225. https://doi.org/10.1016/S1369-703X(02)00221-8

Gonçalves, A. C., Nunes, A. R., Falcão, A., Alves, G., & Silva, L. R. (2021). Dietary effects of anthocyanins in human health: A comprehensive review. *Pharmaceuticals*, *14*(7), 690. https://doi.org/10.3390/ph14070690

Guo, H., & Ling, W. (2015). The update of anthocyanins on obesity and type 2 diabetes: Experimental evidence and clinical perspectives. *Reviews in Endocrine and Metabolic Disorders*, *16*, 1–13. https://doi.org/10.1007/s11154-014-9302-z

Guo, L., Jin, Y., Yang, Y., Liu, J., Liu, C., Zeng, Y., Guo, Q., & Liu, W. (2022). Calcicoptosis induced by purple sweet potato anthocyanins through the nonosmotic regulation of the NFAT5/S100A4-S100A9 pathway in acute lymphoblastic leukemia. *Chemistry & Biodiversity*, *19*(9), e202200447. https://doi.org/10.1002/cbdv.202200447

He, J., Wallace, T. C., Keatley, K. E., Failla, M. L., & Giusti, M. M. (2009). Stability of black raspberry anthocyanins in the digestive tract lumen and transport efficiency into gastric and small intestinal tissues in the rat. *Journal of Agricultural and Food Chemistry*, *57*(8), 3141–3148. https://doi.org/10.1021/jf900567t

Hribar, U., & Poklar Ulrih, N. (2014). The metabolism of anthocyanins. *Current Drug Metabolism*, *15*(1), 3–13. https://doi.org/10.2174/1389200214666131211160308

Hwang, Y. P., Choi, J. H., Han, E. H., Kim, H. G., Wee, J.-H., Jung, K. O., Jung, K. H., Kwon, K.-I., Jeong, T. C., & Chung, Y. C. (2011). Purple sweet potato anthocyanins attenuate hepatic lipid accumulation through activating adenosine monophosphate–activated protein kinase in human HepG2 cells and obese mice. *Nutrition Research*, *31*(12), 896–906. https://doi.org/10.1016/j.nutres.2011.09.026

Ichiyanagi, T., Nashimoto, M., Hayashi, H., & Terahara, N. (2022). Effect of acylated sugar moieties on gastrointestinal absorption of mono-acylated anthocyanins in rats. *BPB Reports*, *5*(3), 24–32. https://doi.org/10.1248/bpbreports.5.3_24

Iqbal, R., Anand, S., Ounpuu, S., Islam, S., Zhang, X., Rangarajan, S., Chifamba, J., Al-Hinai, A., Keltai, M., & Yusuf, S. (2008). Dietary patterns and the risk of acute myocardial infarction in 52 countries: Results of the INTERHEART study. *Circulation*, *118*(19), 1929–1937. https://doi.org/10.1161/CIRCULATIONAHA.107.738716

Jeyaraj, E. J., Vidana Gamage, G. C., Cintrat, J.-C., & Choo, W. S. (2023). Acylated and non-acylated anthocyanins as antibacterial and antibiofilm agents. *Discover Food*, *3*(1), 21. https://doi.org/10.1007/s44187-023-00062-8

Jokioja, J., Yang, B., & Linderborg, K. M. (2021). Acylated anthocyanins: A review on their bioavailability and effects on postprandial carbohydrate metabolism and inflammation. *Comprehensive Reviews in Food Science and Food Safety*, *20*(6), 5570–5615. https://doi.org/10.1111/1541–4337.12836

Kalt, W., Liu, Y., McDonald, J. E., Vinqvist-Tymchuk, M. R., & Fillmore, S. A. (2014). Anthocyanin metabolites are abundant and persistent in human urine. *Journal of Agricultural and Food Chemistry*, *62*(18), 3926–3934. https://doi/abs/10.1021/jf500107j

Kapoor, P., Tiwari, A., Sharma, S., Tiwari, V., Sheoran, B., Ali, U., & Garg, M. (2023). Effect of anthocyanins on gut health markers, Firmicutes–Bacteroidetes ratio and short-chain fatty acids: A systematic review via meta-analysis. *Scientific Reports*, *13*(1), 1729. https://doi.org/10.1038/s41598-023-28764-0

Koh, E. S., Lim, J. H., Kim, M. Y., Chung, S., Shin, S. J., Choi, B. S., Kim, H. W., Hwang, S. Y., Kim, S. W., & Park, C. W. (2015). Anthocyanin-rich Seoritae extract ameliorates renal lipotoxicity via activation of AMP-activated protein kinase in diabetic mice. *Journal of Translational Medicine*, *13*(1), 203. https://doi.org/10.1186/s12967-015-0563-4

Liu, Y., Lin, J., Cheng, T., Liu, Y., & Han, F. (2022). Methylation, hydroxylation, glycosylation and acylation affect the transport of wine anthocyanins in Caco-2 cells. *Foods*, *11*(23), 3793. https://doi.org/10.3390/foods11233793

Mattioli, R., Francioso, A., Mosca, L., & Silva, P. (2020). Anthocyanins: A comprehensive review of their chemical properties and health effects on cardiovascular and neurodegenerative diseases. *Molecules*, *25*(17), 3809. https://doi.org/10.3390/molecules25173809

McDougall, G. J., Fyffe, S., Dobson, P., & Stewart, D. (2007). Anthocyanins from red cabbage–stability to simulated gastrointestinal digestion. *Phytochemistry*, *68*(9), 1285–1294. https://doi.org/10.1016/j.phytochem.2007.02.004

McGhie, T. K., & Walton, M. C. (2007). The bioavailability and absorption of anthocyanins: Towards a better understanding. *Molecular Nutrition & Food Research*, *51*(6), 702–713. https://doi.org/10.1002/mnfr.200700092

Medzhitov, R. (2008). Origin and physiological roles of inflammation. *Nature*, *454*(7203), 428–435. https://doi.org/10.1038/nature07201

Mink, P. J., Scrafford, C. G., Barraj, L. M., Harnack, L., Hong, C.-P., Nettleton, J. A., & Jacobs, D. R., Jr. (2007). Flavonoid intake and cardiovascular disease mortality: A prospective study in postmenopausal women. *The American Journal of Clinical Nutrition*, *85*(3), 895–909. https://doi.org/10.1093/ajcn/85.3.895

Moriya, C., Hosoya, T., Agawa, S., Sugiyama, Y., Kozone, I., Shin-Ya, K., Terahara, N., & Kumazawa, S. (2015). New acylated anthocyanins from purple yam and their antioxidant activity. *Bioscience, Biotechnology, and Biochemistry*, *79*(9), 1484–1492. https://doi.org/10.1080/09168451.2015.1027652.

Moya-Pérez, A., Neef, A., & Sanz, Y. (2015). Bifidobacterium pseudocatenulatum CECT 7765 reduces obesity-associated inflammation by restoring the lymphocyte–macrophage balance and gut microbiota structure in high-fat diet-fed mice. *PLoS One*, *10*(7), e0126976. https://doi.org/10.1371/journal.pone.0126976

Muro, P., Zhang, L., Li, S., Zhao, Z., Jin, T., Mao, F., & Mao, Z. (2024). The emerging role of oxidative stress in inflammatory bowel disease. *Frontiers in Endocrinology*, *15*, 1390351. https://doi.org/10.3389/fendo.2024.1390351

Murphy, K. J., Chronopoulos, A. K., Singh, I., Francis, M. A., Moriarty, H., Pike, M. J., Turner, A. H., Mann, N. J., & Sinclair, A. J. (2003). Dietary flavanols and procyanidin oligomers from cocoa (*Theobroma cacao*) inhibit platelet function. *The American Journal of Clinical Nutrition*, *77*(6), 1466–1473. https://doi.org/10.1093/ajcn/77.6.1466

Mursu, J., Voutilainen, S., Nurmi, T., Tuomainen, T.-P., Kurl, S., & Salonen, J. T. (2008). Flavonoid intake and the risk of Ischaemic stroke and CVD mortality in middle-aged Finnish men: The Kuopio Ischaemic heart disease risk factor study. *British Journal of Nutrition*, *100*(4), 890–895. https://doi.org/10.1017/S0007114508945694

Naseri, R., Farzaei, F., Haratipour, P., Nabavi, S. F., Habtemariam, S., Farzaei, M. H., Khodarahmi, R., Tewari, D., & Momtaz, S. (2018). Anthocyanins in the management of metabolic syndrome: A pharmacological and biopharmaceutical review. *Frontiers in Pharmacology*, *9*, 1310. https://doi.org/10.3389/fphar.2018.01310

Nizamutdinova, I. T., Jin, Y. C., Chung, J. I., Shin, S. C., Lee, S. J., Seo, H. G., Lee, J. H., Chang, K. C., & Kim, H. J. (2009). The anti-diabetic effect of anthocyanins in streptozotocin-induced diabetic rats through glucose transporter 4 regulation and prevention of insulin resistance and pancreatic apoptosis. *Molecular Nutrition & Food Research*, *53*(11), 1419–1429. https://doi.org/10.1002/mnfr.200800526.

Nöthlings, U., Schulze, M. B., Weikert, C., Boeing, H., Van Der Schouw, Y. T., Bamia, C., Benetou, V., Lagiou, P., Krogh, V., & Beulens, J. W. (2008). Intake of vegetables, legumes, and fruit, and risk for all-cause, cardiovascular, and cancer mortality in a European diabetic population. *The Journal of Nutrition*, *138*(4), 775–781. https://doi.org/10.1093/jn/138.4.775

Novotny, J. A., Clevidence, B. A., & Kurilich, A. C. (2012). Anthocyanin kinetics are dependent on anthocyanin structure. *British Journal of Nutrition*, *107*(4), 504–509. https://doi.org/10.1017/S000711451100314X

Oliveira, H., Perez-Gregorio, R., de Freitas, V., Mateus, N., & Fernandes, I. (2019a). Comparison of the in vitro gastrointestinal bioavailability of acylated and non-acylated anthocyanins: Purple-fleshed sweet potato vs red wine. *Food Chemistry*, *276*, 410–418. https://doi.org/10.1016/j.foodchem.2018.09.159

Oliveira, H., Roma-Rodrigues, C., Santos, A., Veigas, B., Brás, N., Faria, A., Calhau, C., de Freitas, V., Baptista, P. V., & Mateus, N. (2019b). GLUT1 and GLUT3 involvement in anthocyanin gastric transport-nanobased targeted approach. *Scientific Reports*, *9*(1), 789. https://doi.org/10.1038/s41598-018-37283-2

Park, M., Yoo, J.-H., Lee, Y.-S., & Lee, H.-J. (2019). Lonicera caerulea extract attenuates non-alcoholic fatty liver disease in free fatty acid-induced HepG2 hepatocytes and in high fat diet-fed mice. *Nutrients*, *11*(3), 494. https://doi.org/10.3390/nu11030494

Pico, J., Corbin, S., Ferruzzi, M. G., & Martinez, M. M. (2019). Banana flour phenolics inhibit trans-epithelial glucose transport from wheat cakes in a coupled in vitro digestion/Caco-2 cell intestinal model. *Food & Function*, *10*(10), 6300–6311. https://doi.org/10.1039/C9FO01679A

Pico, J., & Martínez, M. M. (2019). Unraveling the inhibition of intestinal glucose transport by dietary phenolics: A review. *Current Pharmaceutical Design*, *25*(32), 3418–3433. https://doi.org/10.2174/1381612825666191015154326

Prior, R. L., Wilkes, S. E., Rogers, T. R., Khanal, R. C., Wu, X., & Howard, L. R. (2010). Purified blueberry anthocyanins and blueberry juice alter development of obesity in mice fed an obesogenic high-fat diet. *Journal of Agricultural and Food Chemistry*, *58*(7), 3970–3976. http://dx.doi.org/10.1021/jf902852d

Prior, R. L., Wu, X., Gu, L., Hager, T. J., Hager, A., & Howard, L. R. (2008). Whole berries versus berry anthocyanins: Interactions with dietary fat levels in the C57BL/6J mouse model of obesity. *Journal of Agricultural and Food Chemistry*, *56*(3), 647–653. http://dx.doi.org/10.1021/jf071993o

Rechner, A. R., & Kroner, C. (2005). Anthocyanins and colonic metabolites of dietary polyphenols inhibit platelet function. *Thrombosis Research*, *116*(4), 327–334. https://doi.org/10.1016/j.thromres.2005.01.002

Rein, D., Paglieroni, T. G., Wun, T., Pearson, D. A., Schmitz, H. H., Gosselin, R., & Keen, C. L. (2000). Cocoa inhibits platelet activation and function. *The American Journal of Clinical Nutrition*, *72*(1), 30–35. https://doi.org/10.1093/ajcn/72.1.30

Ryu, D., & Koh, E. (2022). Stability assessment of anthocyanins from black soybean, grape, and purple sweet potato under in vitro gastrointestinal digestion. *Food Science and Biotechnology*, *31*(8), 1053–1062. https://doi.org/10.1007/s10068-022-01071-6

Scazzocchio, B., Varì, R., Filesi, C., Del Gaudio, I., D'Archivio, M., Santangelo, C., Iacovelli, A., Galvano, F., Pluchinotta, F. R., & Giovannini, C. (2015). Protocatechuic acid activates key components of insulin signaling pathway mimicking insulin activity. *Molecular Nutrition & Food Research*, *59*(8), 1472–1481. https://doi.org/10.1002/mnfr.201400816

Schultze, S. M., Hemmings, B. A., Niessen, M., & Tschopp, O. (2012). PI3K/AKT, MAPK and AMPK signalling: Protein kinases in glucose homeostasis. *Expert Reviews in Molecular Medicine*, *14*, e1. https://doi.org/10.1017/S1462399411002109

Steinert, R. E., Ditscheid, B., Netzel, M., & Jahreis, G. (2008). Absorption of black currant anthocyanins by monolayers of human intestinal epithelial Caco-2 cells mounted in using type chambers. *Journal of Agricultural and Food Chemistry*, *56*(13), 4995–5001. https://doi.org/10.1021/jf703670h

Tena, N., Martín, J., & Asuero, A. G. (2020). State of the art of anthocyanins: Antioxidant activity, sources, bioavailability, and therapeutic effect in human health. *Antioxidants*, *9*(5), 451. https://doi.org/10.3390/antiox9050451

Titta, L., Trinei, M., Stendardo, M., Berniakovich, I., Petroni, K., Tonelli, C., Riso, P., Porrini, M., Minucci, S., & Pelicci, P. (2010). Blood orange juice inhibits fat accumulation in mice. *International Journal of Obesity*, *34*(3), 578–588. http://dx.doi.org/10.1038/ijo.2009.266

Toufektsian, M.-C., de Lorgeril, M., Nagy, N., Salen, P., Donati, M. B., Giordano, L., Mock, H.-P., Peterek, S., Matros, A., & Petroni, K. (2008). Chronic dietary intake of plant-derived anthocyanins protects the rat heart against ischemia-reperfusion injury3. *The Journal of Nutrition*, *138*(4), 747–752. https://doi.org/10.1093/jn/138.4.747

Tsuda, T., Horio, F., Uchida, K., Aoki, H., & Osawa, T. (2003). Dietary cyanidin 3-O-β-D-glucoside-rich purple corn color prevents obesity and ameliorates hyperglycemia in mice. *The Journal of Nutrition*, *133*(7), 2125–2130. https://doi.org/10.1093/jn/133.7.2125

Vitolins, M. Z., Anderson, A. M., Delahanty, L., Raynor, H., Miller, G. D., Mobley, C., Reeves, R., Yamamoto, M., Champagne, C., & Wing, R. R. (2009). Action for health in diabetes (look AHEAD) trial: Baseline evaluation of selected nutrients and food group intake. *Journal of the American Dietetic Association*, *109*(8), 1367–1375. https://doi.org/10.1016/j.jada.2009.05.016

Wallace, T. C. (2011). Anthocyanins in cardiovascular disease. *Advances in Nutrition*, *2*(1), 1–7. https://doi.org/10.3945/an.110.000042

Wang, S. Y., & Lin, H.-S. (2000). Antioxidant activity in fruits and leaves of blackberry, raspberry, and strawberry varies with cultivar and developmental stage. *Journal of Agricultural and Food Chemistry*, *48*(2), 140–146. https://doi.org/10.1021/jf9908345

Wu, T., Qi, X., Liu, Y., Guo, J., Zhu, R., Chen, W., Zheng, X., & Yu, T. (2013a). Dietary supplementation with purified mulberry (Morus australis Poir) anthocyanins suppresses body weight gain in high-fat diet fed C57BL/6 mice. *Food Chemistry*, *141*(1), 482–487. http://dx.doi.org/10.1016/j.foodchem.2013.03.046

Wu, Y.-H. S., Chiu, C.-H., Yang, D.-J., Lin, Y.-L., Tseng, J.-K., & Chen, Y.-C. (2013b). Inhibitory effects of litchi (*Litchi chinensis* Sonn.) flower-water extracts on lipase activity and diet-induced obesity. *Journal of Functional Foods*, *5*(2), 923–929. https://doi.org/10.1016/j.jff.2013.02.002

Xu, J., Su, X., Lim, S., Griffin, J., Carey, E., Katz, B., Tomich, J., Smith, J. S., & Wang, W. (2015). Characterisation and stability of anthocyanins in purple-fleshed sweet potato P40. *Food Chemistry*, *186*, 90–96. https://doi.org/10.1016/j.foodchem.2014.08.123

Zhao, C.-L., Yu, Y.-Q., Chen, Z.-J., Wen, G.-S., Wei, F.-G., Zheng, Q., Wang, C.-D., & Xiao, X.-L. (2017a). Stability-increasing effects of anthocyanin glycosyl acylation. *Food Chemistry*, *214*, 119–128. https://doi.org/10.1016/j.foodchem.2016.07.073

Zhao, Y., Hu, X., Liu, Y., Dong, S., Wen, Z., He, W., Zhang, S., Huang, Q., & Shi, M. (2017b). ROS signaling under metabolic stress: Cross-talk between AMPK and AKT pathway. *Molecular Cancer*, *16*, 1–12. https://doi.org/10.1186/s12943-017-0648-1

Zhu, B., Wang, X., & Li, L. (2010). Human gut microbiome: the second genome of human body. *Protein & Cell*, *1*(8), 718–725. https://doi.org/10.1007/s13238-010-0093-z

PART IV

Applications of Acylated Anthocyanins

Modification and Stabilization of Acylated Anthocyanins

12

Luis Cruz

12.1 GENERAL STRATEGIES TO ENHANCE THE STABILITY OF ANTHOCYANINS

In the recent years, the scientific community has been looking for new strategies to enhance anthocyanins' chemical and chromatic stability, particularly their colored species (flavylium cation and quinoidal bases), to increase their industrial applications.[1] Such methods include (1) chemical and enzyme-catalyzed covalent sugar acylation procedures and (2) non-covalent approaches of anthocyanins.

12.1.1 Covalent Strategy: Chemical and Enzymatic Acylation

Acylation of the sugars in anthocyanins is a widespread natural strategy to stabilize molecules and create purple and blue colors in flowers and vegetables, such as red cabbage (RC), black carrots, and sweet and pigmented potatoes.[2–9]

In recent years, a new series of acylated anthocyanins with aromatic/aliphatic acids have been synthesized to enhance their physico-chemical properties, namely, to improve their liposolubility and their chromatic/chemical stability in hydrophobic environments, envisaging their application in lipid-based foods and cosmetic products.[1]

Chemical-catalyzed esterification approaches to obtain anthocyanin–fatty acid conjugates have been employed by using fatty acid chlorides in acetonitrile/dimethylformamide and triethylamine or *via* 1-Ethyl-3-(3-dimethylaminopropyl)carbodiimide-activated fatty acids (lauric acid); however, the reactions revealed low regioselectivity, and complex ester mixtures were obtained.[10]

On the other hand, esterification and *trans*-esterification reactions of pure anthocyanins and anthocyanin-rich extracts with fatty acids or fatty acid methyl esters (saturated and unsaturated) in tertiary alcohols (e.g., *tert*-amyl alcohol and *tert*-butanol) or in aprotic solvents such as acetone, acetonitrile, and pyridine using *Candida antarctica* lipase B (*Cal*B) as a biocatalyst have emerged as a novel and promising approach to enabling higher regioselectivity and higher yields.[11, 12] The enantioselectivity of enzymatic reactions is related to the high stereospecificity of the active site exhibited by *Cal*B to catalyze the acylation at primary alcohols (preferentially) of the sugar residues of anthocyanins.[13] Other authors have reported a systematic study of reactions of anthocyanins from different plant sources (purple corn-husk,

DOI: 10.1201/9781003519591-16

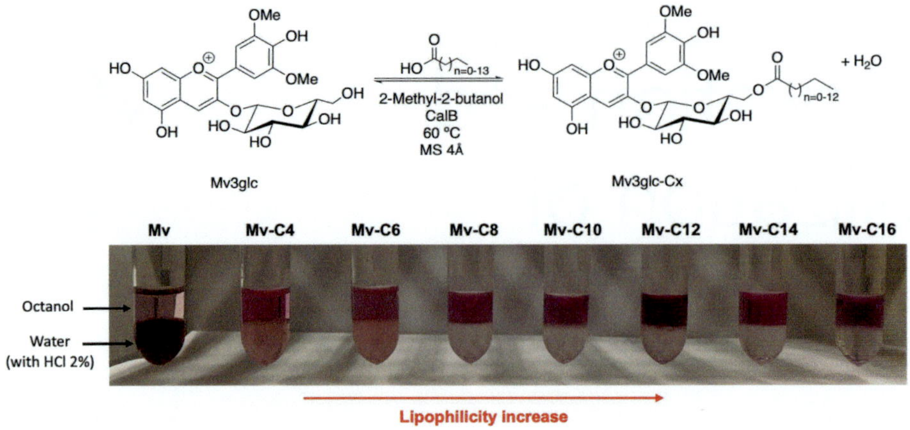

FIGURE 12.1 Enzymatic lipophilization of anthocyanins using *Cal*B as a biocatalyst and partition coefficient determination (log P) of native malvidin-3-glucoside (Mv) and its lipophilic derivatives (Mv-Cx) in an octanol–water biphasic mixture. The anthocyanin content increases in the octanol phase as long as the fatty acid chain length increases.

tiliapo, trueno, plum husk, and bottlebrush flower) with different cinnamic acids (and vinyl esters) using different experimental conditions such as lipase nature, solvents, acyl donor and substrate concentrations, and the content of molecular sieves.[14]

In general, acylated anthocyanins with aromatic and aliphatic acids have shown preferential organic miscibility in an octanol–water biphasic mixture. For example, anthocyanin–fatty acid conjugates showed increased lipophilicity with the chain length and a redshift in their λ_{max} (Figure 12.1). Such conjugates also showed improved antioxidant activity of the native anthocyanin (up to the C8 derivative), exerting a greater protective effect.[15]

12.1.2 Non-Covalent Approaches

Although natural acylated anthocyanins with aromatic acids (caffeic, coumaric, sinapic, and ferulic residues) are intrinsically more stable than nonacylated ones due to the occurrence of intramolecular co-pigmentation phenomena,[16] it is still necessary to investigate new strategies to stabilize the nonacylated ones, which could involve different approaches. Actually, most of the studies reported in the literature so far are associated with the development of stabilization methodologies for nonacylated anthocyanins. Micro- and nanoencapsulation of anthocyanins using biocompatible materials is still the main technique used to protect anthocyanins from the surrounding media.[17], [18] To date, encapsulation that produces capsules includes spray-drying, freeze-drying, ionic gelation, liposome entrapment, nanoencapsulation, yeast encapsulation, phase separation, emulsification, and complexation.[19] The ultimate goal is to produce new systems that improve the stability of anthocyanins to temperature, light, oxygen, and pH variations and that could retain their bioactivity. The typical wall materials may include carbohydrates[20] (maltodextrin, β-cyclodextrin, chitosan, alginate, starch, pectins, and carboxymethylcellulose), proteins[21] (isolated soy protein, whey protein, soy milk, zein, and casein), and plant water-soluble gums[22] (gum arabic and gelatine). Calcium-alginate microparticles have been optimized to encapsulate anthocyanin-rich extract from haskap berries (*Lonicera caerulea*),[23] and the degradation of anthocyanins extracted from black carrots in beverages due to ascorbic acid was reduced through encapsulation in various lecithin liposomal systems.[24] Weiss et al. described the stability of anthocyanin extracts from chokeberry, grape, hibiscus, and purple sweet potato in ω-3-fatty acid-rich oil-in-water emulsions during storage, concluding that extracts with more acylated anthocyanin content undergo less color fading.[25]

12.2 COLOR STABILITY STUDIES OF ACYLATED ANTHOCYANINS

Color stability studies of acylated anthocyanins in lipophilic media toward different factors (pH, temperature, and light) are crucial to understanding their ability to incorporate in oils, lipid-based foods, and/or cosmetic formulations.

The color stability of anthocyanin–fatty acid conjugates was studied in different micelle-like solutions and compared with native dyes. Table 12.1 shows the effect of the environment (micelles vs aqueous solution) and the fatty acid acyl residue on the thermodynamic properties of anthocyanins. Micellar systems such as sodium dodecyl sulfate (SDS) or PEGylated phospholipid-based micelles (PEG-1,2-Distearoyl-sn-glycero-3-phosphoethanolamine) largely increase the color stability of anthocyanins by increasing their equilibrium constants ($pK^{\wedge}a$ and pK'_a) and hydration constant $\left(pK_h\right)$ at high pH values compared with water. Furthermore, anthocyanin–fatty acid conjugates were shown to stabilize the red flavylium cation form, thus extending its color to higher pH values than native anthocyanins in such lipophilic media.

In addition, the effect of acylation and the media resulted in the enhancement of the mole fraction of the purple quinoidal base up to 40% concomitant with the decrease of colorless hemiketal species from 65% to 22% between moderate acid to neutral pH values (Figure 12.2).[12, 27]

Through the determination of the thermodynamic constants of the anthocyanin multistate, a greater color stabilization of the lipophilic cyanidin-3-glucoside-C10 derivative in SDS micelles, compared with that of the parent anthocyanin, was demonstrated by extending the pH domain of the red flavylium cation and the blue quinoidal base by 0.7 units (Figure 12.3). For pH values between 5 and 7, it was possible to observe the greater blue-purple color of the lipophilic-derivative solution, which was probably due to the presence of a higher molar-fraction percentage of the quinoidal base, whereas for the cy3glc solution, hemiketal and chalcones should have been the species preferentially stabilized by the SDS micelles, resulting in a fading of the solution.

TABLE 12.1 Equilibrium Constants Obtained by UV-Vis Spectroscopy for cy3glc, mv3glc, dp3sam, and Their Respective Lipophilic Derivatives in Water and in Micellar Systems of SDS (0.1 M) and PEG-1,2-Distearoyl-sn-glycero-3-phosphoethanolamine (DSPE) (1.3 mM)

COMPOUND	pK_a	$pK^{\wedge}a$	pK'_a	pK_h	K_t	K_i
mv3glc (water)	3.7	2.4	2.3	2.6	0.12	–
mv3glc (SDS)[26]	6.0	–	5.4	6.2	–	–
mv3glc-C6 (SDS)	6.1	5.9	5.7	6.48	0.6	4
mv3glc-C12 (SDS)	6.0	5.75	5.65	6.30	0.5	2
cy3glc (water)	3.89	2.88	2.83	3.01	0.21	0.78
cy3glc (PEG-DSPE)	4.13	3.47	3.36	3.66	0.22	2.0
cy3glc-C12 (PEG-DSPE)	5.09	4.85	4.69	5.34	0.30	4.6
dp3sam (water)	3.70	–	2.92	3.02	–	–
dp3sam (SDS)	5.66	5.37	5.14	5.68	–	–
dp3sam-C8 (SDS)	5.90	5.83	5.82	6.66	–	–

K_a, acidity constant; $K^{\wedge}a$, pseudoequilibrium constant; K'_a, equilibrium constant; K_t, tautomerization constant; K_h, hydration constant; K_i, isomerization constant.

Note: Estimated error ≈ 10%.

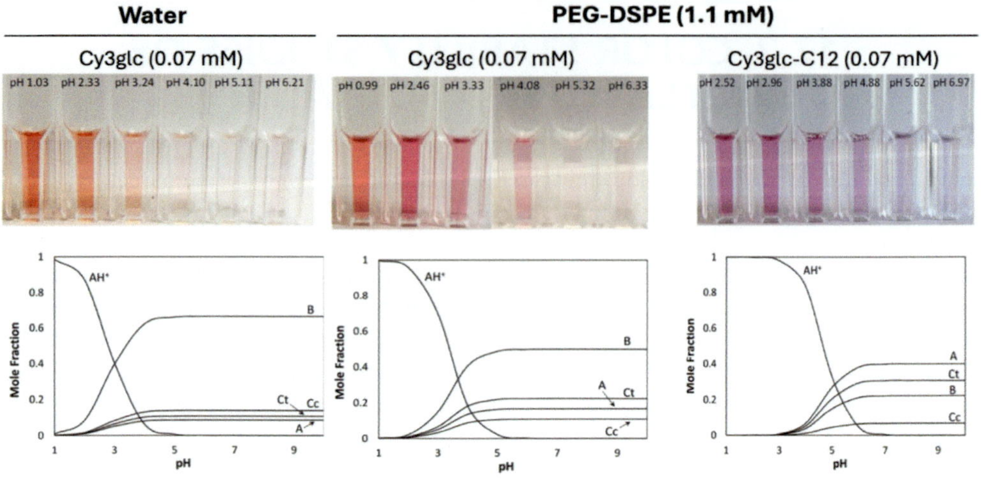

FIGURE 12.2 Mole fraction distribution of cy3glc and cy3glc-C12 (0.07 mM) in water and in PEG-1,2-Distearoyl-sn-glycero-3-phosphoethanolamine at different pH values.

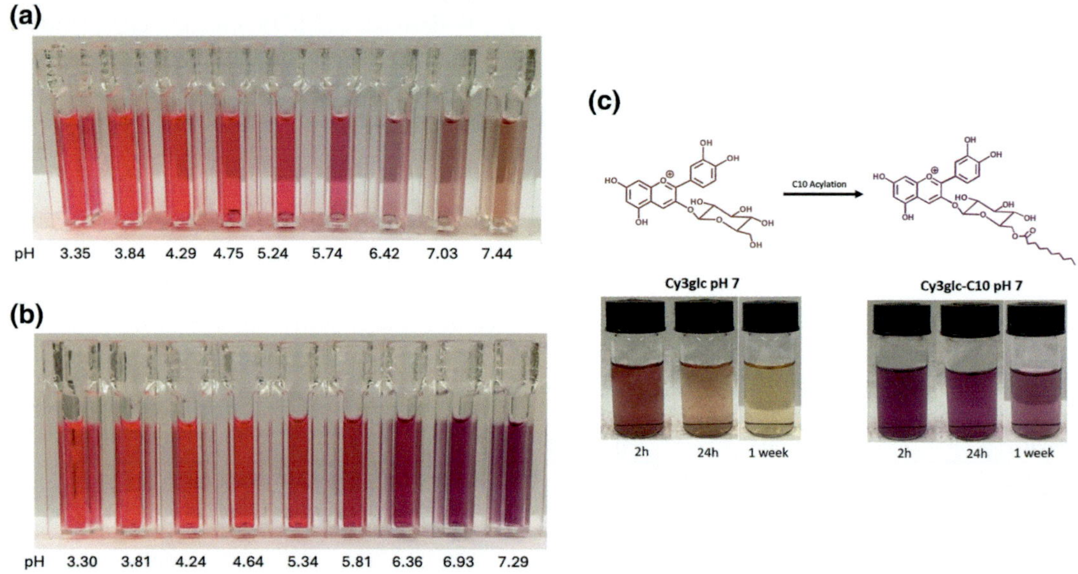

FIGURE 12.3 Aqueous solutions of (a) Cy3glc (5×10^{-5} M) and (b) Cy3glc-C10 (5×10^{-5} M) in 0.1 M SDS between pH 3 and 7; (c) color stability of both anthocyanins in 0.1 M SDS at pH 7 and room temperature overtime.

In a recent work, Oliveira et al. demonstrated that polyacylated anthocyanin-based extracts from hibiscus calyxes (HC), RC, and butterfly pea flower (BPF) maintain their color between pH 3 and 6 over 7 days at room temperature, and showed high resistance to temperature and to bisulfite bleaching. In addition, incorporating such extracts in porcine protein and agar-agar gelatin formulations produced products with stable hues, particularly the blue and purple colors[28] (Figure 12.4).

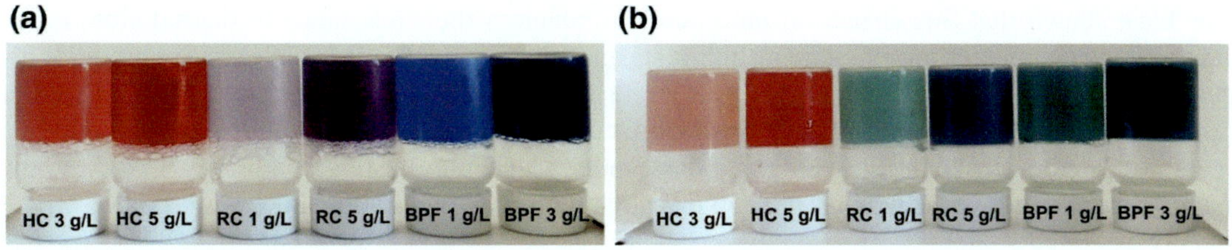

FIGURE 12.4 Gelatins from animal (a) and plant origins (b) with the different acylated anthocyanin-based extracts at two distinct concentrations as natural alternatives to synthetic dyes in the food industry.

12.3 APPLICATIONS

The use of acylated anthocyanins in different food and nonfood products is quite limited because there is a lack of scientific studies about the application of these compounds in different matrices. However, they have been appointed mainly for applications in foods, functional ingredients, indicators/sensors, and cosmetic products. Due to their inherent superior stability rather than the nonacylated ones, it is foreseen that such molecules will gain even more importance in the upcoming years.[29]

12.3.1 Foodstuffs

Acylated anthocyanins are used in the food industry as colorants, functional agents, and indicator sensors. They are found in many plant species, including fruits, berries, vegetables, and tubers, namely, in red potatoes, radishes, RC, black carrots, purple sweet potatoes, purple corn, and BPF, among others.[1, 30]

As colorants, acylated anthocyanins can be used to give foods a purple, blue, or red color. They can be used in jams, fruit preparations, dairy products, ice creams, yogurts, confectionaries, and beverages. As functionalizing agents or functional ingredients, acylated anthocyanins can be used in functional foods to prevent cardiovascular illness, neuronal diseases, cancer, inflammation, and diabetes. Besides their antioxidant capacity, several studies have shown health-promoting properties for acylated anthocyanins, namely, adrenomimetic properties accelerating wound healing, antiproliferative and anti-obesity activities, and therapeutic effects against hyperuricemia and kidney inflammation in hyperuricemic mice. As sensors, acylated anthocyanins can be used mainly as pH and time–temperature indicators to monitor food shelf life in intelligent packaging. Many studies have shown the incorporation of acylated anthocyanins from different sources in films for spoilage monitoring of seafood, meat, fish, and dairy products such as milk and cheese.[31]

12.3.2 Cosmetics

Acylated anthocyanins have garnered significant attention in recent scientific research for their potential in cosmetic applications, owing to their enhanced stability, increased lipophilicity, vibrant colors, and bioactive properties.[32] Acylated anthocyanins are noted for their superior stability against environmental factors such as pH, temperature, and light, which are critical challenges in cosmetic formulations. This stability stems from the intramolecular co-pigmentation effect provided by the acyl groups, typically aromatic acids like *p*-coumaric, caffeic, or ferulic acid, attached to the sugar moieties of anthocyanins.

Studies have shown that this structural modification enhances their resistance to degradation, making them promising candidates for long-lasting colorants and functional ingredients in skincare products (eyeshadows, nail polish, blusher, lipsticks, creams, and sunscreens). For instance, research highlights their use in lipsticks, where acylated anthocyanins from sources like purple sweet potatoes and RC maintain color intensity and shelf stability for over 2 years under accelerated testing conditions.[33] Anthocyanins from red fruit wastes have been valorized and investigated for hair coloration; however, acylated ones are not deeply exploited.[34]

In terms of bioactivity, acylated anthocyanins exhibit potent antioxidant and photoprotective properties, which are highly desirable in cosmetics. They have been shown to scavenge free radicals, reduce reactive oxygen species (ROS) in skin cells, and protect against UV-induced damage. One study demonstrated that acylated anthocyanins, such as cyanidin-3-*O*-glucoside derivatives, incorporated into lipstick formulations provided *in vitro* sun protection factor (SPF) values exceeding 15, indicating their potential as natural UV filters. Additionally, their ability to inhibit tyrosinase – an enzyme involved in melanin production – suggests applications in skin-lightening or anti-melanogenesis products, addressing hyperpigmentation concerns.[35]

The lipophilicity of acylated anthocyanins, improved through acylation with fatty acids (e.g., lauric acid), enhances their solubility in lipid-based cosmetic matrices, overcoming a key limitation of nonacylated anthocyanins. Research on cyanidin-3-*O*-glucoside–lauric acid esters have shown increased lipophilic solubility, enabling their integration into emulsions and creams. Moreover, these compounds have displayed antimicrobial activity against skin pathogens like *Staphylococcus aureus*, suggesting potential use in preservative systems or acne-targeting formulations.[36]

REFERENCES

1. Luo, X. E.; Wang, R.; Wang, J.; Li, Y.; Luo, H.; Chen, S.; Zeng, X. A.; Han, Z. Acylation of Anthocyanins and Their Applications in the Food Industry: Mechanisms and Recent Research Advances. *Foods* 2022, *11* (14), 2166.
2. Gerard, V.; Ay, E.; Morlet-Savary, F.; Graff, B.; Galopin, C.; Ogren, T.; Mutilangi, W.; Lalevee, J. Thermal and Photochemical Stability of Anthocyanins from Black Carrot, Grape Juice, and Purple Sweet Potato in Model Beverages in the Presence of Ascorbic Acid. *J. Agric. Food. Chem.* 2019, *67* (19), 5647–5660. Article. https://doi.org/10.1021/acs.jafc.9b01672.
3. Gras, C. C.; Nemetz, N.; Carle, R.; Schweiggert, R. M. Anthocyanins from Purple Sweet Potato (*Ipomoea batatas* (L.) Lam.) and Their Color Modulation by the Addition of Phenolic Acids and Food-Grade Phenolic Plant Extracts. *Food Chem.* 2017, *235*, 265–274. Article. https://doi.org/10.1016/j.foodchem.2017.04.169.
4. Li, J.; Li, X. D.; Zhang, Y.; Zheng, Z. D.; Qu, Z. Y.; Liu, M.; Zhu, S. H.; Liu, S.; Wang, M.; Qu, L. Identification and Thermal Stability of Purple-Fleshed Sweet Potato Anthocyanins in Aqueous Solutions with Various pH Values and Fruit Juices. *Food Chem.* 2013, *136* (3–4), 1429–1434. Article. https://doi.org/10.1016/j.foodchem.2012.09.054.
5. Oliveira, H.; Basilio, N.; Pina, F.; Fernandes, I.; de Freitas, V.; Mateus, N. Purple-Fleshed Sweet Potato Acylated Anthocyanins: Equilibrium Network and Photophysical Properties. *Food Chem.* 2019, *288*, 386–394. Article. https://doi.org/10.1016/j.foodchem.2019.02.132.
6. Sigurdson, G. T.; Robbins, R. J.; Collins, T. M.; Giusti, M. M. Molar Absorptivities (Epsilon) and Spectral and Colorimetric Characteristics of Purple Sweet Potato Anthocyanins. *Food Chem.* 2019, *271*, 497–504. Article. https://doi.org/10.1016/j.foodchem.2018.07.096.
7. Eichhorn, S.; Winterhalter, P. Anthocyanins from Pigmented Potato (*Solanum tuberosum* L.) Varieties. *Food Res. Int.* 2005, *38* (8–9), 943–948. Article; Proceedings Paper. https://doi.org/10.1016/j.foodres.2005.03.011.
8. Li, A. R.; Xiao, R. S.; He, S. J.; An, X. Y.; He, Y.; Wang, C. T.; Yin, S.; Wang, B.; Shi, X. W.; He, J. R. Research Advances of Purple Sweet Potato Anthocyanins: Extraction, Identification, Stability, Bioactivity, Application, and Biotransformation. *Molecules* 2019, *24* (21), 21. Review. https://doi.org/10.3390/molecules24213816.
9. Zhao, C. L.; Yu, Y. Q.; Chen, Z. J.; Wen, G. S.; Wei, F. G.; Zheng, Q.; Wang, C. D.; Xiao, X. L. Stability-Increasing Effects of Anthocyanin Glycosyl Acylation. *Food Chem.* 2017, *214*, 119–128. https://doi.org/10.1016/j.foodchem.2016.07.073.

10. Zhao, L.-Y.; Chen, J.; Wang, Z.-Q.; Shen, R.-M.; Cui, N.; Sun, A.-D. Direct Acylation of Cyanidin-3-Glucoside with Lauric Acid in Blueberry and Its Stability Analysis. *Int. J. Food Prop.* 2016, *19* (1), 1–12. https://doi.org/10.1080/10942912.2015.1016577; Cruz, L.; Fernandes, V. C.; Araújo, P.; Mateus, N.; de Freitas, V. Synthesis, Characterisation and Antioxidant Features of Procyanidin B4 and Malvidin-3-Glucoside Stearic Acid Derivatives. *Food Chem.* 2015, *174* (0), 480–486. http://dx.doi.org/10.1016/j.foodchem.2014.11.062; Grajeda-Iglesias, C.; Salas, E.; Barouh, N.; Baréa, B.; Figueroa-Espinoza, M. C. Lipophilization and MS Characterization of the Main Anthocyanins Purified from Hibiscus Flowers. *Food Chem.* 2017, *230*, 189–194. https://doi.org/10.1016/j.foodchem.2017.02.140.
11. Cruz, L.; Guimaraes, M.; Araujo, P.; Evora, A.; de Freitas, V.; Mateus, N. Malvidin 3-Glucoside–Fatty Acid Conjugates: From Hydrophilic toward Novel Lipophilic Derivatives. *J. Agric. Food. Chem.* 2017, *65* (31), 6513–6518. https://doi.org/10.1021/acs.jafc.6b05461; De Castro, V. C.; Alves da Silva, P. H.; de Oliveira, E. B.; Desobry, S.; Humeau, C. Extraction, Identification and Enzymatic Synthesis of Acylated Derivatives of Anthocyanins from Jaboticaba (*Myrciaria cauliflora*) Fruits. *Int. J. Food Sci. Technol.* 2014, *49* (1), 196–204. https://doi.org/10.1111/ijfs.12298; Cruz, L.; Benohoud, M.; Rayner, C. M.; Mateus, N.; de Freitas, V.; Blackburn, R. S. Selective Enzymatic Lipophilization of Anthocyanin Glucosides from Blackcurrant (*Ribes nigrum* L.) Skin Extract and Characterization of Esterified Anthocyanins. *Food Chem.* 2018, *266*, 415–419. https://doi.org/10.1016/j.foodchem.2018.06.024; Cruz, L.; Fernandes, I.; Guimaraes, M.; de Freitas, V.; Mateus, N. Enzymatic Synthesis, Structural Characterization and Antioxidant Capacity Assessment of a New Lipophilic Malvidin-3-Glucoside–Oleic Acid Conjugate. *Food Funct.* 2016, *7* (6), 2754–2762. https://doi.org/10.1039/C6FO00466K; Marquez-Rodriguez, A. S.; Guimarães, M.; Mateus, N.; de Freitas, V.; Ballinas-Casarrubias, M. L.; Fuentes-Montero, M. E.; Salas, E.; Cruz, L. Disaccharide Anthocyanin Delphinidin 3-O-Sambubioside from *Hibiscus sabdariffa* L.: Candida Antarctica Lipase B-Catalyzed Fatty Acid Acylation and Study of Its Color Properties. *Food Chem.* 2021, *344*, 128603. https://doi.org/10.1016/j.foodchem.2020.128603; Leonarski, E.; Cesca, K.; de Oliveira, D.; Zielinski, A. A. F. A Review on Enzymatic Acylation as a Promising Opportunity to Stabilizing Anthocyanins. *Crit. Rev. Food Sci. Nutr.* 2023, *63* (24), 6777–6796; https://doi.org/10.1080/10408398.2022.2041541; Li, Z. Y.; Teng, W.; Xie, X.; Bao, Y. W.; Xu, A. H.; Sun, Y. X.; Yang, B. R.; Tian, J. L.; Li, B. Enzymatic Acylation of Cyanidin-3-O-Glucoside with Aromatic and Aliphatic Acid Methyl Ester: Structure–Stability Relationships of Acylated Derivatives. *Food Res. Int.* 2024, *192*. https://doi.org/10.1016/j.foodres.2024.114824.
12. Yang, W.; Kortesniemi, M.; Yang, B.; Zheng, J. Enzymatic Acylation of Anthocyanins Isolated from Alpine Bearberry (*Arctostaphylos alpina*) and Lipophilic Properties, Thermostability, and Antioxidant Capacity of the Derivatives. *J. Agric. Food. Chem.* 2018, *66* (11), 2909–2916. https://doi.org/10.1021/acs.jafc.7b05924.
13. Marton, Z.; Léonard-Nevers, V.; Syrén, P. O.; Bauer, C.; Lamare, S.; Hult, K.; Tranc, V.; Graber, M. Mutations in the Stereospecificity Pocket and at the Entrance of the Active Site of *Candida antarctica* Lipase B Enhancing Enzyme Enantioselectivity. *J. Mol. Catal. B: Enzym.* 2010, *65* (1), 11–17. https://doi.org/10.1016/j.molcatb.2010.01.007.
14. Fernandez-Aulis, F.; Torres, A.; Sanchez-Mendoza, E.; Cruz, L.; Navarro-Ocana, A. New Acylated Cyanidin Glycosides Extracted from Underutilized Potential Sources: Enzymatic Synthesis, Antioxidant Activity and Thermostability. *Food Chem.* 2020, *309*, 125796. https://doi.org/10.1016/j.foodchem.2019.125796.
15. Cruz, L.; Guimarães, M.; Araújo, P.; Évora, A.; de Freitas, V.; Mateus, N. Malvidin 3-Glucoside–Fatty Acid Conjugates: From Hydrophilic toward Novel Lipophilic Derivatives. *J. Agric. Food. Chem.* 2017, *65* (31), 6513–6518. https://doi.org/10.1021/acs.jafc.6b05461.
16. Trouillas, P.; Sancho-García, J. C.; De Freitas, V.; Gierschner, J.; Otyepka, M.; Dangles, O. Stabilizing and Modulating Color by Copigmentation: Insights from Theory and Experiment. *Chem. Rev.* 2016, *116* (9), 4937–4982. https://doi.org/10.1021/acs.chemrev.5b00507.
17. Sharif, N.; Khoshnoudi-Nia, S.; Jafari, S. M. Nano/Microencapsulation of Anthocyanins; a Systematic Review and Meta-Analysis. *Food Res. Int.* 2020, *132*, 18. Review. https://doi.org/10.1016/j.foodres.2020.109077.
18. Yousuf, B.; Gul, K.; Wani, A. A.; Singh, P. Health Benefits of Anthocyanins and Their Encapsulation for Potential Use in Food Systems: A Review. *Crit. Rev. Food Sci. Nutr.* 2016, *56* (13), 2223–2230. Review. https://doi.org/10.1080/10408398.2013.805316.
19. Cai, D.; Li, X.; Chen, J.; Jiang, X.; Ma, X.; Sun, J.; Tian, L.; Vidyarthi, S. K.; Xu, J.; Pan, Z.; et al. A Comprehensive Review on Innovative and Advanced Stabilization Approaches of Anthocyanin by Modifying Structure and Controlling Environmental Factors. *Food Chem.* 2022, *366*, 130611. https://doi.org/10.1016/j.foodchem.2021.130611.
20. Fu, W.; Li, S.; Helmick, H.; Hamaker, B. R.; Kokini, J. L.; Reddivari, L. Complexation with Polysaccharides Enhances the Stability of Isolated Anthocyanins. *Foods* 2023, *12* (9). https://doi.org/10.3390/foods12091846 From NLM.
21. Zhang, L.; Yao, L.; Zhao, F.; Yu, A.; Zhou, Y.; Wen, Q.; Wang, J.; Zheng, T.; Chen, P. Protein and Peptide-Based Nanotechnology for Enhancing Stability, Bioactivity, and Delivery of Anthocyanins. *Adv. Healthc. Mater.* 2023, *12* (25), e2300473. https://doi.org/10.1002/adhm.202300473 From NLM.

22. Chung, C.; Rojanasasithara, T.; Mutilangi, W.; McClements, D. J. Enhancement of Colour Stability of Anthocyanins in Model Beverages by Gum Arabic Addition. *Food Chem.* 2016, *201*, 14–22. https://doi.org/10.1016/j.foodchem.2016.01.051.
23. Celli, G. B.; Ghanem, A.; Brooks, M. S. L. Optimized Encapsulation of Anthocyanin-Rich Extract from Haskap Berries (*Lonicera caerulea* L.) in Calcium-Alginate Microparticles. *J. Berry Res.* 2016, *6* (1), 1–11. Article. https://doi.org/10.3233/jbr-150107.
24. Guldiken, B.; Gibis, M.; Boyacioglu, D.; Capanoglu, E.; Weiss, J. Impact of Liposomal Encapsulation on Degradation of Anthocyanins of Black Carrot Extract by Adding Ascorbic Acid. *Food Funct.* 2017, *8* (3), 1085–1093. Article. https://doi.org/10.1039/c6fo01385f.
25. Klinger, E.; Salminen, H.; Bause, K.; Weiss, J. Stability of Anthocyanin Extracts from Chokeberry, Grape, Hibiscus, and Purple Sweet Potato in w-3-Fatty Acid Rich Oil-in-Water Emulsions. *Food Chem.* 2024, *459*. https://doi.org/10.1016/j.foodchem.2024.140385.
26. Lima, J. C.; Vautier-Giongo, C.; Lopes, A.; Melo, E.; Quina, F. H.; Macanita, A. L. Color Stabilization of Anthocyanins: Effect of SDS Micelles on the Acid-Base and Hydration Kinetics of Malvidin 3-Glucoside (Oenin). *J. Phys. Chem. A* 2002, *106* (24), 5851–5859. Article. https://doi.org/10.1021/jp014081c.
27. Guimarães, M.; Mateus, N.; de Freitas, V.; Cruz, L. Improvement of the Color Stability of Cyanidin-3-Glucoside by Fatty Acid Enzymatic Acylation. *J. Agric. Food. Chem.* 2018, *66* (38), 10003–10010. https://doi.org/10.1021/acs.jafc.8b03536; Mendoza, J.; Pina, F.; Basílio, N.; Guimarães, M.; de Freitas, V.; Cruz, L. Extending the Stability of Red and Blue Colors of Malvidin-3-Glucoside-Lipophilic Derivatives in the Presence of SDS Micelles. *Dyes Pigm.* 2018, *151*, 321–326. https://doi.org/10.1016/j.dyepig.2018.01.007; Cruz, L.; Basílio, N.; de Freitas, V. Color Stabilization of Cyanidin-3-Glucoside-Based Dyes by Encapsulation with Biocompatible PEGylated Phospholipid Micelles. *Dyes Pigm.* 2020, *181*, 108592. https://doi.org/10.1016/j.dyepig.2020.108592.
28. Pereira, A. R.; Fernandes, V. C.; Delerue-Matos, C.; de Freitas, V.; Mateus, N.; Oliveira, J. Exploring Acylated Anthocyanin-Based Extracts as a Natural Alternative to Synthetic Food Dyes: Stability and Application Insights. *Food Chem.* 2024, *461*. https://doi.org/10.1016/j.foodchem.2024.140945.
29. Vidana Gamage, G. C.; Lim, Y. Y.; Choo, W. S. Sources and Relative Stabilities of Acylated and Nonacylated Anthocyanins in Beverage Systems. *J. Food Sci. Technol.* 2022, *59* (3), 831–845. https://doi.org/10.1007/s13197-021-05054-z From NLM.
30. Giusti, M. M.; Wrolstad, R. E. Acylated Anthocyanins from Edible Sources and Their Applications in Food Systems. *Biochem. Eng. J.* 2003, *14* (3), 217–225. https://doi.org/10.1016/S1369-703X(02)00221−8.
31. Pereira, V. A.; de Arruda, I. N. Q.; Stefani, R. Active Chitosan/PVA Films with Anthocyanins from *Brassica oleraceae* (Red Cabbage) as Time–Temperature Indicators for Application in Intelligent Food Packaging. *Food Hydrocoll.* 2015, *43*, 180–188. https://doi.org/10.1016/j.foodhyd.2014.05.014; Moradi, M.; Tajik, H.; Almasi, H.; Forough, M.; Ezati, P. A Novel pH-Sensing Indicator Based on Bacterial Cellulose Nanofibers and Black Carrot Anthocyanins for Monitoring Fish Freshness. *Carbohydr. Polym.* 2019, *222*, 115030. https://doi.org/10.1016/j.carbpol.2019.115030; Choi, I.; Lee, J. Y.; Lacroix, M.; Han, J. Intelligent pH Indicator Film Composed of Agar/Potato Starch and Anthocyanin Extracts from Purple Sweet Potato. *Food Chem.* 2017, *218*, 122–128. http://dx.doi.org/10.1016/j.foodchem.2016.09.050.
32. Li, X.; Liu, C.; Li, Y.; Yuan, K.; Zhang, W.; Cai, D.; Peng, Z.; Hu, Y.; Sun, J.; Bai, W. Bioactivity and Application of Anthocyanins in Skin Protection and Cosmetics: An Extension as a Functional Pigment. *Phytochem. Rev.* 2023, *22* (6), 1441–1467. https://doi.org/10.1007/s11101-023-09890-z.
33. Westfall, A.; Giusti, M. Color Profiles and Stability of Acylated and Nonacylated Anthocyanins as Novel Pigment Sources in a Lipstick Model: A Viable Alternative to Synthetic Colorants. *J. Cosmet. Sci.* 2017, *68* (3), 233–244. PubMed.
34. Rose, P. M.; Cantrill, V.; Benohoud, M.; Tidder, A.; Rayner, C. M.; Blackburn, R. S. Application of Anthocyanins from Blackcurrant (*Ribes nigrum* L.) Fruit Waste as Renewable Hair Dyes. *J. Agric. Food. Chem.* 2018, *66* (26), 6790–6798. https://doi.org/10.1021/acs.jafc.8b01044.
35. Westfall, A.; Sigurdson, G. T.; Giusti, M. M. Antioxidant, UV Protection, and Antiphotoaging Properties of Anthocyanin-Pigmented Lipstick Formulations. *J. Cosmet. Sci.* 2019, *70* (2), 63–76. From NLM.
36. Li, L.; Zhou, P.; Wang, Y.; Pan, Y.; Chen, M.; Tian, Y.; Zhou, H.; Yang, B.; Meng, H.; Zheng, J. Antimicrobial Activity of Cyanidin-3-O-Glucoside–Lauric Acid Ester against *Staphylococcus aureus* and *Escherichia coli*. *Food Chem.* 2022, *383*, 132410. https://doi.org/10.1016/j.foodchem.2022.132410.

Impact of Food Processing

13

Ammu P. Nair, Deepika Kaushik, Emel Oz, Fatih Oz, and Mukul Kumar

13.1 INTRODUCTION

Raw agricultural products undergo a number of processes to enhance their safety, shelf life, nutritional value, and sensory appeal in order to satisfy consumer demands and enable global food distribution. Food processing is the term used for this procedure. These methods transform raw materials into edible products while lowering the risk of spoiling and microbiological contamination. These consist of high-pressure processing, fermentation, drying, freezing, and thermal processing. Processing can significantly affect the nutritional profile of foods, particularly bioactive compounds like anthocyanins, which are water-soluble polyphenolic pigments that give fruits and vegetables their striking red, purple, and blue hues (Al-Juhaimi et al., 2018). Among other health benefits, anthocyanins, which belong to the flavonoid group, have anti-inflammatory, anti-carcinogenic, and antioxidant properties that protect the heart and improve cognitive function. Processing variables that significantly affect anthocyanin stability include temperature, pH, light, and oxygen exposure. For instance, anthocyanins may be better preserved by nonthermal techniques like high-pressure processing, while they may be broken down and their bioavailability reduced by thermal techniques like pasteurization (Carranza-Gomez et al., 2025). The amount of anthocyanin that is retained depends on the food matrix, processing parameters, and storage conditions.

Water-soluble flavonoid pigments known as anthocyanins of which acylated anthocyanins are a particular subclass are responsible for the striking red, purple, and blue hues present in a wide range of fruits, vegetables, and tubers. These compounds, which are glycosides of anthocyanidins, are characterized by a core flavylium cation structure (C6–C3–C6 skeleton) to which sugar moieties are attached, usually at the C3 position, and further modified by acylation with organic acids (Cipriano, 2016). The chemical structure of acylated anthocyanins is formed by joining an anthocyanidin aglycone (such as cyanidin, delphinidin, pelargonidin, peonidin, petunidin, or malvidin) to one or more glycosyl units that are esterified with aliphatic or aromatic acids like *p*-coumaric, caffeic, or ferulic acid (Figure 13.1). The acylation enhances the stability of the flavylium cation and reduces color degradation under various environmental conditions, such as pH, light, and temperature, by promoting π-stacking interactions, which protect the chromophore from water's nucleophilic attack.

DOI: 10.1201/9781003519591-17

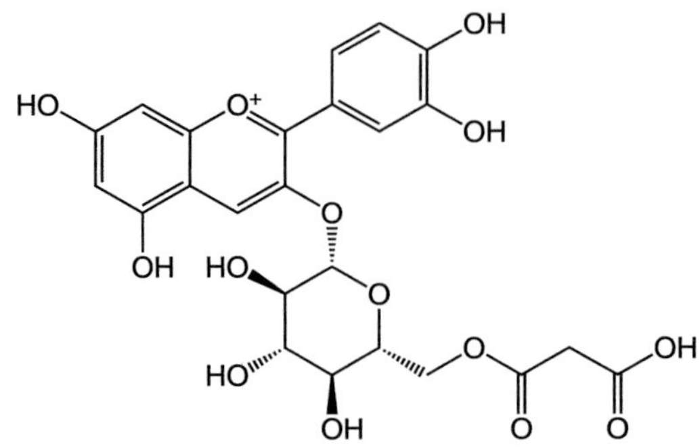

FIGURE 13.1 Structure of acylated anthocyanins – cyanidin-3-(6″-malonylglucoside).

13.2 DISTINCTION BETWEEN ACYLATED AND NONACYLATED ANTHOCYANINS

The red, purple, and blue hues of many fruits and vegetables are caused by a family of water-soluble flavonoid pigments known as anthocyanins. They are separated into acylated and nonacylated forms according to whether or not they have acyl groups attached to their sugar moieties. This structural diversity has a significant impact on their physicochemical properties, stability, bioavailability, and biological activity. In acylated anthocyanins, acyl groups are esterified to their sugar constituents, much like hydroxycinnamic acids (Jeyaraj et al., 2023). This acylation increases their resistance to environmental factors like changes in pH, exposure to light, and heat processing. For instance, compared to their nonacylated counterparts in fruits like berries and grapes, the acylated anthocyanins found in red cabbage are more stable (Table 13.1). Acryl groups promote intramolecular stacking, which increases stability and makes acylated anthocyanins more suitable for applications requiring long shelf life and consistent coloration (Zilic et al., 2019).

However, because of their resistance to degradation, acylated anthocyanins are able to enter the colon undamaged, where they can influence the gut microbiota and support health-promoting initiatives. Additionally, acylated anthocyanins have shown excellent antioxidant properties and inhibitory effects on enzymes such as α-glucosidase and α-amylase, which are important for controlling blood glucose levels after meals (Wang et al., 2023). However, because they lack these acyl modifications, nonacylated

TABLE 13.1 Comparison of Acylated versus Nonacylated Anthocyanins in Terms of Stability

PARAMETER	*NONACYLATED ANTHOCYANINS*	*ACYLATED ANTHOCYANINS*	*REFERENCES*
Light stability	Easily degraded upon light exposure	Improved light resistance due to intramolecular stacking	Jeyaraj et al. (2023)
Thermal stability	Susceptible to thermal degradation	High resistance to heat; stable at elevated temperatures	Zilic et al. (2019)
Light stability	Easily degraded upon light exposure	Improved light resistance due to intramolecular stacking	Jeyaraj et al. (2023)
Storage stability	Shorter shelf life without protective agents	Long shelf life under varied conditions	Wang et al. (2023)
Oxidative stability	More prone to oxidative degradation	Enhanced due to steric protection by acyl groups	Zhao et al. (2017)

anthocyanins are more likely to degrade under similar conditions (Zhao et al., 2017). However, their bioavailability is often higher. Studies have shown that nonacylated anthocyanins are better absorbed in the human gastrointestinal tract, increasing their plasma concentrations after meals. This variation in absorption is caused by the structural simplicity of nonacylated anthocyanins, which facilitates transport across intestinal membranes. Perhaps because of their complex structures that hinder absorption, acylated anthocyanins have a comparatively low bioavailability (Zilic et al., 2019).

13.3 THERMAL PROCESSING

13.3.1 Common Thermal Techniques (Pasteurization, Sterilization, and Blanching)

Thermal processing techniques like pasteurization, sterilization, and blanching are crucial unit procedures in food science that are used to increase food safety, extend shelf life, and maintain quality attributes. Each technique uses heat to achieve specific objectives, such as microbial inactivation, enzyme denaturation, or preservation of sensory and nutritional qualities, although they differ in their methods and effects on food processing. These processes ensure compliance with safety standards while reducing adverse effects on food quality. The ideas of heat transfer, microbial kinetics, and food matrix interactions are used in their development.

13.3.1.1 Pasteurization

Pasteurization is a heat treatment process that uses high temperatures to kill harmful microorganisms in food, prevent spoilage, and extend its shelf life while preserving its nutritional value and sensory appeal. The procedure uses controlled heat, typically below 100°C (e.g., 72°C for 15 s in high-temperature short-time pasteurization or 63°C for 30 min in low-temperature long-time pasteurization) to target pathogens like *Salmonella* spp. and *Listeria monocytogenes* and achieve a logarithmic reduction in microbial populations (Cunningham, 2017). The heat denatures proteins, breaks down microbial cell membranes, and inhibits metabolic processes, rendering pathogens inactive without significantly altering the food's organoleptic properties. When pasteurizing liquid foods, like milk or juice, heat exchangers ensure even heating and rapid cooling to prevent quality degradation. It has a major impact on food processing, ensuring food safety and enabling a longer shelf life in the refrigerator (typically 2–3 weeks for milk) (Azizi-Lalabadi et al., 2023). However, excessive heat can result in minor flavor changes due to Maillard reactions or volatile compound changes, as well as nutrient losses, such as the deterioration of heat-sensitive vitamins like ascorbic acid. Modern innovations, such as ohmic heating for pasteurization, increase energy efficiency and reduce processing times without sacrificing quality in high-viscosity products (Ahmed & Ramaswamy, 2020).

13.2.1.2 Sterilization

Sterilization is typically performed at temperatures above 100°C (e.g., 121°C for 15 min in canning) to achieve a 12-log decrease of resistant spores such as *Clostridium botulinum*. It breaks down microbial DNA, proteins, and cell structures using either dry heat or moist heat (steam or hot water). Retorts and other high-pressure equipment are used in the process to maintain water temperatures high enough to totally deactivate microorganisms (Russell et al., 2023). In food processing, sterilization significantly extends the shelf life of foods (up to several years for canned goods), but due to prolonged exposure to high temperatures, the nutritional and sensory qualities of foods frequently deteriorate. For instance, protein denaturation, vitamin deterioration, and textural changes (like softening in vegetables) are commonly brought on by heat-induced chemical reactions like hydrolysis or oxidation. One innovation that has been highlighted in recent studies is high-pressure thermal sterilization (HPTS), which uses heat and high pressure to reduce processing times and minimize quality losses, particularly in low-acid foods (Marshall et al., 2018).

13.3.1.3 Blanching

Blanching, a mild heat pretreatment, is primarily used to deactivate enzymes and reduce microbial burdens prior to further processing such as drying, canning, or freezing. Blanching, which is typically accomplished by immersing food in steam or hot water (80–100°C) for 1–10 min, targets enzymes like peroxidase and polyphenol oxidase that cause browning and off-flavors (Ranjan et al., 2017). Heat is used in the process to denaturize enzyme proteins, which harms their active sites and prevents catalysis. Blanching also reduces surface microbial contamination and enhances color retention by stabilizing pigments like chlorophyll. In food processing, blanching is necessary to preserve quality for later storage; e.g., it prevents frozen vegetables from oxidatively becoming rancid. However, over-blanching can result in nutrient leaching (including water-soluble vitamins) and cell wall deterioration, which can soften texture. Newer blanching methods, such as infrared or microwave blanching, offer reduced nutrient losses and improved energy efficiency compared to traditional methods (Pandey et al., 2019). Blanching is therefore a preprocessing step that maintains food quality while enhancing the efficiency of later procedures.

13.4 IMPACT ON ANTHOCYANIN STABILITY: DEGRADATION PATHWAYS

Anthocyanins, a family of flavonoid pigments that give plants their striking red, purple, and blue hues, are valued for their antioxidant properties and potential health benefits like cardioprotective, anti-inflammatory, and anti-carcinogenic effects. Anthocyanins break down primarily through two processes: hydrolytic cleavage and oxidative degradation. Anthocyanins undergo deglycosylation at high temperatures or in acidic environments, converting anthocyanidin glycosides into aglycones (Zhao et al., 2021). These aglycones subsequently decompose into phenolic derivatives such as benzoic acid, chalcones, and diketones. Plant vacuoles undergo oxidative degradation in the presence of hydrogen peroxide (H_2O_2), which is often catalyzed by class III peroxidases. This process is particularly accelerated by exposure to high temperatures or light stress (Enaru et al., 2021). By cleaving bonds, this process produces colorless compounds. For example, too much H_2O_2 in vacuoles causes oxidation mediated by peroxidase, which drastically lowers the amount of anthocyanin. Temperature is a crucial factor. According to research, thermal processing follows first-order reaction kinetics, which means that higher temperatures accelerate ring-opening and deglycosylation reactions, reducing the half-life of anthocyanins (Oancea, 2021). In a similar vein, pH affects anthocyanin structure; it is most stable at pH 2–3, while neutral or alkaline environments promote the formation of unstable chalcone forms. Light exposure, particularly ultraviolet (UV) radiation, causes photodegradation, which exacerbates color loss, while oxygen encourages auto-oxidation, which further compromises stability. By stabilizing anthocyanin structures through π-stacking interactions, co-pigmentation with compounds like flavonoids or phenolic acids can slow down degradation. Additionally, new processing methods like microencapsulation and nonthermal methods like pulsed electric fields have been studied to increase stability by protecting anthocyanins from environmental stresses (Chen et al., 2020). For instance, using biopolymers for microencapsulation enhances bioavailability and gastrointestinal retention while decreasing degradation during digestion.

13.5 CASE STUDIES ON THE IMPACT OF PROCESSING METHODS

Tan et al. (2023) compared the physical characteristics, anthocyanin profile, total polyphenol content, and antioxidant activity of RC (*Brassica oleracea* L. var. *capitata* f. *rubra*) with the five processing methods of boiling, steaming, stir-frying, microwaving, and freeze-drying. The goal of the controlled experimental analysis was to elucidate the effects of thermal and nonthermal processing methods on the nutritional and functional properties of this cruciferous vegetable, which is valued for its rich phytochemical composition. Physical attributes like color and texture were assessed using standardized experimental procedures. According to the findings, freeze-drying preserved the pigment's vibrant red color and structural integrity the best, while thermal processes like boiling severely damaged it. The Folin–Ciocalteu assay showed that samples that were freeze-dried had the highest total polyphenol content, indicating minimal loss of phenolic compounds, as opposed to boiling, which caused significant leaching into the cooking water. An analysis of the anthocyanin profile using high-performance liquid chromatography (HPLC) showed that nonthermal processing enhanced the retention of cyanidin derivatives, which are significant contributors to the bioactivity of RC. Antioxidant activity was assessed using the 2,2-diphenyl-1-picrylhydrazyl (DPPH) and 2,2′-azino-bis-(3-ethylbenzothiazoline-6-sulfonic acid) (ABTS) assays, and the findings followed a similar pattern, with freeze-dried samples having a greater ability to scavenge radicals. According to these findings, freeze-drying is the most effective method of preserving the nutritional value of RC because the bioactive compounds in the vegetable are extremely sensitive to processing conditions.

In their study, Enaru et al. (2021) investigated the effects of various food processing techniques on anthocyanins, a family of flavonoid chemicals known for their antioxidant and health-promoting properties and which give many fruits, vegetables, and derived products their vibrant colors. The study systematically evaluates how various processing techniques, such as thermal treatments like boiling and pasteurization, and nonthermal processes like high-pressure processing and ultrasound, drying, and fermentation, affect the stability, chemical structure, and bioactivity of anthocyanins. The authors note that significant degradation of anthocyanins is often caused by heat-induced chemical processes, such as oxidation and hydrolysis, which alter their molecular structure and reduce their antioxidant activity. Conversely, nonthermal methods like high-pressure processing usually preserve anthocyanin content and bioactivity better by minimizing heat exposure. The study also examines how processing parameters such as pH, temperature, and exposure time impact anthocyanin stability, emphasizing how important it is to optimize these variables to maintain the anthocyanins' nutritional and functional properties. The study discusses the implications for food product development and suggests that innovative processing techniques may enhance the preservation of these bioactive components in functional meals. In contrast to boiling, which resulted in significant leaching into the cooking water, the Folin–Ciocalteu test revealed that samples that were freeze-dried had the highest total polyphenol content, suggesting little loss of phenolic compounds. High-performance liquid chromatography (HPLC) analysis of the anthocyanin profile revealed that nonthermal processing improved the retention of cyanidin derivatives, which are important factors in RC's bioactivity. The DPPH and ABTS tests were used to measure antioxidant activity, and the results showed a similar pattern, with freeze-dried samples being better at scavenging radicals. These results indicate that because RC's bioactive compounds are highly sensitive to processing conditions, freeze-drying is the best way to preserve the vegetable's nutritional value.

The study by Luo et al. (2022) examined the acylation of anthocyanins and their potential applications in the food industry, with a focus on the underlying processes and recent advancements. Anthocyanins, a class of flavonoid pigments valued for their antioxidant and health-promoting properties, give fruits and vegetables their vibrant colors. However, their instability under various light, temperature, and pH conditions limits their industrial value. Acrylation, the process of adding acyl groups to anthocyanins, increases the compounds' stability, solubility, and bioavailability, making them more suitable for use in culinary applications. The chemical mechanisms of acylation, including enzymatic and chemical acylation processes, are thoroughly examined in this study along with their effects on the structure and functionality of anthocyanins. It highlights how acylated anthocyanins are more resistant to deterioration, allowing for their use as natural colorants and practical ingredients in beverages, candies, and other foods. The authors also discuss recent advancements like the use of acylated anthocyanins in encapsulation and smart packaging technologies that improve sensory aspects and extend shelf life. By integrating results from multiple studies, the study demonstrates how acylation may be used to address problems with food processing and preservation while fostering health benefits. However, the study does point out flaws in large-scale production techniques and the need for further research into reasonably priced acylation techniques.

13.6 NONTHERMAL PROCESSING TECHNIQUES

13.6.1 High-Pressure Processing

High-pressure processing (HPP) is a nonthermal method of food preservation that deactivates enzymes and microorganisms while preserving the food's nutritional value, sensory appeal, and functionality by applying isostatic pressures, typically between 100 and 1,000 MPa (Huang et al., 2017). In order for HPP to function, food must first be packed and then uniformly transferred through a liquid medium, usually water, in a pressure vessel under high hydrostatic pressure. Microbial cell membranes permeabilize or rupture as a result of this pressure, rupturing non-covalent bonds such as ionic and hydrogen bonds. As a result, the microbes become dormant, and their cellular functions are compromised. For instance, at room temperature, pressures greater than 300 MPa can rupture vegetative bacterial cells, but higher pressures (600 MPa or more) combined with moderate heat (60–70°C) are required to inactivate resistant bacterial spores (Abera, 2019). Additionally, HPP extends shelf life without using significant heat and by using high hydrostatic pressure that denatures and inactivates enzymes through modifications to the tertiary and quaternary structures of proteins. Unlike thermal processing, HPP minimizes the loss of heat-sensitive nutrients like vitamins and bioactive compounds while maintaining organoleptic qualities like flavor and texture. HPP's uniform pressure distribution ensures consistent treatment regardless of food geometry, making it suitable for a wide range of products, including dairy, meats, and juices (Muntean et al., 2016). Among the factors influencing its efficacy are water activity, pH, and food matrix composition; a lower water activity results in a less effective microbial inactivation process. Because HPP can achieve pasteurization or sterilization effects with minimal quality loss, it is positioned as a sustainable alternative to traditional methods, meeting consumer demand for minimally processed, additive-free foods (Elamin et al., 2015).

13.6.2 Pulsed Electric Fields

A new nonthermal food processing method called pulsed electric fields (PEFs) uses short, high-voltage electric pulses to preserve food matrices, which are typically liquid or semi-liquid, and inactivate microorganisms (Niu et al., 2020). The primary mechanism of PEFs is electroporation, in which high-intensity electric fields (5–50 kV/cm) damage the cell membranes of microorganisms. This process creates hydrophilic holes in the lipid bilayer, increasing membrane permeability and allowing intracellular contents to escape. This lowers metabolic activity and deactivates microorganisms. PEF effectiveness is influenced by critical parameters such as electric field strength, pulse duration, frequency, and the food's physicochemical properties (e.g., conductivity and pH). In contrast to thermal pasteurization, PEF operates at ambient or sub-pasteurization temperatures, preserving heat-sensitive nutrients such as vitamins and bioactive compounds, according to studies on fruit juices and dairy products (Martínez et al., 2023). For instance, compared to samples that have undergone thermal processing, orange juice treated with PEF retains higher levels of flavonoids and phenolic acids. Additionally, PEF enhances mass transfer mechanisms and facilitates the extraction of bioactive compounds, such as polyphenols, from fruit by-products by altering the structural makeup of plant cell membranes. However, PEF is less effective against bacterial spores and enzymes, so low heat and combination methods are needed for sterilization (Raso et al., 2022).

13.6.3 Cold Plasma and Ultrasound

Cold plasma (CP) is the creation of partially ionized gas at room temperature that includes ions, electrons, UV photons, reactive oxygen species (ROS), and reactive nitrogen species (RNS). These reactive species degrade microbial cell membranes through oxidative stress, etching, and electrophoration, resulting in cell lysis and inactivating pathogens such as *Salmonella*, *Escherichia coli*, and *Listeria monocytogenes*, with log reductions up to 5-log that are similar to thermal pasteurization but have minimal impact on food quality (Pan et al., 2020). Furthermore, CP deactivates natural enzymes like polyphenoloxidase, which slows down browning processes and modifies the structures of proteins and starches to enhance functionality. On the other hand, ultrasound produces acoustic cavitation by using high-intensity sound waves (20–100 kHz), which leads to the rupture of microbubbles and the creation of localized hot and cold zones (Namjoo et al., 2022). By weakening microbial cell membranes, producing free radicals, and enhancing mass transfer, this mechanical disruption effectively inactivates pathogens like *Pseudomonas fluorescens* and enzymes like pectin methylesterase without compromising nutritional value. CP and ultrasonography (US) cooperate to enhance microbial cleansing and enzyme inactivation through complementary mechanisms such as increased cell membrane permeability and damage from radicals. Further research is necessary to fully comprehend the interactions between CP reactive species and food ingredients and to optimize US parameters for a range of food matrices (Liao et al., 2020).

13.6.4 Advantages Over Conventional Heat Treatment

Advanced heat treatment technologies like ohmic heating, microwave (MW) heating, radiofrequency (RF) heating, and infrared heating provide significant advantages over conventional heat treatment methods in the food processing industry due to their improved energy efficiency, consistent heating, and improved preservation of nutritional and sensory qualities (Rajan et al., 2023). Unlike traditional methods that rely on conductive and convective heat transmission, advanced procedures generate heat directly inside the food matrix, minimizing temperature gradients and reducing processing times. Ohmic heating, for instance, converts electrical energy into heat by using the electrical resistance of food, enabling rapid and uniform cooking while preserving organoleptic properties and bioactive ingredients (Cascais et al., 2021). Studies have shown that ohmic heating reduces thermal load indicators like hydroxymethylfurfural and produces a 5-log drop in microbial counts at 60°C in just 2.91 min, compared to the long times required for conventional heating. Similar to this, MW and RF heating ensure uniform heat distribution and avoid cold spots, a common issue with conventional methods that could compromise microbiological safety, by permeating food with dielectric properties (Rodríguez et al., 2021). These technologies also enhance nutrient preservation; e.g., ohmic heating sweet whey increases the quantity of bioactive peptides compared to traditional methods. However, without significantly altering the internal structures of food, infrared heating effectively targets surface cleaning, reducing bacterial burdens. Modern methods also reduce processing times and energy consumption, which lessens the carbon footprint and encourages sustainability (Soni et al., 2020). For instance, bread loaves may be heated consistently with a temperature differential of less than 5°C when RF and hot air treatment are combined, improving mold control without compromising quality. Fast volumetric heating, electroporation in ohmic systems, and dielectric heating in MW/RF all work together to lessen nutrient breakdown and for the creation of hazardous

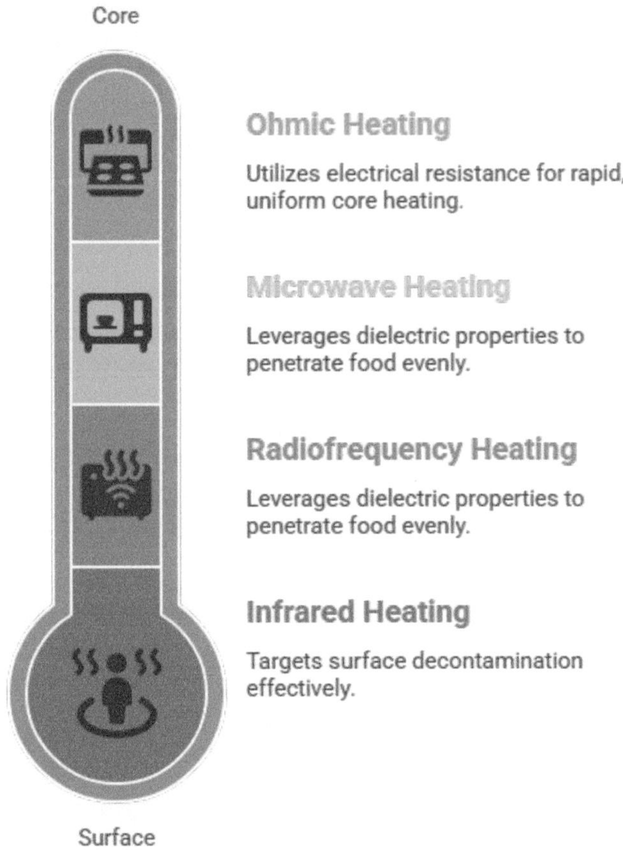

FIGURE 13.2 Advantages of advanced heat treatment technologies.

compounds like acrylamide (Popovich et al., 2017). Additionally, CP alters the structures of proteins and carbohydrates to improve functionality and deactivates natural enzymes like polyphenoloxidase, which slows down browning processes. However, US uses high-intensity sound waves (20–100 kHz) to create acoustic cavitation, which causes microbubbles to burst and creates localized hot and cold zones (Figure 13.2). This mechanical disruption efficiently inactivates pathogens like *Pseudomonas fluorescens* and enzymes like pectin methylesterase without sacrificing nutritional value by weakening microbial cell membranes, generating free radicals, and improving mass transfer (Bilbao-Sainz et al., 2024). Through complementary mechanisms like increased cell membrane permeability and radical damage, CP and US work together to promote microbial cleaning and enzyme inactivation. To completely understand how CP reactive species interact with food ingredients and optimize US parameters for various food matrices, more investigation is needed.

13.7 INFLUENCE OF PH, LIGHT, AND OXYGEN DURING PROCESSING

The quality, safety, and shelf life of food products are greatly influenced by environmental factors such as pH, light, and oxygen. pH affects microbial growth, enzyme activity, and chemical reactions during food preparation. An acidic pH may inhibit the growth of dangerous microorganisms, enhancing food safety, while alkaline conditions may accelerate degradation processes like lipid oxidation, impacting flavor and nutritional value (Goyudianto et al., 2021). Maintaining the pH at the proper level during fermentation, for instance, ensures that dairy products like yogurt have the proper texture and flavor because variations may cause off flavors or denaturation of proteins. Vitamins (like riboflavin in milk) and pigments are among the light-sensitive substances that are broken down by photochemical processes brought on by light exposure, particularly UV and visible light. This causes nutritional loss and color fading (Yin et al., 2024). Because light-induced oxidation can reduce shelf life and alter sensory qualities, this is essential for transparent packaging. Oxygen, a key catalyst for oxidative processes, influences the rancidity of lipids and the degradation of antioxidants, reducing the nutritional value and aesthetic appeal of foods like oils and nuts. Product quality is maintained and oxidative spoiling is decreased by regulating oxygen levels during processes like vacuum packing or modified environment packaging. Deaeration and other techniques are required in beverage processing to increase shelf life because oxygen can exacerbate the microbiological development of aerobic spoilage organisms (Fellows, 2022). Examples of how these parameters interact include processes like thermal treatment, where pH affects the heat stability of proteins and oxygen availability affects the flavor and color of products from the Maillard reaction.

13.7.1 Role of Environmental Conditions in Anthocyanin Degradation

Primary environmental factors such as pH, temperature, light, oxygen, and the presence of co-pigments or enzymes play critical roles in the breakdown of anthocyanins by altering their molecular structure and chemical stability. At low pH (5), they either decompose into chalcones or transform into colorless carbinol pseudobases, which results in color loss. High temperatures, such as those above 35°C, accelerate the breakdown of anthocyanins because they disrupt the conjugated system of chromogenic groups (Kim et al., 2017). According to studies, 43% of the monomeric anthocyanins in blueberry puree are lost when it is pasteurized for 3 min at 95°C. Light exposure, particularly UV radiation, promotes oxidative reactions that lead to photodegradation, while high oxygen levels exacerbate degradation through reactions with ROS and the formation of polymeric pigments (Landi et al., 2015). Enzymatic activity, especially from class III peroxidases and β-glucosidases, further aids in the breakdown of plants. This is evident in *Brunfelsia* flowers, where anthocyanins are actively broken down by vacuolar enzymes as they develop. Additionally, ascorbic acid accelerates the breakdown of anthocyanins in the presence of oxygen by forming complexes that destabilize the pigment structure. These environmental stressors work together to reduce the stability of anthocyanins, which impacts their bioactivity and application in food systems (Yuan et al., 2023).

13.7.2 Protective Role of Acyl Groups in Acidic Media

Similar to acetals or ketals, which cleave in mildly acidic environments, acyl groups are ideal for protecting sensitive functions during reactions involving strong acids like trifluoroacetic acid (TFA) because they are less likely to hydrolyze. The protected molecule's reactivity is further controlled by acyl groups' capacity to donate electrons, preventing unfavorable reactions in acidic environments. The synthesis of peptides and carbohydrates frequently uses this protective mechanism because acyl groups preserve chemoselectivity under extremely acidic conditions (Fenger et al., 2021). Deprotection typically requires basic or reductive conditions because the acyl-protected molecule is not affected by acidic reagents, demonstrating its utility in synthetic processes. This stability is crucial for preserving the protected functional group until selective MAG1 deprotection takes place under basic or reductive conditions because acidic reagents are commonly used in multistep organic synthesis (Figure 13.3). As

Characteristic	Stability in Acidic Media	Hydrolysis Proneness	Reactivity Modulation	Deprotection Conditions
Acyl Groups	High	Low	Yes	Basic/Reductive
Silyl Ethers	Low	N/A	N/A	N/A
Acetals/Ketals	Low	High	N/A	N/A

FIGURE 13.3 Acyl-protecting groups in organic synthesis.

an illustration of the group's resilience in acidic environments, pivaloyl-protected galactose derivatives were found to be effective in acidic glycosylation processes (Sonawane et al., 2020). This is because the electron-rich pivaloyl group stabilized dioxolenium-type intermediates, improving stereoselectivity. Because acyl groups are less likely to hydrolyze than acetals or ketals, which cleave in mildly acidic environments, they are ideal for protecting sensitive functions during processes involving strong acids like TFA. The capacity of acyl groups to donate electrons further controls the reactivity of the protected molecules and prevents adverse reactions in acidic environments. Because acyl groups ensure chemoselectivity in extremely acidic environments, this defense mechanism is frequently employed in the synthesis of peptides and carbohydrates. Acidic reagents cannot affect the acyl-protected molecules because deprotection typically requires basic or reductive conditions, underscoring its utility in synthetic techniques (Yoshida et al., 2017).

13.8 STRATEGIES TO PRESERVE BIOACTIVITY

13.8.1 Use of Protective Matrices (Encapsulation and Biopolymers)

Stability, bioavailability, and controlled release are enhanced when sensitive bioactive substances, microorganisms, or therapeutic agents are encapsulated using protective matrices, especially using biopolymers to shield them from environmental stresses. Biopolymers, such as alginate, chitosan, starch, and gelatin, form a barrier around the core material to shield it from damaging factors like moisture, pH fluctuations, UV radiation, and enzymatic degradation (Detsi et al., 2020). The active ingredient is trapped inside a biopolymeric matrix through methods such as ionic gelation, emulsification, or layer-by-layer assembly. For instance, alginate, a polysaccharide derived from algae, creates hydrogels by ionic crosslinking with calcium ions. This creates a porous but stable matrix that permits gradual release while protecting encapsulated agents, such as plant-growth-promoting rhizobacteria (PGPR), from stressors associated with the soil. Similarly, the biodegradable polymer chitosan forms robust capsules that enhance microbial viability in acidic environments by electrostatically connecting with anionic biopolymers like alginate using its cationic characteristics (Gheorghita et al., 2021). This electrostatic interaction, along with the biopolymer's biocompatibility and biodegradability, ensures minimal toxicity and environmental impact. Additionally, layer-by-layer construction, as exemplified by chitosan and carboxymethyl cellulose, protects probiotics such as *Lactobacillus acidophilus* during gastrointestinal transit. These techniques create coverings with nanolayers that stop enzymes from penetrating. The controlled release mechanism, which relies on diffusion, matrix breakdown, or environmental cues (like pH changes), guarantees targeted distribution (Zabot et al., 2022).

13.8.2 Optimization of Process Parameters (Temperature, Time, and Atmosphere)

Optimizing process variables such as temperature, time, and environment is necessary to maintain the bioactivity of anthocyanins, flavonoid pigments valued for their antioxidant, anti-inflammatory, and anticarcinogenic properties. Temperature significantly affects the stability of anthocyanins because high temperatures can speed up breakdown through oxidative processes or enzymatic activity (Singh et al., 2019). Research indicates that lower temperatures (e.g., 20–25°C) improve the retention of anthocyanins in plant cell cultures. The highest yields for *Perilla frutescens* suspensions were achieved at 20°C, where anthocyanin synthesis increased by 1.8 times compared to higher temperatures. Similarly, since prolonged exposure may cause glycosidic linkages to hydrolyze or degrade thermally, lowering bioactivity, extraction time must be carefully controlled (Song et al., 2022). For instance, using a pH 1 solvent to optimally extract anthocyanins from buni fruit (*Antidesma bunius*) at 68°C for 30 min increased the total monomeric anthocyanin content (555.17 ± 13.34 mg/L) and antioxidant activity. The composition of the atmosphere, especially the amount of oxygen, is also very important because anthocyanins are susceptible to oxidative destruction. Processing in low-oxygen environments, like nitrogen or argon atmospheres, greatly increases stability by lowering oxidative processes (Bedin et al., 2017). Anthocyanin loss is less in vacuum or inert gas environments than in oxygen-rich ones, according to studies. Additionally, novel techniques like RF-assisted enzymatic extraction have demonstrated enhanced anthocyanin yields (50.87 mg/100 g) and preserved bioactivity by reducing processing time and heat exposure (Pravin et al., 2015).

13.8.3 Product Formulation with Stabilizing Agents (Citric Acid and Metal Chelators)

The safety, quality, and shelf life of food products during processing depend on the use of stabilizing agents in product formulation, such as citric acid and metal chelators. Stabilizing chemicals are used to reduce the adverse effects of food processing techniques like freezing, HPP, and thermal treatments that could lead to chemical, physical, and enzymatic deterioration. Citric acid, a weak organic acid, lowers oxidative processes and functions as a pH regulator and antioxidant synergist by chelating metal ions like iron and copper, which encourage lipid peroxidation and food color deterioration (Nangare et al., 2021). Metal chelators, such as polyphosphates or ethylenediaminetetraacetic acid (EDTA), bind pro-oxidant metal ions to prevent oxidative rancidity and nutritional loss in processed foods, such as drinks, canned goods, and emulsions (Allen, 2015). During thermal processing, high temperatures may speed up Maillard reactions or lipid oxidation, changing the flavor and reducing nutrients. Stabilizing agents lessen these effects by maintaining sensory characteristics and structural integrity. For instance, after fruit juices are pasteurized, citric acid increases the stability of ascorbic acid, preserving the nutritional value (Kumar et al., 2023; Liu et al., 2022). Similarly, metal chelators stop enzymatic browning in cut fruits and vegetables by eliminating copper ions required by polyphenol oxidase. By controlling ionic connections, stabilizing chemicals help maintain protein activity; HPP, while helpful for microbial inactivation, can degrade protein structures and cause textural changes. By regulating water activity, stabilizers such as citric acid maintain the integrity of cell membranes; however, freezing can result in the formation of ice crystals, which harms cells and ruins the texture of frozen foods (Marques et al., 2020). The effectiveness of these agents is influenced by concentration, food matrix, and processing conditions; therefore, cautious formulation is required to avoid overuse, which could alter sensory attributes or regulatory compliance.

An investigation into the stability of anthocyanins extracted from purple sweet potatoes (*Ipomoea batatas* L.) in relation to different concentrations of maltodextrin as an encapsulating agent was carried out by Arisanti et al. in 2020. The study assessed the effects of varying maltodextrin concentrations (with DE 10–15) on encapsulation efficiency, total anthocyanin content (TAC), moisture content, and particle size using MW-assisted extraction followed by microencapsulation (Xiao et al., 2022). Anthocyanin stability was evaluated by stress testing the microcapsules under different pH (3, 5, and 7) and temperature (35°C, 75°C, and 115°C) conditions over time. Higher concentrations of maltodextrin were found to significantly improve anthocyanin stability and encapsulation efficiency, especially in alkaline and high-temperature environments. This suggests that maltodextrin functions as an effective protective matrix that inhibits anthocyanin degradation. Maltodextrin's capacity to create a thick encapsulation layer that protects anthocyanins from environmental stressors is responsible for this improvement. The results corroborate earlier research showing maltodextrin is an effective carrier for sensitive bioactive substances, which has also been documented in studies involving various anthocyanin sources and encapsulating agents (Arisanti et al., 2020; Mahdavi et al., 2016). These findings demonstrate the usefulness of maltodextrin in food technology applications, where anthocyanins, which are prized for their antioxidant qualities, are extremely unstable in the absence of adequate protection. Thus, this case study demonstrates how maintaining anthocyanin functionality and expanding their use in food systems exposed to different processing and storage conditions depend on optimizing wall material concentration in microencapsulation processes.

13.9 CONCLUSION

In conclusion, the type of technique employed has a significant impact on the stability of anthocyanins during food processing; nonthermal methods provide better preservation, while thermal methods typically encourage degradation. Anthocyanin resistance to environmental stressors like heat, pH, light, and oxygen is further determined by structural features, especially acylation. Foods rich in anthocyanins can have their nutritional and functional values preserved and degradation effectively reduced by using stabilizing agents, encapsulation, and controlled processing conditions. HPP and PEFs are two examples of nonthermal innovations that show promise for preserving anthocyanins without sacrificing food quality. Overall, to fully utilize the health advantages and business potential of anthocyanins in the food industry, a careful balancing act between appropriate processing methods and protective formulations is required. Optimizing anthocyanin retention and advancing their use in sustainable, functional food systems will require ongoing research and technological developments.

REFERENCES

Abera, G. (2019). Review on high-pressure processing of foods. *Cogent Food & Agriculture*, *5*(1), 1568725.

Ahmed, J., & Ramaswamy, H. S. (2020). Microwave pasteurization and sterilization of foods. In *Handbook of food preservation* (pp. 713–732). CRC Press.

Al-Juhaimi, F., Ghafoor, K., Özcan, M. M., Jahurul, M. H. A., Babiker, E. E., Jinap, S., Sahena, F., Sharifudin, M. S., & Zaidul, I. S. M. (2018). Effect of various food processing and handling methods on preservation of natural antioxidants in fruits and vegetables. *Journal of Food Science and Technology*, *55*(10), 3872–3880.

Allen, K. E. (2015). Metal chelators as antioxidants for food preservation. In *Handbook of antioxidants for food preservation* (pp. 79–104). Woodhead Publishing.

Arisanti, C. I. S., Sukawati, C. B. A. C., Prasetia, I. G. N. J. A., & Wirasuta, I. M. A. G. (2020). Stability of anthocyanins encapsulated from purple sweet potato extract affected by maltodextrin concentration. In *Macromolecular symposia* (vol. 391, no. 1, p.1900127). Wiley.

Azizi-Lalabadi, M., Moghaddam, N. R., & Jafari, S. M. (2023). Pasteurization in the food industry. In *Thermal processing of food products by steam and hot water* (pp. 247–273). Woodhead Publishing.

Bedin, K. C., de Azevedo, S. P., Leandro, P. K., Cazetta, A. L., & Almeida, V. C. (2017). Bone char prepared by CO2 atmosphere: Preparation optimization and adsorption studies of Remazol Brilliant Blue R. *Journal of Cleaner Production*, *161*, 288–298.

Bilbao-Sainz, C., Olsen, C., Chiou, B. S., Rubinsky, B., Wu, V. C., & McHugh, T. (2024). Benefits of isochoric freezing for carrot juice preservation. *Journal of Food Science*, *89*(3), 1324–1336.

Carranza-Gomez, M., Valle-Guadarrama, S., Domínguez-Puerto, R., Sandoval-Castilla, O., & Guerra-Ramírez, D. (2025). Bioactive compounds and pigmenting potential of *Vaccinium corymbosum* extracts separated with aqueous biphasic systems aided by centrifugation. *Processes*, *13*(4), 1072.

Cascais, M., Monteiro, P., Pacheco, D., Cotas, J., Pereira, L., Marques, J. C., & Gonçalves, A. M. (2021). Effects of heat treatment processes: Health benefits and risks to the consumer. *Applied Sciences*, *11*(18), 8740.

Chen, J. Y., Du, J., Li, M. L., & Li, C. M. (2020). Degradation kinetics and pathways of red raspberry anthocyanins in model and juice systems and their correlation with color and antioxidant changes during storage. *LWT*, *128*, 109448.

Cipriano, P. D. A. (2016). *Improved extraction of acylated anthocyanins from purple sweet potato (Ipomoea batatas) for enhanced anti-inflammatory activity and their metabolite production during porcine fecal digestion* (Doctoral dissertation, Texas A&M University).

Cunningham, R. E. (2017). Egg-product pasteurization. In *Egg science and technology* (pp. 289–321). CRC Press.

Detsi, A., Kavetsou, E., Kostopoulou, I., Pitterou, I., Pontillo, A. R. N., Tzani, A., . . . Zoumpoulakis, P. (2020). Nanosystems for the encapsulation of natural products: The case of chitosan biopolymer as a matrix. *Pharmaceutics*, *12*(7), 669.

Elamin, W. M., Endan, J. B., Yosuf, Y. A., Shamsudin, R., & Ahmedov, A. (2015). High pressure processing technology and equipment evolution: A review. *Journal of Engineering Science & Technology Review*, *8*(5).

Enaru, B., Drețcanu, G., Pop, T. D., Stănilă, A., & Diaconeasa, Z. (2021). Anthocyanins: Factors affecting their stability and degradation. *Antioxidants*, *10*(12), 1967.

Fellows, P. J. (2022). *Food processing technology: Principles and practice*. Woodhead Publishing.

Fenger, J. A., Roux, H., Robbins, R. J., Collins, T. M., & Dangles, O. (2021). The influence of phenolic acyl groups on the color of purple sweet potato anthocyanins and their metal complexes. *Dyes and Pigments*, *185*, 108792.

Gheorghita, R., Anchidin-Norocel, L., Filip, R., Dimian, M., & Covasa, M. (2021). Applications of biopolymers for drugs and probiotics delivery. *Polymers*, *13*(16), 2729.

Goyudianto, B. A., Meliana, C., Muliani, D., Sadeli, Y. E., & Ratnasari, N. R. P. (2021). The stability of phycocyanin, phycoerythrin, and astaxanthin from algae towards temperature, pH, light, and oxygen as a commercial natural food colorant. *Indonesian Journal of Life Sciences*, 28–42.

Huang, H. W., Wu, S. J., Lu, J. K., Shyu, Y. T., & Wang, C. Y. (2017). Current status and future trends of high-pressure processing in food industry. *Food Control*, *72*, 1–8.

Jeyaraj, E. J., Vidana Gamage, G. C., Cintrat, J. C., & Choo, W. S. (2023). Acylated and non-acylated anthocyanins as antibacterial and antibiofilm agents. *Discover Food*, *3*(1), 21.

Kim, S., Hwang, G., Lee, S., Zhu, J. Y., Paik, I., Nguyen, T. T., . . . Oh, E. (2017). High ambient temperature represses anthocyanin biosynthesis through degradation of HY5. *Frontiers in Plant Science*, *8*, 1787.

Kumar, A., Kumari, P., Gupta, K., Singh, M., & Tomer, V. (2023). Recent advances in extraction, techno-functional properties, food and therapeutic applications as well as safety aspects of natural and modified stabilizers. *Food Reviews International*, *39*(4), 2233–2276.

Landi, M., Tattini, M., & Gould, K. S. (2015). Multiple functional roles of anthocyanins in plant–environment interactions. *Environmental and Experimental Botany*, *119*, 4–17.

Liao, X., Cullen, P. J., Muhammad, A. I., Jiang, Z., Ye, X., Liu, D., & Ding, T. (2020). Cold plasma–based hurdle interventions: New strategies for improving food safety. *Food Engineering Reviews*, *12*, 321–332.

Liu, S., Sun, H., Ma, G., Zhang, T., Wang, L., Pei, H., Li, X., & Gao, L. (2022). Insights into flavor and key influencing factors of Maillard reaction products: A recent update. *Frontiers in Nutrition*, *9*, 973677.

Luo, X. E., Wang, R., Wang, J., Li, Y., Luo, H., Chen, S., Zeng, X. A., & Han, Z. (2022). Acylation of anthocyanins and their applications in the food industry: Mechanisms and recent research advances. *Foods*, *11*(14), 2166.

Mahdavi, S. A., Jafari, S. M., Assadpoor, E., & Dehnad, D. (2016). Microencapsulation optimization of natural anthocyanins with maltodextrin, gum Arabic and gelatin. *International Journal of Biological Macromolecules*, *85*, 379–385.

Marques, C., Sotiles, A. R., Farias, F. O., Oliveira, G., Mitterer-Daltoe, M. L., & Masson, M. L. (2020). Full physicochemical characterization of malic acid: Emphasis in the potential as food ingredient and application in pectin gels. *Arabian Journal of Chemistry*, *13*(12), 9118–9129.

Marshall, J. W., Schmitt-Kopplin, P., Schuetz, N., Moritz, F., Roullier-Gall, C., Uhl, J., . . . Taylor, A. J. (2018). Monitoring chemical changes during food sterilisation using ultrahigh resolution mass spectrometry. *Food Chemistry*, *242*, 316–322.

Martínez, J. M., Abad, V., Quílez, J., Raso, J., Cebrián, G., & Álvarez-Lanzarote, I. (2023). Pulsed electric fields (PEF) applications in the inactivation of parasites in food. *Trends in Food Science & Technology*, *138*, 470–479.

Muntean, M. V., Marian, O., Barbieru, V., Cătunescu, G. M., Ranta, O., Drocas, I., & Terhes, S. (2016). High pressure processing in food industry–characteristics and applications. *Agriculture and Agricultural Science Procedia*, *10*, 377–383.

Namjoo, M., Moradi, M., Dibagar, N., & Niakousari, M. (2022). Cold plasma pretreatment prior to ultrasound-assisted air drying of cumin seeds. *Food and Bioprocess Technology*, *15*(9), 2065–2083.

Nangare, S., Vispute, Y., Tade, R., Dugam, S., & Patil, P. (2021). Pharmaceutical applications of citric acid. *Future Journal of Pharmaceutical Sciences*, *7*, 1–23.

Niu, D., Zeng, X. A., Ren, E. F., Xu, F. Y., Li, J., Wang, M. S., & Wang, R. (2020). Review of the application of pulsed electric fields (PEF) technology for food processing in China. *Food Research International*, *137*, 109715.

Oancea, S. (2021). A review of the current knowledge of thermal stability of anthocyanins and approaches to their stabilization to heat. *Antioxidants*, *10*(9), 1337.

Pan, Y., Zhang, Y., Cheng, J. H., & Sun, D. W. (2020). Inactivation of Listeria monocytogenes at various growth temperatures by ultrasound pretreatment and cold plasma. *LWT*, *118*, 108635.

Pandey, O. P., Mishra, B. K., & Misra, A. (2019). Comparative study of green peas using with blanching & without blanching techniques. *Information Processing in Agriculture*, *6*(2), 285–296.

Popovich, V. A., Borisov, E. V., Popovich, A. A., Sufiiarov, V. S., Masaylo, D. V., & Alzina, L. (2017). Impact of heat treatment on mechanical behaviour of Inconel 718 processed with tailored microstructure by selective laser melting. *Materials & Design*, *131*, 12–22.

Pravin, T., Sadhasivam, M., & Raghuraman, S. (2015). Optimization of process parameters of Al-10% Cu compacts through powder metallurgy. *Applied Mechanics and Materials*, *813*, 603–607.

Rajan, T. V., Sharma, C. P., & Sharma, A. (2023). *Heat treatment: Principles and techniques*. PHI Learning Pvt. Ltd.

Ranjan, S., Dasgupta, N., Walia, N., Thara Chand, C., & Ramalingam, C. (2017). Microwave blanching: An emerging trend in food engineering and its effects on *Capsicum annuum* L. *Journal of Food Process Engineering*, *40*(2), e12411.

Raso, J., Heinz, V., Alvarez, I., & Toepfl, S. (2022). *Pulsed electric fields technology for the food industry*. Cham, Switzerland: Springer International Publishing.

Rodríguez, L. M. N., Arias, R., Soteras, T., Sancho, A., Pesquero, N., Rossetti, L., . . . Szerman, N. (2021). Comparison of the quality attributes of carrot juice pasteurized by ohmic heating and conventional heat treatment. *LWT*, *145*, 111255.

Russell, C., Hansen, T., & Paulsson, M. (2023). Sterilisation methods. In *Practical pharmaceutics: An international guideline for the preparation, care and use of medicinal products* (pp. 731–748). Cham: Springer International Publishing.

Singh, S., Chakraborty, J. P., & Mondal, M. K. (2019). Optimization of process parameters for torrefaction of *Acacia nilotica* using response surface methodology and characteristics of torrefied biomass as upgraded fuel. *Energy*, *186*, 115865.

Sonawane, R. B., Sonawane, S. R., Rasal, N. K., & Jagtap, S. V. (2020). Chemoselective *O*-formyl and *O*-acyl protection of alkanolamines, phenoxyethanols and alcohols catalyzed by nickel (ii) and copper (ii)-catalysts. *Green Chemistry*, *22*(10), 3186–3195.

Song, S. P., Yang, C., Jiang, C. Z., Wu, Y. M., Guo, R., Sun, H., . . . Zhang, X. K. (2022). Increasing ionic conductivity in Li0. 33La0. 56TiO3 thin-films via optimization of processing atmosphere and temperature. *Rare Metals*, *41*(1), 179–188.

Soni, A., Smith, J., Thompson, A., & Brightwell, G. (2020). Microwave-induced thermal sterilization – a review on history, technical progress, advantages and challenges as compared to the conventional methods. *Trends in Food Science & Technology*, *97*, 433–442.

Tan, S., Lan, X., Chen, S., Zhong, X., & Li, W. (2023). Physical character, total polyphenols, anthocyanin profile and antioxidant activity of red cabbage as affected by five processing methods. *Food Research International*, *169*, 112929.

Wang, Z., Zhang, L., Li, Y., Liu, Q., & Yuan, C. (2023). Non-acylated and acylated anthocynins in red wines of different ages: Color contribution and evaluation. *Journal of Food Composition and Analysis*, *115*, 104951.

Xiao, Z., Xia, J., Zhao, Q., Niu, Y., & Zhao, D. (2022). Maltodextrin as wall material for microcapsules: A review. *Carbohydrate Polymers*, *298*, 120113.

Yin, S., Jin, W., Xi, T., Zhou, X., He, Z., Meng, X., . . . Li, X. (2024). Factors affect the oxygen production of *Chlorella pyrenoidosa* in a bacterial–algal symbiotic system: Light intensity, temperature, pH and static magnetic field. *Process Safety and Environmental Protection*, *184*, 492–501.

Yoshida, K., Oyama, K. I., & Kondo, T. (2017). Structure of polyacylated anthocyanins and their UV protective effect. *Recent Advances in Polyphenol Research*, 171–192.

Yuan, Y., Tian, Y., Gao, S., Zhang, X., Gao, X., & He, J. (2023). Effects of environmental factors and fermentation on red raspberry anthocyanins stability. *LWT*, *173*, 114252.

Zabot, G. L., Schaefer Rodrigues, F., Polano Ody, L., Vinícius Tres, M., Herrera, E., Palacin, H., . . . Olivera-Montenegro, L. (2022). Encapsulation of bioactive compounds for food and agricultural applications. *Polymers*, *14*(19), 4194.

Zhao, C. L., Yu, Y. Q., Chen, Z. J., Wen, G. S., Wei, F. G., Zheng, Q., Wang, C. D., & Xiao, X. L. (2017). Stability-increasing effects of anthocyanin glycosyl acylation. *Food chemistry*, *214*, 119–128.

Zhao, Y. W., Wang, C. K., Huang, X. Y., & Hu, D. G. (2021). Anthocyanin stability and degradation in plants. *Plant Signaling & Behavior*, *16*(12), 1987767.

Zilic, S., Dodig, D., Vancetović, J., Grcic, N., Peric, V., Titan, P., & Maksimović, V. (2019). Composition of anthocyanins in colored grains and the relationship of their non-acylated and acylated derivatives. *Polish Journal of Food and Nutrition Sciences*, *69*(2), 137–146.

14 Encapsulation Strategies for Enhancing Stability, Bioavailability, and Functionality of Acylated Anthocyanins

Ana Rita Pereira, Nuno Mateus, Victor de Freitas, and Ana Fernandes

14.1 INTRODUCTION

Acylated anthocyanins are promising anthocyanin molecules, primarily due to their improved chemical stability, particularly pH, color, and thermal stability, compared to nonacylated anthocyanins. Although they can be found naturally in edible sources, including radishes, red potatoes, red cabbage, black carrots, and purple sweet potatoes (1), acylated anthocyanins may be obtained by chemical (2, 3) and enzymatic acylation (4, 5). The improved performance of acylated anthocyanins can be attributed to the structural modifications in the flavylium nucleus. As polyphenols, anthocyanins contain multiple hydroxyl groups in their molecular structure, including their glycosyl moieties, allowing them to readily interact with various acids to form ester bonds (6). Typically, anthocyanin glycosyl groups undergo esterification with organic acids, while inorganic acids are uncommon (7). The enhanced stability of acylated anthocyanins arises from multiple protective mechanisms, comprising the stacking of their aromatic and planar pyrrole ring, which creates a sandwich-like structure that shields the anthocyanin nucleus from nucleophilic attack (8, 9), hence preventing their degradation (10, 11). Intra- and intermolecular co-pigmentation, as well as self-association, further reinforces this stability. However, mono- and diacylated anthocyanins in aqueous solutions under neutral conditions still showed color loss (11), indicating some acylated anthocyanins' vulnerability in this pH range. Besides the physical effects of acylation, a decrease in polarity, an increase in molecular size, and changes in the spatial structure can be observed (7, 12). Some studies have demonstrated that anthocyanins are fully degraded in a human colon model through a combination of

DOI: 10.1201/9781003519591-18

extract and dimethylol dimethyl-hydantoin, was emulsified under controlled stirring and sonication conditions. In a second study, Span 80 was used as the sole surfactant at varying concentrations (20%, 25%, and 30%), while the rest of the formulation remained unchanged. Both studies assessed physical stability over 28 days and determined antioxidant activity using the 2,2-diphenyl-1-picrylhydrazyl (DPPH) assay. The most stable formulations were obtained with 0.75% Span 60/PEG 400 or 20% Span 80, achieving pH values of 5.8–5.9, suitable for skin application. These formulations exhibited particle sizes between 111.1 and 152.4 nm and maintained antioxidant activity with an IC_{50} value of approximately 38.25 μg/mL, achieving around 81% free radical inhibition. Findings suggest optimized microemulsions effectively preserve anthocyanins and could be promising carriers for skin-related applications. Zang et al. developed and optimized W/O/W purple potato anthocyanin nanoparticles (PPA-NPs) to enhance the bioavailability and bioactivity of anthocyanins (54). The NPs exhibited a high encapsulation efficiency (93.11%), a particle size of 278.36 nm, and a stable zeta potential of 40.44 ± 0.75 mV, which were evaluated over time. *In vivo* experiments in rats showed a 220.36% increase in anthocyanin absorption compared to direct PPA administration. Antioxidant markers improved significantly (Figure 14.1), with increased glutathione (GSH) and superoxide dismutase (SOD) levels and reduced malondialdehyde (MDA) and protein carbonyl content after 30 days of PPA-NP ingestion, indicating lower oxidative stress (54). Additionally, gut microbiota analysis revealed a beneficial modulation, with increased levels of *Lactobacillus*. These findings suggest that PPA-NPs effectively enhance anthocyanin stability, bioavailability, and health benefits, particularly in antioxidant defense and gut microbiota regulation (54). Ravanfar et al. developed solid lipid nanoparticles (SLNs) to improve the stability of anthocyanins from red cabbage under GI conditions

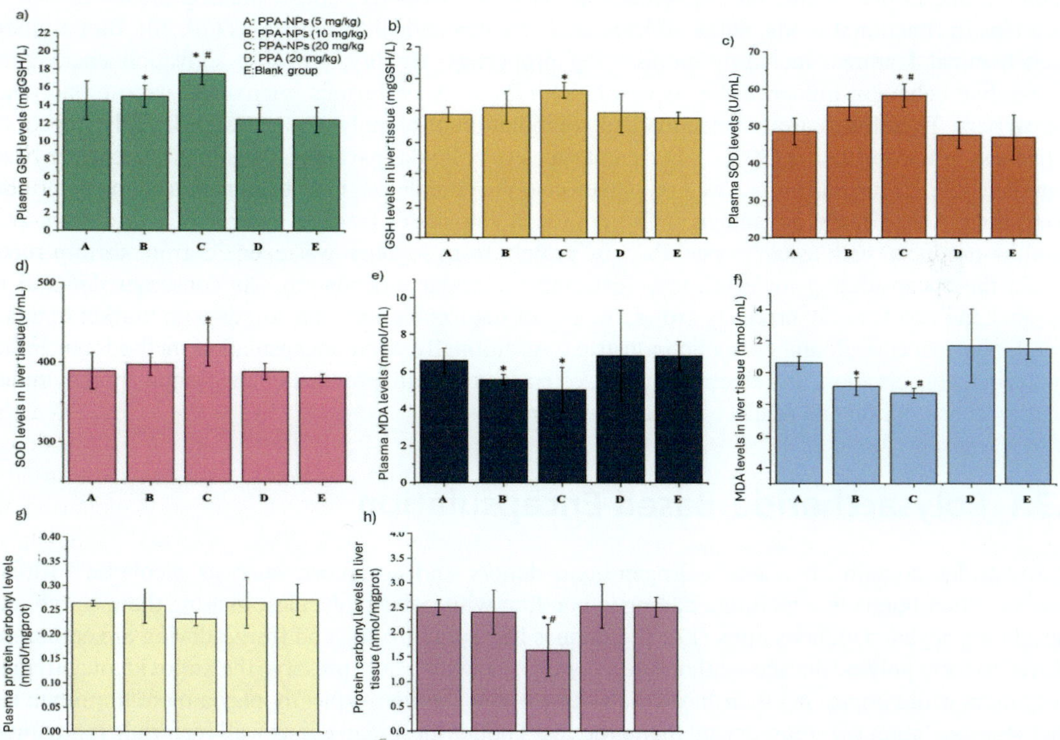

FIGURE 14.1 Evaluation of the antioxidant markers after 30 days with PPA-loaded NPs in W/O/W emulsions (PPA-NPs): content of GSH in the plasma (a) and liver tissue (b); content of SOD in the plasma (c) and (d) liver tissue; content of MDA in the plasma (e) and liver tissue (f), and content of protein carbonyls in the plasma (g) and liver tissue (h). * indicates $p < 0.05$ compared to the blank group. # indicates $p < 0.05$ compared to the PPA group. Adapted from (54).

(55). The SLNs were obtained by diluting a W/O microemulsion, where the aqueous phase containing anthocyanin extract was gradually added to a lipid phase composed of palmitic acid, Span 85, and egg lecithin at 60°C. Ethanol was included as a cosurfactant, and the mixture was stirred before dispersion in an aqueous stabilizer (Pluronic F-127). Optimization of formulation parameters using a pseudo-ternary phase diagram, along with Plackett–Burman and Box–Behnken designs, resulted in spherical SLNs with a particle size of 455 nm and an encapsulation efficiency of 89.2%. The SLNs demonstrated greater stability at pH 3 (simulating gastric conditions) than at pH 5 and 7.4 (intestinal conditions), while temperatures below 25°C helped prevent anthocyanin degradation compared to 40 and 50°C. These results suggest that SLNs can be effective carriers for increased anthocyanins' apparent short-term accelerated stability against environmental stressors such as pH and high temperatures.

14.3 BIOPOLYMER-BASED CARRIERS

Biopolymer-based carriers, such as polysaccharides (e.g., chitosan (CS) and pectin (PC)) and proteins (e.g., whey protein and gelatin), have emerged as an effective and sustainable strategy for encapsulating acylated anthocyanins, enhancing their stability, bioavailability, and controlled release (20). These carriers offer biocompatibility and biodegradability, minimizing environmental impact compared to synthetic carriers. Additionally, their tailored physicochemical properties allow them to protect anthocyanins from degradation due to pH fluctuations, enzymatic activity, and oxidative stress, making them valuable for applications in functional foods, nutraceuticals, and pharmaceutical formulations (20, 30). Biopolymers' physicochemical features, including emulsifying properties, rigidity, porosity, activation energy, and surface-active behavior, influence the encapsulation efficiency of various micro- and nano-biopolymer-based carriers. Therefore, careful selection and combination of biopolymers can significantly impact the effectiveness of encapsulation (30, 56). Different types of colloidal particles, including biopolymer-based NPs or nanogels, inclusion complexes, and polyelectrolyte complexes, have been prepared to encapsulate anthocyanins. A significant advantage of this approach also arises from its potential for upcycling agricultural by-products, such as citrus peel PC, soy protein from soybean waste, or CS from shrimp shells, to obtain the encapsulating material, thus fostering a circular bioeconomy. As consumer demand for clean-label and eco-friendly products grows, the use of natural biopolymers aligns with market trends in the food, pharmaceutical, and cosmetic industries. Additionally, their encapsulation methods, including spray-drying, coacervation, and electrospinning, often require mild processing conditions, thereby reducing energy consumption and reliance on harmful chemicals.

14.3.1 Polysaccharide-Based Encapsulation

Polysaccharides contain abundant hydrogen-bond donors and acceptors, such as alcoholic hydroxyl groups and ether bonds that facilitate hydrogen bonding with polar molecules, making them suitable for encapsulating acylated anthocyanins (19). Techniques like spray-drying and freeze-drying are commonly employed to form polysaccharide–anthocyanin complexes, which help preserve the antioxidant activity of anthocyanins while improving their bioavailability. Polysaccharides typically play a more dominant role in microencapsulation *via* spray-drying than proteins. Their widespread use as wall materials is attributed to their high holding capacity of volatile compounds, low viscosity, high solubility, and effective emulsifying properties (30, 56). Spray-drying is one of the most established and widely used microencapsulation techniques in industrial settings, achieving encapsulation efficiencies of 80%–90% for anthocyanins (57). This method offers key advantages such as fast processing, cost-effectiveness, and scalability, making it highly suitable for large-scale applications. However, the elevated temperatures required can lead to the degradation of heat-sensitive anthocyanins, potentially reducing their functionality. Additionally, the process often results in particles with irregular morphology and broad size distribution, complicating control over their final characteristics (58). Another limitation is the material adhesion to the drying chamber walls, leading to product loss and decreased yield. Despite its benefits, these challenges can restrict its application in food formulations (30). Freeze-drying offers an alternative to spray-drying for preserving water-soluble and thermosensitive bioactive compounds, particularly anthocyanins. This technique operates at low temperatures and in a vacuum environment, which reduces the risk of degradation during the process. However, it is more energy-intensive and time-consuming, typically taking 24–48 h, making it a costlier alternative. Additionally, a porous structure is created that allows oxygen to easily reach anthocyanins, which could promote oxidation (58, 59). There is no significant difference in the encapsulation efficiency compared to spray-drying; however, considering the morphological properties (size and shape), the methodologies showed different features (56). Spray-drying produces spherical and concave particles with a broad size range (from microns to submicrons), typically smooth and crack-free, with higher inlet temperatures promoting hollow structures with lower density. In contrast, freeze-dried powders have irregular shapes, porous structures, and lower density due to the sublimation process (56, 58, 59). Spray-dried particles are generally finer and more uniform, whereas freeze-dried ones tend to be larger, fragmented, and spongy. Additionally, spray-dried powders consist of separate particles, while freeze-drying results in aggregated structures (58, 59). Electrospraying and electrospinning are emerging techniques that use high-voltage electric fields to create particles and fibers, respectively. The main difference between them is the concentration of the biopolymer solution, which determines the final morphology of the material – higher concentrations for fiber formation and lower concentrations favor spherical micro- or NP production (30, 56). The final characteristics of fibers and particles depend on several factors, including processing conditions, solution properties, and environmental parameters. While it has shown promise in pharmaceutical applications, its use in the food sector remains limited due to challenges such as low encapsulation efficiency, particle aggregation, and the need for food-grade biopolymer formulations. Additionally, the high costs associated with optimizing process parameters and ensuring scalability hinder its widespread adoption.

Various polysaccharides have been explored as encapsulating agents due to their biocompatibility, film-forming ability, and protective properties. Examples include starch, CS, maltodextrin (MD), PC, cellulose, and natural gums, each offering distinct functional attributes that influence the stability and release of encapsulated acylated anthocyanins.

14.3.1.1 Maltodextrin

Numerous studies have explored MD for encapsulating acylated anthocyanins as polysaccharide-based carriers. MD is widely used in the industry as a cost-effective, low-viscosity, and water-soluble carrier derived from the partial hydrolysis of starch. Its ability to encapsulate hydrophilic anthocyanins is primarily driven by hydrogen bonding and matrix isolation (19). During processes like spray-drying or freeze-drying, MD initially interacts with anthocyanins through hydrogen bonds, and as drying progresses, it forms a molecular grid structure that effectively entraps the anthocyanins, enhancing their stability and functionality. Ersus and Yurdagel (60) investigated the spray-drying of black carrot anthocyanins using various MDs, including Stardri 10 (10DE), Glucodry 210 (20–23DE), and MDX 29 (28–31DE) as carriers, at different inlet/outlet air temperatures, with a constant feed solid content (20%). The anthocyanin content of the powders microencapsulated with Glucodry 210 (630 mg anthocyanin/100 g dry matter of powder) was 28.5% higher than that of MDX 29 (482 mg anthocyanin/100 g dry matter of powder) and Stardri 10 (499 mg anthocyanin/100 g dry matter of powder). The study also highlights the influence of outlet air temperature on the retention of anthocyanin content during the spray-drying process. A decrease in outlet temperature resulted in higher anthocyanin preservation (36.83% for MDX 29, 16.05% for Glucodry 210, and 7.9% for SD 10), particularly for MDs with higher dextrose equivalents (DEs). These findings show that higher DE MDs are more sensitive to higher outlet air temperatures because lower-molecular-weight MDs contain shorter chains, and oxidation reactions of aldehydes at the open sides of the molecules may lead to structural deformations during heating processes (60). Scanning electron microscopic (SEM) micrographs showed particle sizes ranging from approximately 3 μm to 20 μm, and all capsules spray-dried with MDs with different DEs exhibited smooth spherical shapes. Antioxidant activity assays showed that as the anthocyanin content increases, the EC_{50} value decreases. The anthocyanin content of encapsulated powders decreased by 33% after 64 days of storage at 25°C, while only an 11% loss was observed at 4°C, highlighting a temperature-dependent degradation. The samples retained their pink color at 4°C with a half-life of anthocyanins of 25 months, whereas they turned brown at 25°C with a drop in half-life to 10 months. Additionally, exposure to light at 25°C reduced the half-life to 9 months, emphasizing the combined effect of temperature and light on anthocyanin degradation (60). The microencapsulation of red cabbage (*Brassica oleracea* L. var. *capitata* L. f. *rubra*) anthocyanin-rich extract by spray-drying proves to be an effective method for producing a natural food colorant (61). Different concentrations of MD and gum arabic (GA) were evaluated as encapsulating agents. The study showed that the concentrated red cabbage extract (CE) contained 1.6 times more anthocyanins than the raw extract (RE), with the microencapsulated microparticles from CE achieving up to 85 mg of anthocyanins per 100 g. The highest yield of microparticles (67%) was obtained with formulations containing higher MD concentrations. The resulting microparticles exhibited uniformity in size, spherical shape, and smooth surfaces, making them suitable for food applications. Additionally, the combination of MD and GA improved the thermal stability of the microparticles, with the glass transition temperature ranging between 37 and 90°C, significantly higher than the storage temperature. These findings highlight the potential of microencapsulated red cabbage anthocyanin-rich extract as a promising natural colorant for the food industry (61). Building on the previous study that utilized spray-drying for red cabbage anthocyanin-rich extract microencapsulation, the current work explores a different technique, drum drying, to encapsulate the same anthocyanin-rich extract (62). This method, combining MD DE 20 (MD20) and GA at various ratios, achieved encapsulation efficiencies ranging from 93.65% to 98.85%, with the highest efficiency observed for the MD20:AG 20:80 formulation. This combination also resulted in the highest anthocyanin content (42.26 ± 0.58 mg C3G equivalents/100 g DW) and excellent solubility (over 90%) (62). The encapsulated extract exhibited a mixed crystalline-amorphous structure, with increased crystallinity at higher AG concentrations. When incorporated into the Thai dessert "Aluar," the encapsulated extract improved storage stability, reduced lipid oxidation, increased antioxidant activity, and extended shelf life, demonstrating the potential of drum drying as a viable method for preserving anthocyanins and developing functional food ingredients (62). A recent study explored the encapsulation of anthocyanins extracted from dried BPFs (*Clitoria ternatea* L.) using foam-mat drying, with monoglycerides (MG) as foaming agents and MD as a stabilizer (63). The optimal formulation, containing 3.5% MG and 20% MD dried at 70°C, retained up to 210 mg of anthocyanins per 100 g of product with an encapsulation efficiency of 98%. The results showed that increasing MG content enhanced foam porosity, reducing drying time, while higher MD levels improved foam stability and anthocyanin retention. The study also demonstrated that increasing the concentration of MG or MD contributed to the thermal stability of anthocyanins during processing. More than 90% of anthocyanins were retained in BPF powders within 120 min, particularly in formulations containing MG levels above 3.5% and MD concentrations exceeding 15% (63). These results highlight the protective role of coating agents such as MG and MD in encapsulating anthocyanins, effectively shielding them from thermal degradation. The antibacterial activity of encapsulated BPF powder was evaluated, showing that powders with 2.5% and 3.5% MG effectively inhibited *Bacillus subtilis*. However, no inhibition was observed for *Staphylococcus aureus*, *Escherichia coli*, or *Pseudomonas aeruginosa* (63). These results suggest that MG enhances the encapsulation efficiency and antibacterial properties of BPF powders, indicating potential applications in the food and cosmetic industries. Vergara et al. (64) explored the microencapsulation of anthocyanin extract from purple-fleshed cultivated potatoes (purple potato extract (PPE)) using spray-drying with MD as an encapsulating agent. The optimal formulation achieved an encapsulation efficiency of 86% for an inlet air temperature of 130°C and a PPE:MD ratio of 1:4, demonstrating effective retention of anthocyanins (64). The degradation rate constant of anthocyanins in purple potato extract encapsulated in maltodextrin (PPE-MD) was significantly lower than that in the free extract, indicating enhanced stability. Moreover, microencapsulation increased anthocyanin bioaccessibility by 20%, highlighting its protective effect for *in vitro* GI digestion (Figure 14.2). The study also confirmed that

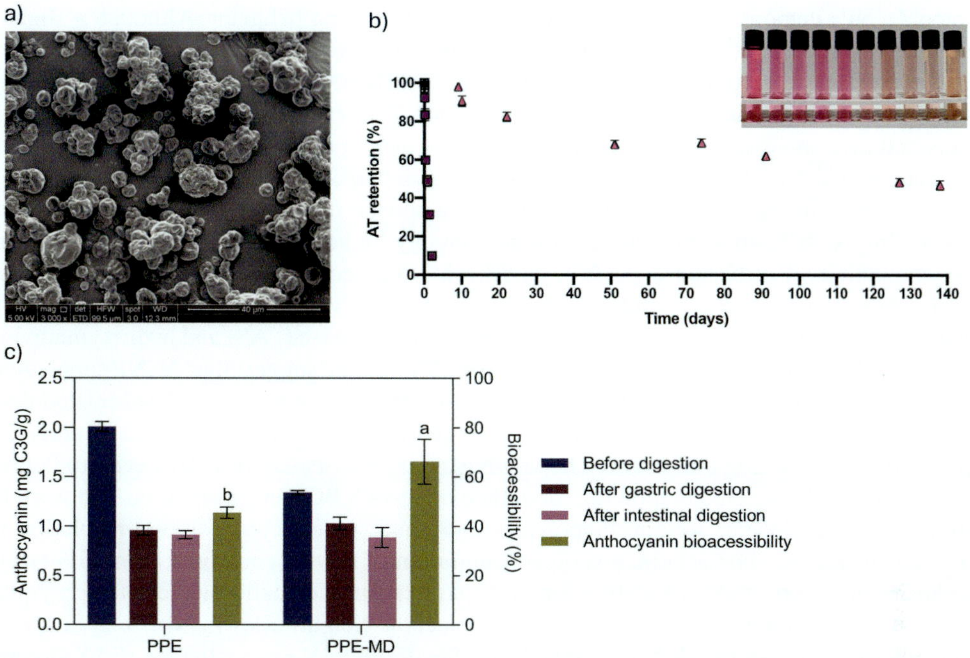

FIGURE 14.2 (a) SEM photograph PPE-MD. (b) Evolution of anthocyanin retention from PPE and PPE-MD during storage at 60°C. (PPE (□) and PPE-MD (Δ)). The inset photograph shows the visual degradation of non-encapsulated anthocyanins during the time-course of the storage stability assay. (c) Bioaccessibility of anthocyanins from PPE and PPE-MD after *in vitro* GI digestion. Adapted from (64).

the encapsulated anthocyanins remained stable during storage. These findings suggest that anthocyanins from purple potatoes, when encapsulated, can serve as a stable natural colorant and functional ingredient with improved antioxidant and anti-inflammatory properties (64).

14.3.1.2 Chitosan and Derivatives

CS is a widely used biopolymer in biomedical applications due to its exceptional biocompatibility, biodegradability, nontoxicity, and bioactive properties. Its non-antigenic nature, high permeability, and ability to interact with biological systems make it valuable for drug delivery, wound healing, and tissue engineering (19). Anthocyanins extracted from purple-fleshed sweet potato were successfully encapsulated using an emulsification-crosslinking technique, where CS served as the polymer and sodium tripolyphosphate (NaTPP) acted as the crosslinker agent (65). The optimal conditions for obtaining high-quality microcapsules included a stirring intensity of 1,600 rpm, a CS concentration of 3% (w/v), a pH of 3.0 for both the CS and NaTPP solutions, and a core-to-CS ratio of 150% (w/w). SEM analysis confirmed the formation of spherical yet irregular microparticles, while FTIR analysis demonstrated interactions among anthocyanins, CS, and NaTPP. Hydrogen bonds are established between anthocyanins and protonated CS, while tripolyphosphate ions contribute to ionic/electrostatic interactions, enhancing both encapsulation efficiency and the antioxidant activity of anthocyanins (65). Chatterjee et al. described the development, characterization, and antioxidant activity of black carrot anthocyanin-loaded CS nanoparticles (ACNPs) using the ionic gelation technique (66). ACNPs exhibited a small size of 274 nm and a polydispersity index (PDI) of 0.376, with a bimodal particle size distribution and a spherical surface morphology with a coacervate structure. FTIR analysis revealed physicochemical interactions of anthocyanins with CS, namely, hydrogen bonding interactions between the anthocyanin NPs and the polymeric CS matrix. The encapsulation process achieved an efficiency of 70%. The viscosity profile of encapsulated ACNPs demonstrated Newtonian and shear-thickening behavior. Although a slight reduction in the *in vitro* antioxidant activity of anthocyanins was observed after encapsulation, as determined by DPPH, ferric reducing antioxidant power (FRAP), and ABTS assays, their *in vivo* antioxidant capacity was enhanced, as shown by serum antioxidant evaluations (66). A related study reports the hypolipidemic effects and mechanisms of black carrot ACNP supplementation in male Wistar rats, in the context of high-fat and alcohol-enriched diets (67). ACNPs were also prepared using the conventional ionotropic gelation method, containing 1% (w/v) of CS solution at a concentration of 1 g/L. The findings revealed that rats receiving ACNP supplementation exhibited notably higher catalase and superoxide dismutase (SOD) enzyme activity. Furthermore, ACNPs significantly reduced serum cholesterol and triglyceride levels, diminishing lipid-induced oxidative stress in serum and liver tissues. Western blotting analyses indicated the normalization of key enzymes such as fatty acid synthase (FAS) and hydroxymethylglutaryl Co-A reductase (HMGCR) involved in lipid metabolism. The results suggest that ACNP supplementation can effectively mitigate dyslipidemia, offering a promising strategy for managing lipid-related disorders. The encapsulation of anthocyanins in biopolymers is an evolving research frontier with potential applications in preventing and treating diet-induced hepato-regenerative diseases (67). Atay et al. (68) developed CS/gelatin microparticles by electrospraying for anthocyanin-rich BCE encapsulation. The work focuses on the effects of CS molecular weight, biopolymer ratio, and solvent composition on solution properties and capsule formation. For gelatin, a solid concentration of 8 wt% combined with 20% (v/v) acetic acid was chosen as the optimal formulation for successful capsule formation *via* electrospraying, while regarding CS, the maximum solubility varied with molecular weight, allowing for 5, 2, and 1.25 wt% for the low, intermediate, and high Mw, respectively. Small-angle X-ray scattering (SAXS) confirmed the formation of nanoscale compacted structures due to the strong electrostatic interactions between the positively charged amino groups of CS and the negatively charged residues in gelatin (68). Electrospray structures incorporating 17wt% of anthocyanin-rich BCE were formulated using the ideal CS/gelatin ratios, and their encapsulation efficiency (exceeding 75% in all cases) and release behavior in two food simulants (10% ethanol and 3% acetic acid) were assessed. A faster and higher release occurred in the acidic medium due to increased swelling of the microstructures. Notably, higher CS content in the formulation reduced anthocyanin diffusion, highlighting the role of polymer composition in modulating release properties (68).

14.3.1.3 Other Common Polysaccharides

Various polysaccharides have been explored as carrier materials for acylated anthocyanin encapsulation. Research on agave fructans as carrier agents for blue corn anthocyanins (rich in cyanidin-3-(6-malonyl-glucoside)) *via* spray-drying demonstrated their effectiveness (69). Blue corn extract was combined with different fructan concentrations (6%, 8%, 10%, and 12% w/v) and dried at 150°C. Encapsulation with 6% (w/v) fructans resulted in powders with favorable physical, thermal, and morphological properties, along with high bioactive compound retention. The glass transition temperature (>50°C) indicated strong stability during storage. The findings suggest that using 6% (w/v) agave fructans optimizes anthocyanin (29.27%), polyphenol (73.03%), and antioxidant activity (42.57%) retention (69), making the encapsulated powders a promising natural food pigment with additional functional benefits, including antioxidant properties and soluble fiber content. García-Tejeda et al. (70) evaluated phosphorylated starch as an encapsulating agent loaded with purple maize anthocyanins (acylated with malonyl and succinyl moieties) using spray-drying under varying inlet temperatures (130–170 °C) and polysaccharide concentrations (10%–20%). The phosphorylation, performed *via* reactive extrusion with sodium tripolyphosphate (STPP), led to reduced crystallinity, lower viscosity, and increased solubility – properties favorable for spray-drying. FTIR analysis confirmed anthocyanin retention in the microcapsules, while scanning electron microscopy revealed smoother particles at higher starch concentrations. The highest encapsulation yield (49.11%) was achieved when the concentration of phosphorylated starch in the emulsion was 20%. These findings highlight phosphorylated starch as an effective carrier for anthocyanin encapsulation, enhancing their stability and potential applications in food systems (70). Different polysaccharides commonly used as encapsulating agents, including MD, AG, xanthan gum, and gellan gum, were evaluated for their effectiveness in protecting the bioactive compounds of ultrasound-assisted red cabbage extract during freeze-drying (71). Comparatively, MD proved to be the most effective encapsulating agent, better preserving the phenolic content and antioxidant activity (74.15%, significantly higher than the range of 3.88%–13.81% observed for other polysaccharides) after freeze-drying. The encapsulation efficiency (EE) of anthocyanins varied between 88.75% and 96.63%, with the lowest values observed for formulations containing gellan gum (88.75%) and xanthan gum (91.52%). In contrast, higher EE values were achieved when MD (96.63%) and AG (95.42%) were used as encapsulating agents. This enhancement could be linked to the ability of AG to form a protective barrier around the core material, reducing exposure to environmental factors, while MD likely contributed to the development of an amorphous matrix that stabilized the encapsulated anthocyanins (71). More importantly, this study uses sorption isotherms to assess the stability of the encapsulated red cabbage extract, helping predict its durability and quality under various environmental conditions. The isotherm analysis revealed a behavior characteristic of hygroscopic materials, where water absorption capacity increases as water activity increases. Differential enthalpy and entropy values indicated that as the moisture content increased from 2% to 14%, the isosteric heat (qst) and entropy decreased, suggesting weaker water adsorption forces at higher moisture levels (71). These data are essential for understanding the thermal stability of the encapsulates. The qst value decreased from 27.518 kJ/mol to 12.505 kJ/mol with increasing moisture, reflecting the reduced energy required for water adsorption under storage conditions. Regarding isokinetic energy and Gibbs free energy, the values found (364.88 J/mol and –1.596 kJ/mol, respectively) suggest that the encapsulate has relatively high thermodynamic stability, which is an indicator that the anthocyanin encapsulates would remain stable even under challenging storage conditions (71). Combining different polysaccharides (MD, GA, xanthan gum, and PC) and their blends with red cabbage anthocyanins by freeze-drying showed improved stability against oxygen, sunlight, and high temperature, and storage of acylated anthocyanins. Particularly, the MD and PC mixture exhibited efficient thermal and shelf-life stability, and delayed anthocyanin release throughout the simulated GI digestion (72).

Cyclodextrins, particularly β-cyclodextrin (β-CD), are naturally occurring cyclic oligosaccharides known for their biocompatibility and biodegradability, being widely used in food and pharmaceutical applications. Their ability to form inclusion complexes with sensitive compounds such as anthocyanins offers effective protection against oxidation and thermal degradation, while also helping mask undesirable tastes and odors (73). Feng et al. (74) encapsulated purple corn cob anthocyanins (PCCAs) using β-CD to enhance the stability and organoleptic and sensory qualities of purple corn flour (PCF) solid drinks. Encapsulating the anthocyanins with β-CD at a 2:1 wall-to-core ratio significantly reduced the astringency and sourness of the drink. Interestingly, this did not affect the functional properties of the anthocyanins when compared to their unencapsulated form, even at higher concentrations. Structurally, the resulting microcapsules were irregular in shape, and structural analysis demonstrated hydrogen bond interaction between PPCA and β-CD. Roasting the PCF at 175°C for 4 min was found to preserve a high level of anthocyanins and resulted in a better flavor profile (74). Overall, the use of β-CD proved to be an effective way to enhance both the stability and sensory quality of anthocyanin-rich beverages, supporting its use as a valuable tool in developing more appealing functional foods.

14.4 PROTEIN-BASED ENCAPSULATION

Food-grade proteins, such as whey protein isolate (WPI), casein, and soy protein, are promising alternatives to polysaccharides commonly used for anthocyanin encapsulation as drug delivery applications. The increasing attention in the food industry is mainly due to its health benefits. However, protein-related factors must be carefully optimized, as they significantly influence the rheological properties and sensory characteristics of anthocyanin-enriched food and beverage products (75).

14.4.1 Whey Protein Isolate

Whey protein, a by-product of cheese production, has gained attention as an effective encapsulating material for nutraceuticals due to its high nutritional value, availability, and excellent techno-functional properties, including resistance to pepsin digestion. Its composition includes β-lactoglobulin, α-lactalbumin, bovine serum albumin, immunoglobulins, and lactoferrin. Additionally, heat treatment can modify its structure by exposing hydrophobic segments, enhancing its ability to interact with bioactive compounds, and improving encapsulation efficiency (76). Wang and co-workers reported that electrostatic interactions between the positive charge of anthocyanins and the negative charge of protein contribute to the formation of the conjugate (77). Whey protein concentrate (WPC)–xylose (D) conjugates were explored as carriers for encapsulating PSPAs through antisolvent precipitation, resulting in an average particle size of 579.6 nm and a PDI of 0.4, indicating a homogeneous system. The conjugates significantly enhanced acylated anthocyanins' stability, with retention rates of 88.84%–88.98% at 4°C and improved thermal

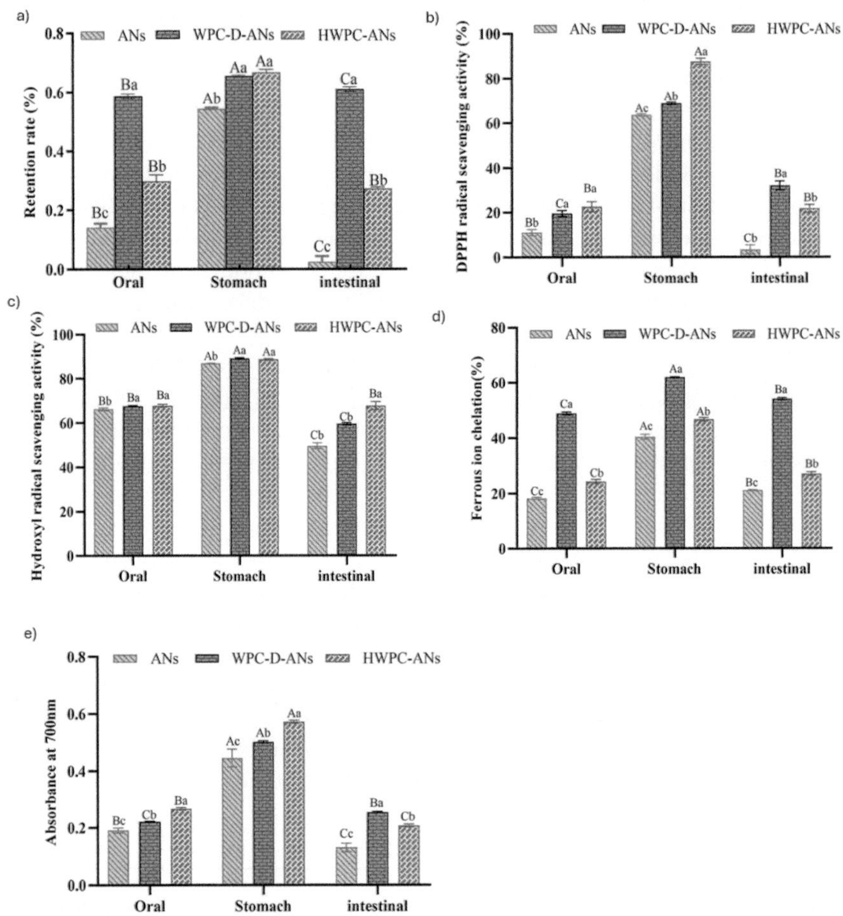

FIGURE 14.3 (a) Retention rates, (b) DPPH radical scavenging activity (%), (c) hydroxyl radical scavenging activity (%), (d) Fe^{2+} chelating activity (%), and (d) reducing power during simulated GI digestion of free anthocyanins (ANs), anthocyanins in WPC– D conjugates, and heat treatment of WPC-loaded anthocyanin nanoparticles (HWPC-ANs). Different uppercase and lowercase letters indicate significant differences between and within groups at the $p < 0.05$ level. Adapted from (77).

protection, reducing degradation at 60–90°C by about 2%. WPC–D increased anthocyanin retention during *in vitro* digestion, particularly in the intestinal phase, where 61.03% was retained compared to 2.50% in the control (Figure 14.3). WPC–D conjugates significantly improved anthocyanin stability under storage, thermal, light, and oxidative conditions while increasing bioavailability and antioxidant potential (77). In addition to the electrostatic interactions, Zhang et al. also observed hydrogen bonding between encapsulated PSPAs and sweet potato protein hydrolysates (SPPHs) (78). The results demonstrated that SPPHs with a molecular weight of 8,000 Da significantly enhanced the strength of sodium alginate (SA) gels through hydrogen bonding and other interactions, resulting in a denser gel structure and higher PSPA encapsulation efficiency (87.27%) compared to SA gels without SPPHs (61.73%). Additionally, after commercial sterilization at 98°C and storage at 37°C for 21 days, the PSPA concentration in SA-SPPH gels was 2.5 times higher than that in SA-only gels, demonstrating improved stability. The incorporation of SPPH facilitated crosslinking with calcium ions, promoted gelation, enhanced water retention, and slowed the release of PSPA, highlighting its potential for hydrogel-based food applications with improved stability and nutritional value.

14.5 POLYSACCHARIDE–PROTEIN-BASED ENCAPSULATION

Polysaccharide–protein encapsulation systems are emerging as innovative carriers for anthocyanins, offering excellent biocompatibility and functional versatility. Due to anthocyanins' structures and interactions with proteins and polysaccharides, tailored anthocyanin–biopolymer assemblies enhance stability, solubility, and controlled release properties. These complexes exhibit improved structural integrity and antioxidant capacity, making them promising candidates for food and pharmaceutical applications. Moreover, protein–polysaccharide conjugates can significantly enhance emulsification and gelation properties, outperforming individual biopolymers in stabilizing complex formulations (79). Micro- and nanoencapsulation of acylated anthocyanins with different biopolymer combinations has been reported. Tao et al. prepared red cabbage anthocyanin-loaded composite NPs by zein protein and anionic polysaccharides (PC, chondroitin sulfate, and λ-carrageenan) self-assembly at pH 4 (80). Spherical NPs were obtained with a particle size ranging from 102.6–360.4 nm and a negative zeta potential. Encapsulation efficiency depended on the polysaccharide type and the raw material adding sequence, with values between 68.03% and 80.55%. Anthocyanin-loaded zein/polysaccharide NPs coated with PC were the best option to encapsulate acylated anthocyanins, keeping good environmental stability (pH 2–8, 0–100 mM of NaCl) and storage stability (4°C for 4 weeks and 40°C for 18 days) (80). Dai et al. investigated the formation of the PPCA complex using PC, WPI, and CS to enhance digestive stability and α-glucosidase inhibitory activity (81). WPI–PC–PCCA had the highest encapsulation efficiency (58.84%) followed by the CS–PC–PCCA complex (48.13%), which demonstrated a medium particle size (200–300 nm), the lowest hygroscopicity (10.23 ± 0.28%), the lowest solubility (10.57 ± 1.26%), and the highest zeta potential (28.20 ± 1.14), indicating high stability (81). Hydrogen bonding and electrostatic interactions were the driving forces observed in the complex formation. In this case, the polysaccharide–polysaccharide system, CS–PC–PCCA, demonstrated the strongest α-glucosidase inhibition and the highest stability during simulated digestion, highlighting its potential for developing bioactive ingredients for glycemic control (81). Tamaroh et al. evaluated the effectiveness of spray-drying for encapsulating anthocyanin extract from purple yam (*Dioscorea alata* L.) using different ratios of combined MD and WPI (82). The formulation with a higher proportion of WPI (1:3) provided the best retention of bioactive compounds, as indicated by its higher antioxidant activity (65.16% radical scavenging activity), total phenolic content (776.25 mg gallic acid equivalent/100 g), and anthocyanin concentration (60.83 mg/100 g). The particle size of NPs was between 300.5 and 304.2 nm, with a negative zeta potential and an irregularly round shape with several depressions. Deng et al. (83) investigated the microencapsulation of purple corn anthocyanins (PCAs) also using spray-drying with MD and its combination with WPI or GA as wall materials. The highest encapsulation efficiencies (93.4%–94.6%) were obtained at specific PCA-to-wall material ratios, namely, 1:1 PCA to MD; 2:3 PCA to MD–GA and 1:1 PCA to MD–WPI (83). All microcapsules demonstrated lower water content and hygroscopicity, together with a good water solubility. Incorporating MD-encapsulated PCA enhanced its stability, particularly in the presence of Ca^{2+}, vitamin C, or Fe^{2+}. MD–PCA exhibited higher stability at 25°C, MD–GA–PCA showed superior resistance to temperatures at 40°C and light exposure, and MD–WPI–PCA was more stable under high humidity (75%) and during GI digestion. After 2 h of gastric digestion, the retention of MD–PCA, MD–GA–PCA, and MD–WPI–PCA was 89.40%, 85.73%, and 92.95%, respectively. However, after 2 h of intestinal digestion, these values decreased to 35.36%, 30.33%, and 43.12%, respectively, indicating that MD–WPI offers better protection for PCA throughout the GI digestion (83).

Complex coacervation is a microencapsulation technique that involves phase separation through the interaction of oppositely charged colloids, typically proteins and polysaccharides. The pH of the reaction is a critical factor in complex coacervation, as it influences the charge and solubility of the polymers, which in turn affects the formation and stability of the coacervates. A study of coacervate systems of potato protein isolate (PPI) combined with polysaccharides (GA, guar gum, inulin, PC, and soluble fiber) to encapsulated anthocyanins from black carrot concentrate shows that the solubility of PPI is highest at pH 2 (76.79%) and lowest at pH 12 (6.79%) (84). The antioxidant activity of anthocyanins was best preserved when PPI was combined with soluble fiber, followed by the high antioxidant levels in GA (8.64 μmol Trolox/g) and PC (8.36 μmol Trolox/g) microcapsules. The stability of anthocyanins was pH-dependent, with the highest stability observed at pH 1. The encapsulation efficiency ranged from 64.00% to 81.20%, with the highest efficiency for the PPI–GA microcapsules, due to their film-forming and emulsifying properties. FTIR analysis confirmed the integration of proteins and polysaccharides, and SEM imaging revealed more uniform and compact shapes (84). A study exploring the nanoencapsulation of anthocyanin-rich red cabbage extract using complex coacervation with WPI and high-methoxyl pectin showed that pH 4 was the best condition to blend the biopolymers and induce NP formation (85). The NPs showed pH sensitivity, progressively breaking from pH 4 to 7 (Figure 14.4a). The process generated monodisperse NPs (PDI < 0.200) with an average size of 380 nm and a negative charge (–17 mV). The encapsulation efficiency was only 30%, influenced by the anthocyanins' chemical structure, with diacylated and C3-triglucoside forms being better retained than monoacylated C3-diglucosides. Antioxidant activity decreased for loaded anthocyanins compared to the free extract (Figure 14.4b and c), probably due to the interaction between biopolymers and anthocyanins that limited the scavenging capacity of the latter

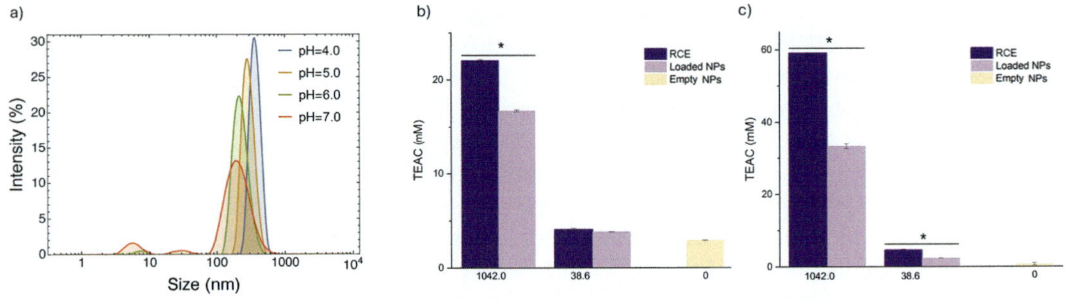

FIGURE 14.4 (a) Influence of pH on dynamic light scattering intensity signals (%) of particle distributions of the nano-coacervates of RCE (1,042.0 mg/L); antioxidant capacity measured by ABTS (b) and FRAP (c) of NPs (at 38.6 mg/L and 1,042.0 mg/L RCE), RCE (at the same ACN concentration), and empty NPs. * denotes significant differences ($p < 0.05$) between the extract and coacervates at the same ACN concentration. Adapted from (85).

(85). A combination of GA and inulin with pea protein to encapsulate red cabbage anthocyanin extract showed the potential of this coacervation system for developing stable natural colorants for food applications. It achieved high encapsulation efficiencies (≥ 95%), with the highest system reaching 98.92%, with the encapsulated anthocyanins retaining bioactivity, demonstrating antioxidant properties and α-amylase inhibition (22.48% at 0.5 mg/mL). The coacervates exhibited solubilities between 70.12% and 75.84% and porous morphology, confirmed by FTIR analysis.

14.6 INORGANIC MATERIAL-BASED ENCAPSULATION SYSTEMS

Inorganic material-based encapsulation systems have gained increasing attention for stabilizing and delivering anthocyanins. Materials such as halloysite nanotubes (HNTs), silica NPs, and layered double hydroxides offer structural rigidity, controlled porosity, and tunable surface chemistry, providing an effective means to protect anthocyanins from degradation while enhancing their bioavailability. These systems can also enable targeted and stimulus-responsive release, particularly in functional foods, pharmaceuticals, and cancer therapy. A study developed a multifunctional nanoplatform based on natural anthocyanins from black carrot chelated with gadolinium ions, designed for dual-modal photoacoustic imaging (PAI) and magnetic resonance imaging (MRI), as well as photothermal therapy (PTT) for tumors (86). The NPs exhibited suitable particle size (40 nm) and zeta potential (–14 mV), good dispersibility, and physiological stability. *In vitro* cytotoxicity and staining assays demonstrated excellent biocompatibility and a strong photothermal effect. *In vivo*, the NPs demonstrated high contrast enhancement for MRI and PAI, making them efficient as dual-modal imaging agents regardless of the administration route (intravenous or intratumoral). After intratumoral injection, the results showed complete tumor elimination without damaging normal tissue. Anthocyanins extracted from black carrots ensured biocompatibility and biodegradability, validated by safety assessments. This study highlights the clinical potential of this anthocyanin-based nanomedicine platform for precise and efficient cancer diagnosis and therapy (86). Hamedi and Koosha also explored encapsulated black carrot anthocyanin extract in biomedical applications, focusing on pH-responsive drug delivery inorganic systems using HNTs as biocompatible and cost-effective carriers for targeted anthocyanin delivery (87). Under vacuum conditions, anthocyanins were successfully loaded into the lumen of HNTs, as confirmed by transmission electron microscopy imaging (Figure 14.5a), with a loading efficiency of 4%. X-ray diffraction analysis revealed shifts in crystalline reflections, indicating structural modifications upon anthocyanin encapsulation. The *in vitro* release study showed that approximately 46% of the black carrot anthocyanins loaded into HNTs were released under GI conditions (Figure 14.5b). Notably, the encapsulated anthocyanins exhibited a potent antiproliferative activity against MCF-7 breast cancer and HT-29 colon cancer cell lines compared to free anthocyanins (Figure 14.5c). The release profile of anthocyanins from HNTs was pH-dependent, making this system particularly suited for targeting the acidic microenvironment of cancer cells (87).

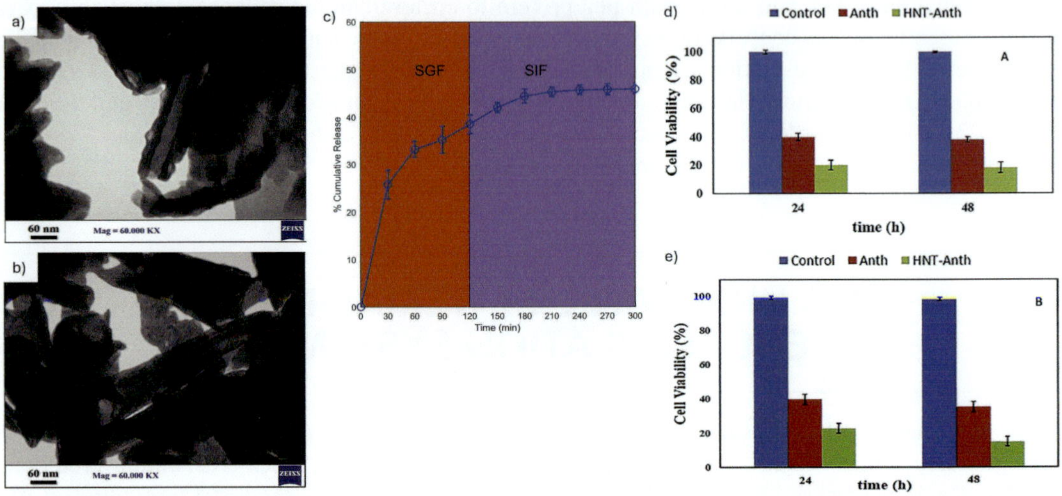

FIGURE 14.5 TEM images of (a) neat HNTs and (b) anthocyanin-loaded HNTs; (c) release of black carrot anthocyanins encapsulated in HNTs under simulated gastric fluid (SGF) and simulated intestinal fluid (SIF) conditions. Cell viability of (d) MCF-7 and (e) HT-29 cancerous cell lines treated with anthocyanins and anthocyanin-loaded HNTs.

Adapted from Ref. 87.

14.7 CONCLUSIONS AND PERSPECTIVES

The stability and bioavailability of anthocyanins are of growing scientific interest due to their human health benefits. Encapsulation strategies developed for food and pharmaceutical applications can enhance the stability of acylated anthocyanins, protecting them from degradation due to environmental factors such as light, heat, and pH changes. This extended shelf life reduces food and pharmaceutical waste while minimizing the need for synthetic stabilizers. Furthermore, encapsulation systems can improve the solubility and bioavailability of anthocyanins, ensuring efficient utilization of these bioactive compounds and reducing unnecessary losses.

Among the different encapsulation systems reported in this chapter, lipid-based systems, including liposomes, niosomes, and emulsified delivery systems, enhance anthocyanin structural integrity, bioavailability, and intestinal uptake by protecting against degradation, facilitating membrane permeation, and optimizing release kinetics, making them effective carriers for both topical and systemic applications. Biopolymer carriers, including protein- and polysaccharide-based particles, are widely employed due to their natural origin, biocompatibility, and capacity to create stable matrices that shield anthocyanins from degradation caused by environmental factors. Besides the commonly used MD, CS derivatives and whey protein have also demonstrated potential in enhancing acylated anthocyanin bioavailability and facilitating their targeted delivery for therapeutic applications in humans. Spray-drying and freeze-drying are the most common methods for acylated anthocyanins loaded in biopolymer carriers due to their high encapsulation efficiency, even at an industrial scale. Coacervation – typically through protein–polysaccharide interactions – has gained attention for its ability to form structured micro- and nanocomplexes that provide enhanced protection and controlled release properties. This technique enables the development of delivery systems that stabilize anthocyanins and regulate their release, optimizing their functionality in food, pharmaceutical, and cosmetic formulations. Meanwhile, inorganic encapsulation systems, such as silica NPs and HNTs, provide a different level of protection through their rigid structures, ensuring sustained release and pH-dependent delivery. Despite the advantages of the encapsulation approach, some challenges remain. Some biopolymers may need chemical modifications to enhance their functionality, which may involve non-sustainable solvents or crosslinking agents. Additionally, scalability and cost-effectiveness can be limiting factors for widespread industrial adoption. On the other hand, the encapsulation system chosen depends on the intended application, physicochemical characteristics of anthocyanins, and the desired release profile. While these carriers offer promising benefits for targeted delivery of acylated anthocyanins, enhancing their stability in the GI tract, boosting bioavailability, and enabling sustained release, their safety and potential toxicity need to be assessed. Although many of these complexes are biodegradable and can be efficiently removed through normal metabolic pathways, their high biocompatibility also facilitates their movement within biological systems, which may increase the possibility of nonspecific interactions, potentially leading to harmful accumulation at the cellular level.

Future research should focus on improving the bioavailability of encapsulated acylated anthocyanins. While encapsulation enhances stability, the challenge remains in optimizing the absorption and metabolic fate of these bioactive compounds. Developing emerging strategies for specific targeted release in different biological systems, accompanied by model and *in vitro* human cell studies, holds great potential for further enhancing anthocyanin bio-efficacy.

ACKNOWLEDGMENTS

This work received financial support from the PT national funds (FCT/MECI, Fundação para a Ciência e Tecnologia and Ministério da Educação, Ciência e Inovação) through the project UID/50006/2025 DOI 10.54499/UID/50006/2025-Laboratório Associado para a Química Verde - Tecnologias e Processos Limpos. Ana Fernandes (AF) acknowledges Fundação para a Ciência e Tecnologia (FCT) for the CEEC contract CEECIND/00029/2018/CP1545/CT0010 https://doi.org/10.54499/CEECIND/00029/2018/CP1545/CT0010).

REFERENCES

1. Giusti MM, Wrolstad RE. Acylated Anthocyanins from Edible Sources and Their Applications in Food Systems. *Biochemical Engineering Journal*. 2003;14(3):217–225.
2. Zhao L-Y, Chen J, Wang Z-Q, Shen R-M, Cui N, Sun A-D. Direct Acylation of Cyanidin-3-Glucoside with Lauric Acid in Blueberry and Its Stability Analysis. *International Journal of Food Properties*. 2016;19(1):1–12.
3. Cruz L, Fernandes VC, Araújo P, Mateus N, de Freitas V. Synthesis, Characterisation and Antioxidant Features of Procyanidin B4 and Malvidin-3-Glucoside Stearic Acid Derivatives. *Food Chemistry*. 2015;174:480–486.
4. Cruz L, Benohoud M, Rayner CM, Mateus N, de Freitas V, Blackburn RS. Selective Enzymatic Lipophilization of Anthocyanin Glucosides from Blackcurrant (*Ribes nigrum* L.) Skin Extract and Characterization of Esterified Anthocyanins. *Food Chemistry*. 2018;266:415–419.
5. Cruz L, Guimarães M, Araújo P, Évora A, de Freitas V, Mateus N. Malvidin 3-Glucoside–Fatty Acid Conjugates: From Hydrophilic toward Novel Lipophilic Derivatives. *Journal of Agricultural and Food Chemistry*. 2017;65(31):6513–6518.
6. Wang Y, Julian McClements D, Chen L, Peng X, Xu Z, Meng M, et al. Progress on Molecular Modification and Functional Applications of Anthocyanins. *Critical Reviews in Food Science and Nutrition*. 2024;64(31):11409–11427.
7. Zhao C-L, Yu Y-Q, Chen Z-J, Wen G-S, Wei F-G, Zheng Q, et al. Stability-Increasing Effects of Anthocyanin Glycosyl Acylation. *Food Chemistry*. 2017;214:119–128.

8. Zeng F, Zeng H, Ye Y, Zheng S, Zhuang Y, Liu J, et al. Preparation of Acylated Blueberry Anthocyanins through an Enzymatic Method in an Aqueous/Organic Phase: Effects on Their Colour Stability and pH-Response Characteristics. *Food & Function.* 2021;12(15):6821–6829.
9. Yang W, Kortesniemi M, Ma X, Zheng J, Yang B. Enzymatic Acylation of Blackcurrant (Ribes nigrum) Anthocyanins and Evaluation of Lipophilic Properties and Antioxidant Capacity of Derivatives. *Food Chemistry.* 2019;281:189–196.
10. Fenger J-A, Moloney M, Robbins RJ, Collins TM, Dangles O. The Influence of Acylation, Metal Binding and Natural Antioxidants on the Thermal Stability of Red Cabbage Anthocyanins in Neutral Solution. *Food & Function.* 2019;10(10):6740–6751.
11. Pereira AR, Fernandes VC, Delerue-Matos C, de Freitas V, Mateus N, Oliveira J. Exploring Acylated Anthocyanin-Based Extracts as a Natural Alternative to Synthetic Food Dyes: Stability and Application Insights. *Food Chemistry.* 2024;461:140945.
12. Escribano-Bailón MT, Santos-Buelga C, Rivas-Gonzalo JC. Anthocyanins in Cereals. *Journal of Chromatography A.* 2004;1054(1):129–141.
13. Shehata E, Day-Walsh P, Kellingray L, Narbad A, Kroon PA. Spontaneous and Microbiota-Driven Degradation of Anthocyanins in an In Vitro Human Colon Model. *Molecular Nutrition & Food Research.* 2023;67(19):e2300036.
14. Wang Z-C, Tanzeela N, Sun L, Fang Z, Yan Y, Li D, et al. Effect of In Vitro Gastrointestinal Digestion on the Composition and Bioactivity of Anthocyanins in the Fruits of Cultivated *Lycium ruthenicum* Murray. *CyTA – Journal of Food.* 2019;17(1):552–562.
15. Zhao L, Temelli F, Chen L. Encapsulation of Anthocyanin in Liposomes Using Supercritical Carbon Dioxide: Effects of Anthocyanin and Sterol Concentrations. *Journal of Functional Foods.* 2017;34:159–167.
16. Jokioja J, Yang B, Linderborg KM. Acylated Anthocyanins: A Review on Their Bioavailability and Effects on Postprandial Carbohydrate Metabolism and Inflammation. *Comprehensive Reviews in Food Science and Food Safety.* 2021;20(6):5570–5615.
17. Ayvaz H, Cabaroglu T, Akyildiz A, Pala CU, Temizkan R, Ağçam E, et al. Anthocyanins: Metabolic Digestion, Bioavailability, Therapeutic Effects, Current Pharmaceutical/Industrial Use, and Innovation Potential. *Antioxidants (Basel).* 2022;12(1).
18. Ryu D, Sung Y, Hong J, Koh E. Cellular Uptake of Anthocyanins Extracted from Black Soybean, Grape, and Purple Sweet Potato Using INT-407 Cells. *Food Science and Biotechnology.* 2021;30(10):1383–1391.
19. Fu Z, Ju H, Xu G-S, Wu Y-C, Chen X, Li H-J. Recent Development of Carrier Materials in Anthocyanins Encapsulation Applications: A Comprehensive Literature Review. *Food Chemistry.* 2024;439:138104.
20. Rashwan AK, Karim N, Xu Y, Xie J, Cui H, Mozafari MR, et al. Potential Micro-/Nano-Encapsulation Systems for Improving Stability and Bioavailability of Anthocyanins: An Updated Review. *Critical Reviews in Food Science and Nutrition.* 2023;63(19):3362–3385.
21. Islam Shishir MR, Karim N, Gowd V, Zheng X, Chen W. Liposomal Delivery of Natural Product: A Promising Approach in Health Research. *Trends in Food Science & Technology.* 2019;85:177–200.
22. Reza Mozafari M, Johnson C, Hatziantoniou S, Demetzos C. Nanoliposomes and Their Applications in Food Nanotechnology. *Journal of Liposome Research.* 2008;18(4):309–327.
23. Lee C, Na K. Anthocyanin-Loaded Liposomes Prepared by the pH-Gradient Loading Method to Enhance the Anthocyanin Stability, Antioxidation Effect and Skin Permeability. *Macromolecular Research.* 2020;28(3):289–297.
24. Mozafari MR. Liposomes: An Overview of Manufacturing Techniques. *Cellular & Molecular Biology Letters.* 2005;10(4):711–719.
25. Maja L, Željko K, Mateja P. Sustainable Technologies for Liposome Preparation. *The Journal of Supercritical Fluids.* 2020;165:104984.
26. Muñoz-Shuguli C, Vidal CP, Cantero-López P, Lopez-Polo J. Encapsulation of Plant Extract Compounds Using Cyclodextrin Inclusion Complexes, Liposomes, Electrospinning and Their Combinations for Food Purposes. *Trends in Food Science & Technology.* 2021;108:177–186.
27. Mendoza J, Basílio N, Pina F, Kondo T, Yoshida K. Rationalizing the Color in Heavenly Blue Anthocyanin: A Complete Kinetic and Thermodynamic Study. *The Journal of Physical Chemistry B.* 2018;122(19):4982–4992.
28. Oliveira H, Basílio N, Pina F, Fernandes I, de Freitas V, Mateus N. Purple-Fleshed Sweet Potato Acylated Anthocyanins: Equilibrium Network and Photophysical Properties. *Food Chemistry.* 2019;288:386–394.
29. Ossman T, Fabre G, Trouillas P. Interaction of Wine Anthocyanin Derivatives with Lipid Bilayer Membranes. *Computational and Theoretical Chemistry.* 2016;1077:80–86.
30. Mohammadalinejhad S, Kurek MA. Microencapsulation of Anthocyanins – Critical Review of Techniques and Wall Materials. *Applied Sciences.* 2021;11(9).

31. Guldiken B, Gibis M, Boyacioglu D, Capanoglu E, Weiss J. Ascorbic Acid-Induced Degradation of Liposome-Encapsulated Acylated and Non-Acylated Anthocyanins of Black Carrot Extract. *Journal of the Science of Food and Agriculture*. 2021;101(13):5707–5714.
32. Guldiken B, Gibis M, Boyacioglu D, Capanoglu E, Weiss J. Impact of Liposomal Encapsulation on Degradation of Anthocyanins of Black Carrot Extract by Adding Ascorbic Acid. *Food and Function*. 2017;8(3):1085–1093.
33. Ghareaghajlou N, Hallaj-Nezhadi S, Ghasempour Z. Nano-Liposomal System Based on Lyophilization of Monophase Solution Technique for Encapsulating Anthocyanin-Rich Extract from Red Cabbage. *Dyes and Pigments*. 2022;202:110263.
34. Cheng M, Zhang X, Cao J, Zheng X, Zhang Z. Caco-2 Cell Transport of Purple Sweet Potato Anthocyanins-Phospholipids Complex. *Journal of Food Science and Technology*. 2018;55(1):304–312.
35. Joshua DL, Shengmin S, Anthony YHL, Chung SY. Metabolism of Dietary Polyphenols and Possible Interactions with Drugs. *Current Drug Metabolism*. 2007;8(5):499–507.
36. Jorge AF, Grijalvo S, Pais A, Eritja R. Lipid Nanocarriers for Oligonucleotide Delivery to the Brain. In: Group TF, editor. *Nanoparticles for Brain Drug Delivery*. 1st ed. Jenny Standford Publishing, 2021, p. 31.
37. Abdelkader H, Alani AWG, Alany RG. Recent Advances in Non-Ionic Surfactant Vesicles (Niosomes): Self-Assembly, Fabrication, Characterization, Drug Delivery Applications and Limitations. *Drug Delivery*. 2014;21(2):87–100.
38. Rajera R, Nagpal K, Singh SK, Mishra DN. Niosomes: A Controlled and Novel Drug Delivery System. *Biological and Pharmaceutical Bulletin*. 2011;34(7):945–953.
39. Azeem A, Anwer MK, Talegaonkar S. Niosomes in Sustained and Targeted Drug Delivery: Some Recent Advances. *Journal of Drug Targeting*. 2009;17(9):671–689.
40. Hamishehkar H, Rahimpour Y. Niosomes as Carrier in Dermal Drug Delivery. In: Sezer AD, editor. *Recent Advances in Novel Drug Carrier Systems*. Rijeka: IntechOpen, 2012.
41. Priprem A, Damrongrungruang T, Limsitthichaikoon S, Khampaenjiraroch B, Nukulkit C, Thapphasaraphong S, et al. Topical Niosome Gel Containing an Anthocyanin Complex: A Potential Oral Wound Healing in Rats. *AAPS PharmSciTech*. 2018;19(4):1681–1692.
42. Fidan-Yardimci M, Akay S, Sharifi F, Sevimli-Gur C, Ongen G, Yesil-Celiktas O. A Novel Niosome Formulation for Encapsulation of Anthocyanins and Modelling Intestinal Transport. *Food Chemistry*. 2019;293:57–65.
43. McClements DJ. *Food Emulsions: Principles, Practices, and Techniques*. 2nd ed. Boca Raton: CRC Press, 2004.
44. Muschiolik G. Multiple Emulsions for Food Use. *Current Opinion in Colloid & Interface Science*. 2007;12(4):213–220.
45. Garti N, Bisperink C. Double Emulsions: Progress and Applications. *Current Opinion in Colloid & Interface Science*. 1998;3(6):657–667.
46. Nimaming N, Sadeghpour A, Murray BS, Sarkar A. Hybrid Particles for Stabilization of Food-Grade Pickering Emulsions: Fabrication Principles and Interfacial Properties. *Trends in Food Science & Technology*. 2023;138:671–684.
47. Tadros T, Izquierdo P, Esquena J, Solans C. Formation and Stability of Nano-Emulsions. *Advances in Colloid and Interface Science*. 2004;108–109:303–318.
48. McClements DJ. Emulsion Design to Improve the Delivery of Functional Lipophilic Components. *Annual Review of Food Science and Technology*. 2010;1:241–269.
49. Wooster TJ, Golding M, Sanguansri P. Impact of Oil Type on Nanoemulsion Formation and Ostwald Ripening Stability. *Langmuir*. 2008;24(22):12758–12765.
50. Liu J, Tan Y, Zhou H, Muriel Mundo JL, McClements DJ. Protection of Anthocyanin-Rich Extract from pH-Induced Color Changes Using Water-in-Oil-in-Water Emulsions. *Journal of Food Engineering*. 2019;254:1–9.
51. Koirala P, Sriprablom J, Winuprasith T. Anthocyanin-Rich Butterfly Pea Petal Extract Loaded Double Pickering Emulsion Containing Nanocrystalline Cellulose: Physicochemical Properties, Stability, and Rheology. *Foods*. 2023;12(22).
52. Desnita R, Wahdaningsih S, Hervianti S. Span 60 as Surfactant of Topical Microemulsion of Purple Sweet Potato (*Ipomoea batatas* L.) Ethanol Extract and Antioxidant Activity Test Using DPPH Method. *International Journal of PharmTech Research*. 2016;9:198–203.
53. Desnita R, Veronika M, Wahdaningsih S. Topical Microemulsion's Formulation of Purple Sweet Potato (*Ipomoea batatas* L.) Ethanol Extract as Antioxidant by Using Various Concentration of Span 80. *International Journal of PharmTech Research*. 2016;9:234–239.
54. Zang J, Xiao P, Liu Z, Liu Y, Zhang Q, Chen J, et al. Preparation and Characterization of W/O/W Purple Potato Anthocyanin Nanoparticles: Antioxidant Effects and Gut Microbiota Improvement in Rats. *Journal of Food Science*. 2024;89(12):9901–9922.

55. Ravanfar R, Tamaddon AM, Niakousari M, Moein MR. Preservation of Anthocyanins in Solid Lipid Nanoparticles: Optimization of a Microemulsion Dilution Method Using the Placket–Burman and Box–Behnken Designs. *Food Chemistry*. 2016;199:573–580.
56. Sharif N, Khoshnoudi-Nia S, Jafari SM. Nano/Microencapsulation of Anthocyanins; a Systematic Review and Meta-Analysis. *Food Research International*. 2020;132:109077.
57. Samborska K, Boostani S, Geranpour M, Hosseini H, Dima C, Khoshnoudi-Nia S, et al. Green Biopolymers from by-Products as Wall Materials for Spray Drying Microencapsulation of Phytochemicals. *Trends in Food Science & Technology*. 2021;108:297–325.
58. Ozkan G, Franco P, De Marco I, Xiao J, Capanoglu E. A Review of Microencapsulation Methods for Food Antioxidants: Principles, Advantages, Drawbacks and Applications. *Food Chemistry*. 2019;272:494–506.
59. Rezvankhah A, Zahra E-D, Askari G. Encapsulation and Delivery of Bioactive Compounds Using Spray and Freeze-Drying Techniques: A Review. *Drying Technology*. 2020;38(1–2):235–258.
60. Ersus S, Yurdagel U. Microencapsulation of Anthocyanin pigments of Black Carrot (*Daucus carota* L.) by Spray Drier. *Journal of Food Engineering*. 2007;80(3):805–812.
61. Machado MH, Almeida ADR, Maciel MVDOB, Vitorino VB, Bazzo GC, da Rosa CG, et al. Microencapsulation by Spray Drying of Red Cabbage Anthocyanin-Rich Extract for the Production of a Natural Food Colorant. *Biocatalysis and Agricultural Biotechnology*. 2022;39:102287.
62. Sakulnarmrat K, Wongsrikaew D, Konczak I. Microencapsulation of Red Cabbage Anthocyanin-Rich Extract by Drum Drying Technique. *LWT*. 2021;137:110473.
63. Maneeratanachot S, Chetpattananondh P, Kungsanant S. Encapsulation of Anthocyanin from Butterfly Pea Flowers (*Clitoria ternatea* L.) Extract Using Foam-Mat Drying. *Food and Bioproducts Processing*. 2024;145:105–115.
64. Vergara C, Pino MT, Zamora O, Parada J, Pérez R, Uribe M, et al. Microencapsulation of Anthocyanin Extracted from Purple Flesh Cultivated Potatoes by Spray Drying and Its Effects on In Vitro Gastrointestinal Digestion. *Molecules [Internet]*. 2020;25(3).
65. Laila U, Rochmadi R, Pudjiraharti S. Microencapsulation of Purple-Fleshed Sweet Potato Anthocyanins with Chitosan-Sodium Tripolyphosphate by Using Emulsification-Crosslinking Technique. *Journal of Mathematical and Fundamental Sciences*. 2019;51:29–46.
66. Chatterjee NS, Dara PK, Perumcherry Raman S, Vijayan DK, Sadasivam J, Mathew S, et al. Nanoencapsulation in Low-Molecular-Weight Chitosan Improves In Vivo Antioxidant Potential of Black Carrot Anthocyanin. *Journal of the Science of Food and Agriculture*. 2021;101(12):5264–5271.
67. Sreerekha PR, Dara PK, Vijayan DK, Chatterjee NS, Raghavankutty M, Mathew S, et al. Dietary Supplementation of Encapsulated Anthocyanin Loaded-Chitosan Nanoparticles Attenuates Hyperlipidemic Aberrations in Male Wistar Rats. *Carbohydrate Polymer Technologies and Applications*. 2021;2:100051.
68. Atay E, Fabra MJ, Martínez-Sanz M, Gomez-Mascaraque LG, Altan A, Lopez-Rubio A. Development and Characterization of Chitosan/Gelatin Electrosprayed Microparticles as Food Grade Delivery Vehicles for Anthocyanin Extracts. *Food Hydrocolloids*. 2018;77:699–710.
69. Sánchez-Madrigal MÁ, Quintero-Ramos A, Amaya-Guerra CA, Meléndez-Pizarro CO, Castillo-Hernández SL, Aguilera-González CJ. Effect of Agave Fructans as Carrier on the Encapsulation of Blue Corn Anthocyanins by Spray Drying. *Foods [Internet]*. 2019;8(7).
70. García-Tejeda YV, Salinas-Moreno Y, Hernández-Martínez ÁR, Martínez-Bustos F. Encapsulation of Purple Maize Anthocyanins in Phosphorylated Starch by Spray Drying. *Cereal Chemistry*. 2016;93(2):130–137.
71. Pusty K, Dash KK, Tiwari A, Balasubramaniam VM. Ultrasound Assisted Extraction of Red Cabbage and Encapsulation by Freeze-Drying: Moisture Sorption Isotherms and Thermodynamic Characteristics of Encapsulate. *Food Science Biotechnology* 2023;32(14):2025–2042.
72. Sendri N, Singh S, Bhatt S, Gupta M, Bhandari P. Insight into the Influence of Oxygen, Sunlight and Temperature on the Stability and Color Attributes of Red Cabbage Anthocyanins and In Vitro Gastrointestinal Behaviour. *Food Chemistry Advances*. 2023;3:100359.
73. Fernandes A, Sousa A, Azevedo J, Mateus N, de Freitas V. Effect of Cyclodextrins on the Thermodynamic and Kinetic Properties of Cyanidin-3-*O*-Glucoside. *Food Research International*. 2013;51(2):748–755.
74. Ying F, Xingwei H, Ren G, Qiao J, Zhang Z, Jiang Z, Wang H, Zhu Y. Effects of Roasting on Purple Corn Flour Quality and Addition of Anthocyanins and Their Compounds with Cyclodextrin on Purple Corn Flour Solid Drink Quality. *LWT - Food Science and Technology*. 2025;225:117946.
75. Wang Y, Jian Z, Zhang L. Anthocyanin-Dietary Proteins Interaction and Its Current Applications in Food Industry. *Food Reviews International*. 2023;39(6):3301–3313.
76. Singh TK, Øiseth SK, Lundin L, Day L. Influence of Heat and Shear Induced Protein Aggregation on the In Vitro Digestion Rate of Whey Proteins. *Food & Function*. 2014;5(11):2686–2698.

77. Wang F, Zhang S, Luo Q, Xie Q, Xu H, Deng G. Glycated Whey Protein Concentrate for Purple Sweet Potato Anthocyanins Delivery: Physicochemical Stability and In Vitro Digestion Characteristics. *Journal of Functional Foods*. 2024;114:106059.
78. Zhang M, Feng Y, Xiao J, Sun C, Tu J, Niu L. Sweet Potato Protein Hydrolysates Solidified Calcium-Induced Alginate Gel for Enhancing the Encapsulation Efficiency and Long-Term Stability of Purple Sweet Potato Anthocyanins in Beads. *Food Chemistry: X*. 2024;23:101775.
79. Paliya BS, Sharma VK, Sharma M, Diwan D, Nguyen QD, Aminabhavi TM, et al. Protein-Polysaccharide Nanoconjugates: Potential Tools for Delivery of Plant-Derived Nutraceuticals. *Food Chemistry*. 2023;428:136709.
80. Tao W, Ferreira IMPLVO, He J, de Freitas V, Mateus N, Oliveira H, et al. Tailored Zein-Polysaccharide Nanoparticles for Anthocyanin Encapsulation: Insights into Preparation and Characterization. *Food Hydrocolloids*. 2025:111365.
81. Dai J, Ruan Y, Feng Y, Li B. Physical Properties, α-Glucosidase Inhibitory Activity, and Digestive Stability of Four Purple Corn Cob Anthocyanin Complexes. *Foods*. 2022;11(22).
82. Tamaroh S, Sari YP. Antioxidant and Physical Characteristics of Anthocyanin Extract from Purple Yam (*Dioscorea alata* L.) Nanoencapsulation: Effect of Maltodextrin and Whey Protein Isolate Ratios. *Current Research in Nutrition and Food Science*. 2024;12(1):245–253.
83. Deng W, Li X, Ren G, Bu Q, Ruan Y, Feng Y, et al. Stability of Purple Corn Anthocyanin Encapsulated by Maltodextrin, and Its Combinations with Gum Arabic and Whey Protein Isolate. *Foods*. 2023;12(12).
84. Aktaş H, Custodio-Mendoza J, Szpicer A, Pokorski P, Samborska K, Kurek MA. Polysaccharide-Potato Protein Coacervates for Enhanced Anthocyanin Bioavailability and Stability. *International Journal of Biological Macromolecules*. 2024;282:136829.
85. Fierri I, De Marchi L, Chignola R, Rossin G, Bellumori M, Perbellini A, et al. Nanoencapsulation of Anthocyanins from Red Cabbage (*Brassica oleracea* L. var. *capitata* f. *rubra*) through Coacervation of Whey Protein Isolate and Apple High Methoxyl Pectin. *Antioxidants [Internet]*. 2023;12(9).
86. Guo C, Sun J, Dong J, Cai W, Zhao X, Song B, et al. A Natural Anthocyanin-Based Multifunctional Theranostic Agent for Dual-Modal Imaging and Photothermal Anti-Tumor Therapy. *Journal of Materials Chemistry B*. 2021;9(36):7447–7460.
87. Hamedi S, Koosha M. Designing a pH-Responsive Drug Delivery System for the Release of Black-Carrot Anthocyanins Loaded in Halloysite Nanotubes for Cancer Treatment. *Applied Clay Science*. 2020;197:105770.

Technological Applications of Acylated Anthocyanins

15

Innovation Across Industries

Asma Ayub, Muhammad Azlan Habib, Muhammad Naveed, Dureshahwar, and Waqar Majeed

15.1 INTRODUCTION

Anthocyanins (ACNs), which are water-soluble natural colorants that belong to the class of flavonoids that exhibit bioactive qualities because of their antioxidant qualities, are derived from the Greek words *Anthos*, which means flower, and *kyaneos*, which means dark blue (Tarone et al., 2020a; Luzardo-Ocampo et al., 2021). ACN-rich foods have several health benefits, including protection against cardiovascular diseases (CVDs), obesity, diabetes, and eye disease, and have antimicrobial, anticancer, and neuroprotective properties (Tena et al., 2020; Gonçalves et al., 2021; Alam et al., 2021).

ACNs, which include cyanidin (Cy), delphinidin (Dp), malvidin (Mv), pelargonidin (Pg), peonidin (Pn), and petunidin (Pt), are glycosylated derivatives of flavylium cations (2-phenylbenzopyrilium) with various methoxyl or hydroxyl substitutes on the rings, separated by an oxygen-containing six-membered heterocyclic ring (Rodriguez-Amaya et al., 2019). ACNs are naturally occurring plant-derived pigments that have found extensive use in the food industry, where they serve as natural colorants, antioxidants, preservatives, and functional ingredients (Xiong et al., 2019). Despite their valuable bioactive properties, these compounds possess certain limitations that restrict their broader application. Being water-soluble, ACNs exhibit poor solubility in organic solvents and are highly sensitive to environmental conditions such as elevated temperatures, exposure to light, and extreme pH levels (Fenger et al., 2019). These characteristics can lead to degradation and instability during processing and storage. ACNs are widely distributed in nature and are commonly found in various plant sources, including fruits, vegetables, flowers, leaves, cereals, grains, roots, and tubers (Tena et al., 2020; Arruda et al., 2021).

Acylation occurs either naturally in plants or can be achieved synthetically or enzymatically. The process introduces acyl groups typically aromatic (e.g., *p*-coumaric, ferulic, and caffeic acids) or aliphatic (e.g., malonic acid) to the sugar residues attached to the anthocyanidin core. This modification promotes intra- and intermolecular stacking, resulting in improved stability, particularly in aqueous and thermally stressed environments (Matsufuji et al., 2007).

 DOI: 10.1201/9781003519591-19

15.1.1 Importance of Acylation in Enhancing Functional Properties

Acylated ACNs have emerged as promising alternatives to their nonacylated counterparts because of their enhanced functional properties, which arise from structural modifications introduced through acyl group attachment (Cruz et al., 2016; Luo et al., 2018; Yang et al., 2019). These modifications can occur naturally in plants or be achieved synthetically through chemical or enzymatic acylation *in vitro*. Acylation improves the physicochemical characteristics of ACNs, particularly by increasing their stability and solubility in organic solvents compared to nonacylated forms. For instance, Yang et al. (2019) reported that enzymatically acylated blackcurrant ACNs exhibited greater thermal stability than their native counterparts. The acylated derivatives demonstrated higher half-lives and activation energies, indicating that the acylation process effectively enhanced their resistance to thermal degradation.

15.1.2 Scope and Objectives

This chapter explores the innovative and expanding applications of acylated ACNs across diverse technological fields. Acylated ACNs, known for their enhanced stability and bioactivity compared to their non-acylated counterparts, are increasingly being investigated for their roles in pharmaceuticals, cosmetics, food technology, and materials science. The chapter highlights recent advances in extraction, formulation, and functional deployment of acylated ACNs, while examining their potential to replace or complement synthetic compounds in various industrial applications. The chapter discusses the chemical structure and unique properties of acylated ACNs that contribute to their technological relevance, and current and emerging applications of acylated ACNs in the pharmaceutical industry, including drug delivery, anti-inflammatory agents, and antioxidants, and explores their cosmetic uses, such as in natural colorants, ultraviolet (UV) protection, and antiaging formulations.

15.2 STRUCTURAL AND FUNCTIONAL SIGNIFICANCE

15.2.1 Chemical Structure and Mechanism of Acylation

Acylation refers to the hydroxyl groups with aliphatic or aromatic acyl donors and represents a widespread modification among plant secondary metabolites. This modification plays a crucial role in altering the physical, chemical, and biological properties of plant-derived compounds (Cunningham & Edwards, 2008). In plants, acylated ACNs are synthesized *in vivo* through the phenylpropanoid biosynthetic pathway, which originates from phenylalanine (Appelhagen et al., 2018). The biosynthetic route leading from phenylalanine to ACNs involves three key stages: the general phenylalanine pathway, the early steps of the flavonoid pathway, and the final stage dedicated to ACN formation (Shi & Xie, 2014). Typically, the biosynthesis of acylated ACNs is derived from three primary anthocyanidin precursors cyanidin, pelargonidin, and delphinidin – which undergo acyl modification to yield structurally diverse and more stable pigment forms.

15.2.2 Stability and Solubility Enhancements

Chemical acylation is an effective strategy to enhance the stability of ACNs. Through this process, hydroxyl groups are converted into ester groups, which are inherently more stable and less reactive. As a result, acylated ACNs, such as acylated cyanidin-3-glucoside, exhibit significantly greater stability compared to their nonacylated counterparts. In addition to improved stability, chemical acylation contributes to enhanced lipophilicity, thereby broadening the potential application of ACNs in lipid-based systems (Ishihara & Nakajima, 2003; Tommasini et al., 2004). Similarly, enzymatic acylation offers an alternative approach for synthesizing acylated ACNs, utilizing biocatalysts to achieve greater fat solubility and functional versatility (Cruz et al., 2017).

15.2.3 Improved Antioxidant and Bioactive Properties

Epidemiological studies have demonstrated a strong association between ACN consumption and a reduced risk of various chronic and degenerative diseases. Owing to their potent antioxidant activity, acylated ACNs are considered promising functional bioactive compounds capable of contributing to the prevention of cardiovascular disorders, neurodegenerative diseases, cancer, inflammation, and diabetes (He & Giusti, 2010; Jing et al., 2008; Mattioli et al., 2020). During normal metabolic processes, the human body generates reactive oxygen species (ROS) and other free radicals. When present in excess, these reactive molecules can damage biomolecules such as carbohydrates, lipids, proteins, DNA, and RNA, ultimately contributing to the onset of various diseases, including cancer, Alzheimer's disease, autoimmune deficiencies, and diabetes (Li et al., 2019). The health-promoting effects of ACNs and their acylated derivatives are largely attributed to their superior antioxidant capacity and ability to effectively scavenge free radicals, thereby mitigating oxidative stress and preventing cellular damage.

15.3 PHARMACEUTICAL APPLICATIONS

15.3.1 Antioxidant and Anti-Inflammatory Therapeutics

Plant-based diets rich in bioactive phytochemicals including phenolics, ACNs, and vitamins have been strongly associated with a reduced risk of various human diseases (Zhang et al., 2020). ACNs and their derivatives exert their therapeutic and protective effects primarily through antioxidant and anti-inflammatory mechanisms, as both oxidative stress and inflammation are central contributors to the development of numerous pathological conditions. Excessive levels of pro-oxidants can induce oxidative stress, leading to damage of key cellular macromolecules such as lipids, proteins, and nucleic acids, which in turn may cause cytotoxicity, genotoxicity, or carcinogenesis. The antioxidant activity of ACNs mitigates oxidative stress through multiple pathways, including the direct scavenging of ROS and the enhancement of endogenous antioxidant defense systems. Specifically, ACNs can stimulate the activity of critical antioxidant enzymes such as superoxide dismutase (SOD), catalase (CAT) and glutathione peroxidase (GPx), thereby maintaining cellular redox balance and protecting against oxidative damage (Shahidi & Ambigaipalan, 2015).

15.3.2 Cancer Prevention and Adjunct Therapies

ACNs are a subclass of flavonoids within the broader group of polyphenolic compounds. Their polyphenolic structure plays a vital role in determining their antioxidant potential. Various analytical methods have been developed to assess the antioxidant activity of ACNs, primarily based on their underlying mechanisms, such as radical scavenging and reducing power. Common assays include the 2,2-diphenyl-1-picrylhydrazyl (DPPH) free radical scavenging test, ferric reducing antioxidant power (FRAP) assay, lipid peroxidation inhibition assay, and metal ion chelation assay. Antioxidant activity has long been recognized as the primary defense mechanism against the early stages of mutagenesis (Amatori et al., 2016). By effectively scavenging free radicals, ACNs help minimize oxidative DNA damage, thereby reducing the risk of tumorigenesis and contributing to cancer prevention (Kalemba-Drożdż et al., 2020).

15.3.3 Neuroprotective and Cardioprotective Effects

Neurodegenerative diseases, such as Parkinson's disease, Alzheimer's disease, and amyotrophic lateral sclerosis, are clinically characterized by progressive impairment of cognitive and motor functions (Mehdi et al., 2022). Dietary intake of ACNs has been shown to enhance cognitive performance and provide neuroprotective effects, thereby mitigating the progression of neurodegenerative disorders and reducing cognitive decline. These findings highlight the potential of ACNs as functional dietary compounds for the prevention of neurological diseases. Similarly, CVDs – a group of disorders affecting the heart and blood vessels – have gained global attention due to their high prevalence and mortality rates (Krga & Milenkovic, 2019). Numerous studies have demonstrated that the consumption of ACN-rich foods (such as berries and nuts) and ACN extracts or purified ACN supplements can exert beneficial effects in reducing the risk and severity of CVDs (Larsson et al., 2018).

15.3.4 Drug Delivery Systems: Encapsulation and Nanocarriers

Nanoencapsulation technologies facilitate multiple types of interactions such as van der Waals forces, hydrophobic interactions, and hydrogen bonding between nanocarriers and bioactive natural compounds like ACNs. These interactions contribute to improved physicochemical stability, controlled release, and enhanced bioavailability of the encapsulated compounds. Notably, nanocarrier systems also possess the unique capability to cross biological barriers, including the blood–brain barrier, thereby amplifying the therapeutic efficacy of ACNs and other encapsulated molecules (Gonçalves et al., 2022).

15.3.5 Synergistic Use in Functional Formulations

ACNs, natural flavonoid pigments found in fruits and vegetables, exhibit strong antioxidant, anti-inflammatory, and anticarcinogenic activities, making them valuable in pharmaceutical formulations. Synergistic strategies such as nanoencapsulation (e.g., using chitosan nanoparticles or niosomes) have been employed to enhance their therapeutic potential. For example, cyanidin-3-glucoside encapsulated in chitosan nanoparticles demonstrated improved antioxidant activity and gastroprotection (Ozkan et al., 2024). Co-formulation with other bioactive compounds like resveratrol or curcumin further boosts their efficacy in managing chronic conditions such as cancer and CVDs (Lewandowska et al., 2014).

15.4 COSMETIC APPLICATIONS

15.4.1 Role as Natural Pigments in Personal Care Products

Acylated ACNs, owing to their enhanced lipid solubility, can function as efficient natural pigments, particularly when modified through chemical or enzymatic acylation processes. For instance, the enzymatic conjugation of malvidin-3-*O*-glucoside with oleic acid and various saturated fatty acids of differing chain lengths has been shown to impart distinct lipophilic characteristics and potentially augment antioxidant capacity (Cruz et al., 2017). In the cosmetics industry, skincare and makeup products are formulations designed to enhance or improve the appearance and condition of the human body through various application methods, including spreading, sprinkling, or spraying. ACNs exhibit pH-dependent color variability, a property that underlies the wide spectrum of colors observed in nature and can be leveraged to improve the color performance of cosmetic and personal care products. Beyond their vivid pigmentation, ACNs are valued in cosmetic applications for their natural origin, safety profile, and additional bioactivities, including antioxidant and protective effects on the skin.

15.4.2 Antiaging and Skin Protection Properties

Using pure standards of cyanidin- and malvidin-3-*O*-glucosides along with ACN extracts derived from elderberry and red wine, Correia et al. (2021) demonstrated that these compounds possess notable photoprotective and skin beneficial properties. Most of the tested ACNs exhibited UV-filtering capacity, reduced the formation of ROS in human skin keratinocytes and fibroblasts, and inhibited the activity of skin-degrading enzymes such as tyrosinase, elastase, hyaluronidase, and collagenase, without inducing cytotoxic effects. These characteristics suggest that ACN-based formulations can provide dual functionality, offering both photoprotection and antiaging benefits. Furthermore, in a preventive context, ACNs extracted from red grapes and chokeberries have been shown to suppress melanoma cell proliferation, decrease mitochondrial membrane potential, and mitigate oxidative damage while exhibiting no harmful effects on normal skin cells (Diaconeasa et al., 2017).

15.4.3 Photoprotective Effects and UV Shielding

ACNs display distinct UV absorption bands within the 280–320 nm range, which are modulated by factors such as the structure of their aglycones, the nature of sugar conjugation, and the specific acylation patterns (Agati et al., 2013; Mejia-Giraldo et al., 2022). In plants, ACNs are predominantly localized within the epidermal cells, where they accumulate in vacuoles and function as natural sunscreens by shielding tissues from excessive solar radiation. This photoprotective role has inspired growing interest in their potential application as protective agents for human skin (Diaconeasa et al., 2020). Through their UV-absorbing capacity and antioxidant activity, ACNs can effectively mitigate the harmful effects of UV radiation on the epidermis, thereby contributing to skin protection and overall dermal health.

15.4.4 Formulation of Natural and Clean-Label Products

Skincare formulations enriched with antioxidants are increasingly being investigated for their protective and restorative effects on human skin. Although ACNs are promising candidates for such applications due to their potent bioactivity, their practical use is often constrained by instability issues. These water-soluble natural pigments are highly sensitive to environmental factors such as pH fluctuations, temperature variations, and light exposure, which can lead to undesirable color changes and degradation. To overcome these limitations and enhance their applicability, innovative strategies have been explored – one of which involves combining ACNs with clay minerals. This approach improves pigment stability and opens new opportunities for their incorporation into biocosmetic formulations (Silva et al., 2020).

15.5 FOOD AND BEVERAGE INDUSTRY

15.5.1 Heat- and pH-Stable Natural Colorants

ACNs have become popular natural colorants because they modify their appearance through distinct color transformations between acidic red to basic green or yellow color states. Sensor films comprising indicators have been developed by exploiting the pH-triggered color-shifting feature to monitor packaged food properties in real time. The pH-reactive indicator films made with ACNs serve as monitoring tools for fresh milk along with beverages and fish and meat products (Weston et al., 2020). The CS-LRA (Cassava Starch-Lycium Ruthenicum Anthocyanins) films show pH sensitivity by changing their color while immersed in solutions of pH 2–13 (Qin et al., 2019).

The chemical process of acylation allows petanin to preserve its color while maintaining superior intensity and achieving increased stability (Tarone et al., 2020b). Acylated ACNs thrive under thermal, pH, and light conditions better than nonacylated ACNs, which allows their application in pH-responsive color change indicator films along with encapsulation and in bio-based sensors for smart food packaging (Echegaray et al., 2023).

15.5.2 Use in Functional and Fortified Foods

The total concentration of ACNs in red/blue fruits and vegetables shows wide variations not only between plant kinds but also within specific plant groups. This variation is influenced by factors like orchard and farming conditions, harvesting timing, and ripening state. The ACN content of fresh berries exists between 100 and 700 mg/100 g of weight (Neveu et al., 2010). ACNs are found in a wide variety of red, purple, and blue plant foods. The common sources are purple corn, cherries, plums, pomegranates, eggplants, wine grapes, black carrots, red cabbage, and purple cauliflower. Many berries now contain ACNs such as maqui, and hence their production and consumption rates have increased substantially (Igwe et al., 2019). Determining the daily consumption of ACNs remains complicated because insufficient data exist regarding food contents containing ACNs. The health advantages of ACNs extend to their applications within the food, pharmaceutical, and cosmetic sectors (Câmara et al., 2022).

15.5.3 Smart Packaging Indicators and Edible Films

ACN-based intelligent packaging serves as the key contributor to the food application of ACNs among all other applications (Liu et al., 2024). Various indicators such as electrochemical potentiometer biosensors, radiofrequency identification and time–temperature indicators, and ripeness and pH indicators are extensively used to monitor packaged food quality in intelligent packaging systems (Sahraeian et al., 2024). Real-time monitoring of food freshness is a critical indicator since it detects food deterioration that exceeds human visual perception. The industrialization of ACN-based color indicators faces various obstacles primarily due to their instability (Roy & Rhim, 2021). Acylated ACNs demonstrate superior potential in food applications and function as a sustainable food color solution instead of synthetic dyes due to their stability (Pereira et al., 2024).

15.5.4 Nutraceutical Beverages and Supplements

ACNs exist as biologically active compounds that promote human health benefits. ACNs demonstrate a strong impact on human health benefits due to their bioavailability levels. Two key definitions exist regarding bioavailability: total bioavailability and apparent bioavailability (Han et al., 2019).

Systemic absorption of a foreign chemical starts with the gastrointestinal wall passing xenobiotics through first-pass metabolism, which results in both original chemicals and metabolites reaching the bloodstream at total bioavailability rates, and apparent bioavailability determines xenobiotic dose rates reaching blood without alternations. Bioaccessibility is defined as the fraction of an ingested compound (such as anthocyanins, or ACN) that is released from its food matrix within the gastrointestinal tract and becomes available for absorption. It reflects the amount of food matrix-released ACN that can penetrate intestinal and stomach membranes after digestion (Lila et al., 2016).

ACN metabolism and absorption are crucial areas of study to comprehend their therapeutic and health-promoting effects (Cosme et al., 2020). Food components such as alcohol, lipids, and other matrix constituents can significantly influence the structural stability and bioavailability of ACNs, which are also affected by environmental factors like temperature and pH. Interestingly, the co-consumption of ethanol with ACN-rich foods has been shown to enhance the absorption of these compounds across the intestinal membrane, thereby improving their bioavailability and potential physiological effects (Fernandes et al., 2015).

ACNs interact with various food components, including pectic compounds, proteins, and other phenolic substances, during digestion (Lang et al., 2021). According to molecular docking models, ACNs and alfa-casein form complexes, wherein alfa-casein assists the delivery of ACNs into the bloodstream (Koh et al., 2020). ACNs along with their metabolites are widely distributed throughout various body sections, including the brain and eyes, following the digestion process (Ayvaz et al., 2023).

15.6 TEXTILE AND INDUSTRIAL APPLICATIONS

15.6.1 Biobased and Sustainable Textile Dyes

ACNs serve as a potential dye agent for fabric applications. The dyeing and antibacterial behavior was investigated among silk wool and cotton textiles during the research (Jing et al., 2014). The ACN water extracts from clove, pomegranate and peony used to dye fabrics proved effective against Staphylococcus aureus bacteria with high antimicrobial properties. The highest adsorption rate of ACNs occurs when using TiO2 metal oxide at a solution pH of 3–4 (Junger et al., 2017).

The co-pigmentation of ACNs with caffeine produces increased population of red flavylium cations and blue quinoidal bases at moderate pH values (Trouillas et al., 2016). The resulting concentration of colored ACNs reaches its maximum value at the pH condition that corresponds to the highest level of TiO2 dye adsorption (Junger et al., 2019). The observed results describe the formation of insoluble mixed compounds that combine plant-based dyes with inorganic substrates suitable for cosmetic use (Brudzyńska et al., 2021).

15.6.2 Smart Textiles with Responsive Color Change

Smart textiles, commonly referred to as intelligent or stimuli-sensitive textiles, are a branch of material innovation. These textiles exhibit active engagement with their surroundings by allowing programmability that enhances numerous applications, including biomedicine, defense, aerospace, energy storage, fashion, sports equipment, leisure activities, and wireless communication systems (Yilmaz, 2019).

Color-changing responsive textiles can be classified based on the external stimuli that trigger their chromatic transformation. These include (1) photochromic textiles, which undergo color changes upon exposure to electromagnetic radiation; (2) thermochromic fabrics, which respond to variations in environmental temperature; (3) electrochromic textiles, which alter their color when an electrical potential is applied; and (4) ionochromic textiles, which exhibit color shifts resulting from interactions with ionic species that influence their absorption or emission spectra (Karastathi et al., 2024).

15.6.3 Applications in Paper, Ink, and Paint Industries

Acylated ACNs maintain their pigments effectively and resist external conditions, which makes them suitable for paper production, ink manufacturing, and paint applications. These sustainable and unharmful materials respond to current market forces seeking environmentally friendly and harmless materials in these applications. Acylated ACNs show potential as biological dyes for paper applications because they deliver attractive hues and decompose naturally. Acylated ACNs in inks represent eco-friendly dye alternatives to synthetic chemicals, thus decreasing both industrial pollution and human health risks (Xiong et al., 2019).

Acylated ACNs maintain stability, which leads to longer product life and stable print outcomes. The use of natural paints with integrated acylated ACNs provides both indoor safety and nontoxic characteristics by eliminating volatile organic compounds (VOCs) from indoor spaces. The enhanced pigments settle successfully on painted surfaces as they resist fading under light exposure (Luo et al., 2022).

15.7 ENVIRONMENTAL AND TECHNOLOGICAL INNOVATIONS

15.7.1 Biosensors and Diagnostic Devices

Biosensors analyze chemical analytes through electrical, thermal, or optical inputs by enabling specific biochemical reactions using biological molecules, such as enzymes, immune systems, tissues, organelles, or complete cells, as per the literature sources. ACNs display structural transformations when hydrogen ion concentrations change (Lzaod & Dutta, 2022).

The pH-sensing capabilities of biosensors have garnered significant attention because their structures incorporate ACN pigments as their vital components. Researchers used commercially available contact lenses with and without lacreon to create biosensors. *Brassica oleracea* ACNs were extracted, and contact lenses were functionalized by soaking followed by drop casting processes using ACN aqueous solutions. Tests on lacreon contact lenses revealed evenly distributed color changes that were easily observable by human eyes over the physiological pH range from 6.5 to 7.5 (Riaz et al., 2019).

15.7.2 Biopolymer Integration and Sustainable Packaging

Bio-fabrication and tissue engineering allow borders to be crossed for learning about biocompatible materials with suitable fabrication properties through their studies of engineered biological constructs (Blando et al., 2018). Biopolymers represent a crucial field of exploration of living textile applications and biomedicine. The growing interest in food preservation methods is significantly driven by ACNs due to their antioxidant and antimicrobial properties and their ability to delay product degradation (Jiao et al., 2023).

ACNs together with carbon dots (CDs) create active and smart food packaging elements that offer multiple benefits, including defense against bacterial growth and effective blocking of UV radiation (Wagh et al., 2023). Cells incorporating ACNs demonstrated stronger antioxidant characteristics compared to cellulose-based films as per the experimental outcomes. Numerous *in vivo* studies showed on-the-spot analysis capabilities of these films while analyzing freshness of food in pork, fish, and shrimp samples. Research demonstrate that active films in food packaging provide significant benefits for both food preservation and longevity, and for quality assessment (Rosales-Murillo et al., 2024).

15.7.3 Applications for Bioelectronics and Eco-Friendly Materials

The combination of acylated ACNs raises interest in eco-friendly materials and bioelectronics because of their enhanced color intensity and substantial stability features. ACN-based materials are now being explored through recent studies for their multiple applications in eco-friendly materials and bioelectronic devices (Sun et al., 2024).

ACNs are used in a single device for both energy harvesting and pH detection. This dual functionality is due to ACNs' unique properties. A green sensor system based on bacterial cellulose nanofibers loaded with ACNs exists for Al (III) ion detection. These sensors displayed high selectivity and sensitivity along with visible color reactions to detect specific metal ions for potential uses in environmental monitoring (Arghavani et al., 2023).

Biodegradable films containing ACNs are used as indicators of product freshness in food packaging systems. Research has developed biopolymer films with ACN extracts combined with ZnO nanoparticles for smart packaging applications (Cruces et al., 2024). The films possess antimicrobial functionality combined with visual food quality monitoring capacity, which enhances food safety and extension of product shelf life. The diversity of acylated ACNs demonstrates their capability to create durable materials that adapt to their environment for both electronic applications and environmentally friendly uses (Koshy et al., 2023).

15.8 CURRENT CHALLENGES AND FUTURE PERSPECTIVES

15.8.1 Limitations in Extraction, Stability, and Cost

ACNs face challenges when used for pH detection because of their low stability. Laboratory tests show that ACN aglycones break down into pale colorless or brown products, leading to their complete loss of pH sensitivity (Sakulnarmrat & Konczak, 2022). The stability of ACNs depends on their structure, e.g., pelargonidin, cyanidin, and delphinidin show different stability patterns, and numerous environmental factors, including temperature and light exposure. The presence of sugars, ascorbic acid, other phenolic compounds, enzymes, metal ions, and oxygen attached to ACNs affect their stability. The extraction of acylated ACNs from sources such as sweet potato and purple carrot can present a solution since their acyl groups generate steric hindrance that prevents water-based attacks while blocking colorless hemiketal and chalcone formation (Kossyvaki et al., 2022).

15.8.2 Industrial-Scale Production and Standardization

Acylated ACNs are a class of modified plant pigments known for their enhanced stability, antioxidant properties, and strong coloration, and they find applications in the food, cosmetic, and pharmaceutical sectors. Scientists have created a system of genetically modified cell cultures from *Arabidopsis thaliana* and tobacco species for maximizing acylated ACN production. Bioreactors enable sustainable production methods that are replacing traditional extraction strategies in various industries (Ijod et al., 2024). Researchers use microorganisms like *Escherichia coli* and *Saccharomyces cerevisiae* to produce ACNs in the laboratory. The yields from microbial production are high, but the generation of complex acylated derivatives presents a substantial production challenge. Standardized purification steps play an essential role in satisfying safety requirements. Additional research must determine how acylation treatments affect drug metabolism and bioavailability to guarantee effective end-use performance (Lin et al., 2023)

15.8.3 Opportunities for Interdisciplinary Research and Development

Stable natural pigments called acylated ACNs have opened many research opportunities across fields including biochemistry and innovation development. Recent research demonstrates that ACNs can benefit different fields, including food science, materials engineering, pharmacology, and data analytics (Xu et al., 2024).

These natural pigments display environmental sensitivity to heat, pH, and light, which demands unique delivery methods. The strategies work toward extending shelf life and enhancing bioavailability because of their applications in functional foods and nutraceuticals (Jeyaraj et al., 2023). ACNs are used as intelligent food packaging materials because they display pH-dependent color alterations. Currently, researchers are developing smart food packaging systems that utilize ACN-based films to indicate food spoilage. Crops provide essential nutrition and function simultaneously as sustainable natural sources of coloring agents and bioactive components, effectively replacing synthetic food additives (Jokioja et al., 2021).

REFERENCES

Agati, G., Brunetti, C., Di Ferdinando, M., Ferrini, F., Pollastri, S. and Tattini, M. 2013. Functional roles of flavonoids in photoprotection: new evidence, lessons from the past. *Plant Physiol. Biochem.* 72:35–45.

Alam, M.A., Islam, P., Subhan, N., Rahman, M.M., Khan, F., Burrows, G.E., Nahar, L. and Sarker, S.D. 2021. Potential health benefits of anthocyanins in oxidative stress related disorders. *Phytochem. Rev.* 20(4):705–749.

Amatori, S., Mazzoni, L., Alvarez-Suarez, J.M., Giampieri, F., Gasparrini, M., Forbes-Hernandez, T.Y., Afrin, S., Provenzano, A.E., Persico, G., Mezzetti, B., et al. 2016. Polyphenol-rich strawberry extract (PRSE) shows in vitro and in vivo biological activity against invasive breast cancer cells. *Sci. Rep.* 6(1):30917.

Appelhagen, I., Wulff-Vester, A.K., Wendell, M., Hvoslef-Eide, A.K., Russell, J., Oertel, A., Martens, S., Mock, H.P., Martin, C. and Matros, A. 2018. Colour bio-factories: towards scale-up production of anthocyanins in plant cell cultures. *Metab. Eng.* 48:218–232.

Arghavani, S., Mohseni-Shahri, F.S. and Moeinpour, F. 2023. Anthocyanin-loaded bacterial cellulose nanofiber as a green sensor for monitoring the selective naked eye and visual detection of Al (III) ions. *Anal. Sci. Adv.* 4:324–334.

Arruda, H.S., Silva, E.K., Peixoto Araujo, N.M., Pereira, G.A., Pastore, G.M. and Marostica Junior, M.R. 2021. Anthocyanins recovered from agri-food by-products using innovative processes: trends, challenges, and perspectives for their application in food systems. *Molecules* 26(9):2632.

Ayvaz, H., Cabaroglu, T., Akyildiz, A., Pala, C.U., Temizkan, R., Ağçam, E., Ayvaz, Z., Durazzo, A., Lucarini, M., Direito, R. and Diaconeasa, Z. 2023. Anthocyanins: metabolic digestion, bioavailability, therapeutic effects, current pharmaceutical/industrial use, and innovation potential. *Antioxidants* 12(1):48.

Blando, F., Calabriso, N., Berland, H., Maiorano, G., Gerardi, C., Carluccio, M.A. and Andersen, Ø.M. 2018. Radical scavenging and anti-inflammatory activities of representative anthocyanin groupings from pigment-rich fruits and vegetables. *Int. J. Mol. Sci.* 19:1–15.

Brudzyńska, P., Sionkowska, A. and Grisel, M. 2021. Plant-derived colorants for food, cosmetic and textile industries: a review. *Materials* 14:1–18.

Câmara, J.S., Locatelli, M., Pereira, J.A.M., Oliveira, H., Arlorio, M., Fernandes, I., Perestrelo, R., Freitas, V. and Bordiga, M. 2022. Behind the scenes of anthocyanins – from the health benefits to potential applications in food, pharmaceutical and cosmetic fields. *Nutrients* 14:5133.

Correia, P., Oliveira, H., Araújo, P., Brás, N.F., Pereira, A.R., Moreira, J., de Freitas, V., Mateus, N., Oliveira, J. and Fernandes, I. 2021. The role of anthocyanins, deoxyanthocyanins and pyranoanthocyanins on the modulation of tyrosinase activity: an in vitro and in silico approach. *Int. J. Mol. Sci.* 22:6192.

Cosme, P., Rodríguez, A.B., Espino, J. and Garrido, M. 2020. Plant phenolics: bioavailability as a key determinant of their potential health-promoting applications. *Antioxidants* 9.

Cruces, F., García, M.G., Poyatos, E., Vivancos, J.L., Fuentes, A., Martínez-Máñez, R. and Ochoa, N.A. 2024. Super hydrophobic pectin-anthocyanin colorimetric sensor to indicate freshness in bovine meat. *Food Bioprocess. Technol.* 1563–1579.

Cruz, L., Fernandes, I., Guimaraes, M., de Freitas, V. and Mateus, N. 2016. Enzymatic synthesis, structural characterization and antioxidant capacity assessment of a new lipophilic malvidin-3-glucoside-oleic acid conjugate. *Food Funct.* 7:2754–2762.

Cruz, L., Guirnaraes, M., Araujo, P., Evora, A., de Freitas, V. and Mateus, N. 2017. Malvidin 3-glucoside-fatty acid conjugates: from hydrophilic toward novel lipophilic derivatives. *J. Agric. Food Chem.* 65:6513–6518.

Cunningham, O.D. and Edwards, R. 2008. Modifying the acylation of flavonols in *Petunia hybrida*. *Phytochem.* 69:2016–2021.

Diaconeasa, Z., Ayvaz, H., Rugină, D., Leopold, L., Stănilă, A., Socaciu, C., Tăbăran, F., Luput, L., Mada, D.C., Pintea, A., et al. 2017. Melanoma inhibition by anthocyanins is associated with the reduction of oxidative stress biomarkers and changes in mitochondrial membrane potential. *Plant Foods Hum. Nutr.* 72:404–410.

Diaconeasa, Z., Știrbu, I., Xiao, J., Leopold, N., Ayvaz, Z., Danciu, C., Ayvaz, H., Stănilă, A., Nistor, M. and Socaciu, C. 2020. Anthocyanins, vibrant color pigments, and their role in skin cancer prevention. *Biomedicines* 8:336.

Echegaray, N., Guzel, N., Kumar, M., Guzel, M., Hassoun, A. and Lorenzo, J.M. 2023. Recent advancements in natural colorants and their application as coloring in food and in intelligent food packaging. *Food Chem.* 404.

Fenger, J.A., Moloney, M., Robbins, R.J., Collins, T.M. and Dangles, O. 2019. The influence of acylation, metal binding and natural antioxidants on the thermal stability of red cabbage anthocyanins in neutral solution. *Food Funct.* 10:6740–6751.

Fernandes, I., Faria, A., de Freitas, V., Calhau, C. and Mateus, N. 2015. Multiple-approach studies to assess anthocyanin bioavailability. *Phytochem. Rev.* 14:899–919.

Gonçalves, A.C., Falcão, A., Alves, G., Lopes, J.A. and Silva, L.R. 2022. Employ of anthocyanins in nanocarriers for nano delivery: in vitro and in vivo experimental approaches for chronic diseases. *Pharmaceutics* 14:2272.

Gonçalves, A.C., Nunes, A.R., Falcão, A., Alves, G. and Silva, L.R. 2021. Dietary effects of anthocyanins in human health: a comprehensive review. *Pharmaceuticals* 14(7):690.

Han, F., Yang, P., Wang, H., Fernandes, I., Mateus, N. and Liu, Y. 2019. Digestion and absorption of red grape and wine anthocyanins through the gastrointestinal tract. *Trends Food Sci. Technol.* 83:211–224.

He, J. and Giusti, M.M. 2010. Anthocyanins: natural colorants with health-promoting properties. *Annu. Rev. Food Sci. Technol.* 1:163–187.

Igwe, E.O., Charlton, K.E. and Probst, Y.C. 2019. Usual dietary anthocyanin intake, sources and their association with blood pressure in a representative sample of Australian adults. *J. Hum. Nutr. Diet.* 32.

Ijod, G., Nawawi, N.I.M., Anwar, F., Rahim, M.H.A., Ismail-Fitry, M.R., Adzahan, N.M. and Azman, E.M. 2024. Recent microencapsulation trends for enhancing the stability and functionality of anthocyanins: a review. *Food Sci. Biotechnol.* 33:2673–2698.

Ishihara, K. and Nakajima, N. 2003. Structural aspects of acylated plant pigments: stabilization of flavonoid glucosides and interpretation of their functions. *J. Mol. Catal. B Enzym.* 23:411–417.

Jeyaraj, E.J., Vidana Gamage, G.C., Cintrat, J.C. and Choo, W.S. 2023. Acylated and non-acylated anthocyanins as antibacterial and antibiofilm agents. *Discov Food* 3:21.

Jiao, X., Xie, J., Du, H., Bian, X., Wang, C., Zhou, L. and Wen, Y. 2023. Antibacterial smart absorbent pad with Janus structure for meat preservation. *Food Packag. Shelf Life.* https://doi.org/10.1016/j.fpsl.2023.101066.

Jing, P., Bomser, J.A., Schwartz, S.J., He, J., Magnuson, B.A. and Giusti, M.M. 2008. Structure function relationships of anthocyanins from various anthocyanin-rich extracts on the inhibition of colon cancer cell growth. *J. Agric. Food Chem.* 56:9391–9398.

Jing, P., Song, L.H., Shen, S.Q., Zhao, S.J., Pang, J. and Qian, B.J. 2014. Characterization of phytochemicals and antioxidant activities of red radish brines during lactic acid fermentation. *Molecules.* https://doi.org/10.3390/molecules19079675.

Jokioja, J., Yang, B. and Linderborg, K.M. 2021. Acylated anthocyanins: a review on their bioavailability and effects on postprandial carbohydrate metabolism and inflammation. *Comp. Rev Food Sci. Food Safety* 20:5570–5615.

Junger, I.J., Homburg, S.V., Meissner, H., Grethe, T., Pfeiffer, A.S., Fiedler, J., Herrmann, A., Blachowicz, T. and Ehrmann, A. 2017. Influence of the pH value of anthocyanins on the electrical properties of dye-sensitized solar cells. *AIMS Energy* 5.

Junger, I.J., Udomrungkhajornchai, S., Grimmelsmann, N., Blachowicz, T. and Ehrmann, A. 2019. Effect of caffeine copigmentation of anthocyanin dyes on DSSC efficiency. *Materials* 12.

Kalemba-Drożdż, M., Cierniak, A. and Cichoń, I. 2020. Berry fruit juices protect lymphocytes against DNA damage and ROS formation induced with heterocyclic aromatic amine PhIP. *J. Berry Res.* 10(1):95–113. https://doi.org/10.3233/JBR–190429.

Karastathi, N., Salmane, A.K. and Parker, B. 2024. pH-responsive textile with anthocyanin-laden hydrogel yarns: a colorimetric and feasibility study. *Res. Direct.: Biotechnol. Design.* 2:e17.

Koh, J., Xu, Z. and Wicker, L. 2020. Blueberry pectin and increased anthocyanins stability under in vitro digestion. *Food Chem. Food Chem.* 302.

Koshy, J.T., Vasudevan, D., Sangeetha, D. and Prabu, A.A. 2023. Biopolymer based multifunctional films loaded with anthocyanin rich floral extract and ZnO nano particles for smart packaging and wound healing applications. *Polymers (Basel)* 15.

Kossyvaki, D., Contardi, M., Athanassiou, A. and Fragouli, D. 2022. Colorimetric indicators based on anthocyanin polymer composites: a review. *Polymers* 14:4129.

Krga, I. and Milenkovic, D. 2019. Anthocyanins: from sources and bioavailability to cardiovascular-health benefits and molecular mechanisms of action. *J. Agric. Food Chem.* 67:1771–1783.

Lang, Y., Tian, J., Meng, X., Si, X., Tan, H., Wang, Y., Shu, C., Chen, Y., Zang, Z., Zhang, Y., Wang, J. and Li, B. 2021. Effects of α-casein on the absorption of blueberry anthocyanins and metabolites in rat plasma based on pharmacokinetic analysis. *J. Agric. Food Chem.* 69.

Larsson, S.C., Drca, N., Bjorck, M., Back, M. and Wolk, A. 2018. Nut consumption and incidence of seven cardiovascular diseases. *Heart* 104:1615–1620.

Lewandowska, U., Gorlach, S., Owczarek, K., Hrabec, E. and Szewczyk, K. 2014. Synergistic interactions between anticancer chemotherapeutics and phenolic compounds and anticancer synergy between polyphenols. *Adv. Hygiene Exp. Med. Postepy Higieny i Medycyny Doswiadczalnej* 68.

Li, A., Xiao, R., He, S., An, X., He, Y., Wang, C., Yin, S., Wang, B., Shi, X. and He, J. 2019. Research advances of purple sweet potato anthocyanins: extraction, identification, stability, bioactivity, application, and biotransformation. *Molecules* 24:3816.

Lila, M.A., Burton-Freeman, B., Grace, M. and Kalt, W. 2016. Unraveling anthocyanin bioavailability for human health. *Annu. Rev. Food Sci. Technol.* 7.

Lin, Y., Li, C., Shi, L. and Wang, L. 2023. Anthocyanins: modified new technologies and challenges. *Foods* 12:1–27.

Liu, J., Zhao, Y., Zhang, J., Kong, Y., Liu, P., Fang, Y., Cui, M., Pei, T., Zhong, X., Xu, P., Qiu, W., Yang, D., Martin, C. and Zhao, Q. 2024. Production of species-specific anthocyanins through an inducible system in plant hairy roots. *Metab. Eng.* 81.

Luo, C.L., Zhou, Q., Yang, Z.W., Wang, R.D. and Zhang, J.L. 2018. Evaluation of structure and bioprotective activity of key high molecular weight acylated anthocyanin compounds isolated from the purple sweet potato (*Ipomoea batatas* L. cultivar Eshu No.8). *Food Chem.* 241:23–31.

Luo, X., Wang, R., Wang, J., Li, Y., Luo, H., Chen, S., Zeng, X. and Han, Z. 2022. Acylation of anthocyanins and their applications in the food industry: mechanisms and recent research advances. *Foods* 11.

Luzardo-Ocampo, I., Ramírez-Jiménez, A.K., Yañez, J., Mojica, L. and Luna-Vital, D.A. 2021. Technological applications of natural colorants in food systems: a review. *Foods* 10(3):634.

Lzaod, S. and Dutta, T. 2022. Recent advances in the development of oxidoreductase-based biosensors for detection of phenolic antioxidants in food and beverages. *ACS Omega* 7.

Matsufuji, H., Kido, H., Misawa, H., Yaguchi, J., Otsuki, T., Chino, M., Takeda, M. and Yamagata, K. 2007. Stability to light, heat, and hydrogen peroxide at different pH values and DPPH radical scavenging activity of acylated anthocyanins from red radish extract. *J. Agric. Food Chem.* 55:3692–3701.

Mattioli, R., Francioso, A., Mosca, L. and Silva, P. 2020. Anthocyanins: a comprehensive review of their chemical properties and health effects on cardiovascular and neurodegenerative diseases. *Molecules* 25:3809.

Mehdi, A., Lamiae, B., Samira, B., Ramchoun, M., Abdelouahed, K., Tamas, F. and Hicham, B. 2022. Pomegranate (*Punica granatum* L.) attenuates neuroinflammation involved in neurodegenerative diseases. *Foods* 11.

Mejia-Giraldo, J.C., Gallardo, C. and Puertas-Mejia, M.A. 2022. Selected extracts from high mountain plants as potential sunscreens with antioxidant capacity. *Photochem. Photobiol.* 98:211–219.

Neveu, V., Perez-Jiménez, J., Vos, F., Crespy, V., du Chaffaut, L., Mennen, L., Knox, C., Eisner, R., Cruz, J., Wishart, D. and Scalbert, A. 2010. Phenol-explorer: an online comprehensive database on polyphenol contents in foods. *Database (Oxford)*. bap024.

Ozkan, G., Ceyhan, T., Çatalkaya, G., Rajan, L., Ullah, H., Daglia, M. and Capanoglu, E. 2024. Encapsulated phenolic compounds: clinical efficacy of a novel delivery method. *Phytochem. Rev.* 23(3):781–819.

Pereira, A.R., Fernandes, V.C., Delerue-Matos, C., de Freitas, V., Mateus, N. and Oliveira, J. 2024. Exploring acylated anthocyanin-based extracts as a natural alternative to synthetic food dyes: stability and application insights. *Food Chem.* 461:140945.

Qin, Y., Liu, Y., Yong, H., Liu, J., Zhang, X. and Liu, J. 2019. Preparation and characterization of active and intelligent packaging films based on cassava starch and anthocyanins from *Lycium ruthenicum* Murr. *Int. J. Biol. Macromol.* 134.

Riaz, R.S., Elsherif, M., Moreddu, R., Rashid, I., Hassan, M.U., Yetisen, A.K. and Butt, H. 2019. Anthocyanin-functionalized contact lens sensors for ocular pH monitoring. *ACS Omega* 4.

Rodriguez-Amaya, D.B. 2019. Natural food pigments and colorants. In *Bioactive Molecules in Food*, Mérillon, J.-M., Ramawat, K.G., Eds., Springer International Publishing: Cham, Switzerland, pp. 867–901.

Rosales-Murillo, S.S., Sánchez-Bodón, J., Hernández Olmos, S.L., Ibarra-Vázquez, M.F., Guerrero-Ramírez, L.G., Pérez-Álvarez, L. and Vilas-Vilela, J.L. 2024. Anthocyanin-loaded polymers as promising nature-based, responsive, and bioactive materials. *Polymers* 16.

Roy, S. and Rhim, J.W. 2021. Anthocyanin food colorant and its application in pH-responsive color change indicator films. *Crit. Rev. Food Sci. Nutr.* 61.

Sahraeian, S., Abdollahi, B. and Rashidinejad, A. 2024. Biopolymer-polyphenol conjugates: novel multifunctional materials for active packaging. *Int. J. Biol. Macromol.* 280:135714.

Sakulnarmrat, K. and Konczak, I. 2022. Encapsulation of *Melodorum fruticosum* Lour. anthocyanin-rich extract and its incorporation into model food. *LWT*. 153:112546.

Shahidi, F. and Ambigaipalan, P. 2015. Phenolics and polyphenolics in foods, beverages and spices: antioxidant activity and health effects – a review. *J. Funct. Foods* 18:820–897.

Shi, M.Z. and Xie, D.Y. 2014. Biosynthesis and metabolic engineering of anthocyanins in *Arabidopsis thaliana*. *Recent Pat. Biotechnol.* 5:205–209.

Silva, G., da Silva, K., Silva, C., Gonçalves, J. and Quina, F. 2020. Hybrid pigments from anthocyanin analogues and synthetic clay minerals. *ACS Omega* 5:3354.

Sun, W., Dong, J., Li, W., Gao, X., Liu, J. and Nan, D. 2024. An anthocyanin-based eco-friendly triboelectric nanogenerator for pH monitoring and energy harvesting. *Molecules* 29:1–12.

Tarone, A.G., Cazarin, C.B.B. and Marostica Junior, M.R. 2020a. Anthocyanins: new techniques and challenges in microencapsulation. *Food Res. Int.* 133:109092.

Tarone, A.G., Silva, E.K., Barros, H.D.D.F.Q., Cazarin, C.B.B. and Junior, M.R.M. 2020b. High-intensity ultrasound-assisted recovery of anthocyanins from Jabuticaba by-products using green solvents: effects of ultrasound intensity and solvent composition on the extraction of phenolic compounds. *Food Res. Int.* 140:110048.

Tena, N., Martín, J. and Asuero, A.G. 2020. State of the art of anthocyanins: antioxidant activity, sources, bioavailability, and therapeutic effect in human health. *Antioxidants* 9:451.

Tommasini, S., Raneri, D., Ficarra, R., Calabrò, M.L., Stancanelli, R. and Ficarra, P. 2004. Improvement in solubility and dissolution rate of flavonoids by complexation with-cyclodextrin. *J. Pharmaceut. Biomed.* 35:379–387.

Trouillas, P., Sancho-García, J.C., De Freitas, V., Gierschner, J., Otyepka, M. and Dangles, O. 2016. Stabilizing and modulating color by copigmentation: insights from theory and experiment. *Chem. Rev.* 116.

Wagh, R.V., Khan, A., Priyadarshi, R., Ezati, P. and Rhim, J.W. 2023. Cellulose nanofiber-based multifunctional films integrated with carbon dots and anthocyanins from *Brassica oleracea* for active and intelligent food packaging applications. *Int. J. Biol. Macromol.* 233:123567.

Weston, M., Phan, M.A.T., Arcot, J. and Chandrawati, R. 2020. Anthocyanin-based sensors derived from food waste as an active use-by date indicator for milk. *Food Chem.* 326.

Xiong, Y., Zhang, P., Warner, R.D. and Fang, Z. 2019. 3-deoxyanthocyanidin colorant: nature, health, synthesis, and food applications. *Compr. Rev. Food Sci. Food Saf.* 18:1533–1549.

Xu, M., Fang, D., Kimatu, B.M., Lyu, L., Wu, W., Cao, F. and Li, W. 2024. Recent advances in anthocyanin-based films and its application in sustainable intelligent food packaging: a review. *Food Control* 162:110431.

Yang, W., Kortesniemi, M., Ma, X., Zheng, J., Yang, B. 2019. Enzymatic acylation of blackcurrant (*Ribes nigrum*) anthocyanins and evaluation of lipophilic properties and antioxidant capacity of derivatives. *Food Chem.* 281:189–196.

Yilmaz, N.D. 2019. *Smart Textiles: Wearable Nanotechnology*. Wiley-Scrivener, John Wiley & Sons, Inc.

Zhang, H., Xu, Z., Zhao, H., Wang, X., Pang, J., Li, Q., Yang, Y. and Ling, W. 2020. Anthocyanin supplementation improves anti-oxidative and anti-inflammatory capacity in a dose-response manner in subjects with dyslipidemia. *Redox Biol.* 32:101474.

Index

A

B

C

D

S

T

U

V

W

Y

Z